SEARCH ALGORITHMS FOR ENGINEERING OPTIMIZATION

Edited by **Taufik Abrão**

Search Algorithms for Engineering Optimization

http://dx.doi.org/10.5772/45841
Edited by Taufik Abrão

Contributors

Abdelkader Zeblah, Rami Abdelkader, Yoshio Uwano, Bruno Augusto Angélico, Márcio Mendonça, Lúcia Valéria R. De Arruda, Taufik Abrão, Fabio Durand, Alysson Santos, , Larissa Melo, Lucas Garcia, Oleksiy Pogrebnyak, Enrique Guzmán, Juan Gabriel Zambrano Nila, Fernando Ciriaco, Paul Jean E. Jeszensky, Lucas Dias H. Sampaio, Mateus De Paula Marques, Mário Henrique Adaniya, Aleksandar Jevtić, Bo Li, Chung-Ming Kuo, Ivan Casella, Alfeu Sguarezi, Carlos Capovilla, Ernesto Ruppert, José Puma, Hamid Reza Baghaee Majidi

Published by InTech

Janeza Trdine 9, 51000 Rijeka, Croatia

Publishing Process Manager Iva Lipovic

Technical Editor InTech DTP team

Cover InTech Design team

Additional hard copies can be obtained from orders@intechopen.com

Search Algorithms for Engineering Optimization, Edited by Taufik Abrão
p. cm.
ISBN 978-953-51-0983-9

For EU product safety concerns:
IN TECH d.o.o., Prolaz Marije Krucifikse Kozulić 3, 51000 Rijeka, Croatia,
info@intechopen.com or visit our website at intechopen.com.

Meet the editor

Taufik Abrão received his B. S., M. Sc. and Ph.D., all in Electrical Engineering, from the Polytechnic School of the University of Sao Paulo (EPUSP), Brazil, in 1992, 1996, and 2001, respectively. Since March 1997 he has been with the Department of Electrical Engineering, Londrina State University (UEL) Brazil, where he is currently an Associate Professor in communications engineering. In 2012 he was an Academic Visitor at CSPC University of Southampton, UK. Dr. Abrão is involved in editorial board activities of several journals and has served as TCP member in several symposium and conferences. His research interests lie in communications, including multi-user detection and estimation, MC-CDMA and MIMO systems, cooperative communication and relaying, resource allocation, heuristic and convex optimization aspects. He is co-authored of more than 180 research papers published in specialized international journals and conferences.

Contents

Preface

Heuristic Search is an important sub-discipline of optimization theory and finds applications in a vast variety of fields, including life science and engineering. Over the years, search methods have made an increasing number of appearances in engineering systems, primarily because of the capability in providing effective near-optimum solutions with low-complexity, more cost-effective and less time consuming. Heuristic Search is a method that might not always find the best solution but is guaranteed to find a good solution in reasonable time, i.e., by sacrificing completeness it increases efficiency. Search methods have been useful in solving tough engineering-oriented problems that either could not be solved any other way or solutions take a very long time to be computed.

The primary goal of this book is to provide a variety of applications for search methods and techniques in different fields of electrical engineering. By organizing relevant results and applications, the book will serve as a useful resource for students, researchers and practitioners to further exploit the potential of search methods in solving hard non-polynomial optimization problems that arise in advanced engineering technologies, such as image and video processing issues, detection and resource allocation in telecommunication systems, security and harmonic reduction in power generation systems, as well as redundancy optimization problem and search-fuzzy learning mechanisms in industrial applications. To better explore those engineering-oriented search methods, this book is organized in four parts. In Part 1, three search optimization procedures applied to image and video processing are discussed. In Part 2, three specific hard optimization problems that arise in telecommunications systems are solved using guided search procedures: multiuser detection, power-rate allocation, anomaly detection and routing optical channel allocation problems are treaded deploying a collection of guided-search algorithms, such as Ant Colony, Particle Swarm, Genetic, Simulation Annealing, Tabu, Evolutionary Programming, Neighborhood Search and Hyper-Heuristic. Search methods applied to power systems and industrial processes are developed in Part 3: cognitive concepts and methods, such as fuzzy cognitive maps and adaptive fuzzy learning mechanisms are aggregated in order to efficiently model and solve optimization problems found in reliable power generation and industrial applications. Finally, the last chapter is devoted to conceptual and formal aspects of Grover-type quantum search, which constitutes Part 4.

It is our sincere hope that the book will help readers to further explore the potential of search methods in solving efficiently hard-complexity engineering optimization problems.

Taufik Abrão
Electrical Engineering Department,
State University of Londrina (DEEL-UEL),
Londrina, Paraná, Brazil

Section 1

Image Reconstruction

Chapter 1

Search Algorithm for Image Recognition Based on Learning Algorithm for Multivariate Data Analysis

Juan G. Zambrano, E. Guzmán-Ramírez and
Oleksiy Pogrebnyak

Additional information is available at the end of the chapter

http://dx.doi.org/10.5772/ 52179

1. Introduction

An image or a pattern can be recognized using prior knowledge or the statistical information extracted from the image or the pattern. The systems for image recognition and classification have diverse applications, e.g. autonomous robot navigation[1], image tracking radar [2], face recognition [3], biometrics [4], intelligent transportation, license plate recognition, character recognition [5] and fingerprints [6].

The problem of automatic image recognition is a composite task that involves detection and localization of objects in a cluttered background, segmentation, normalization, recognition and verification. Depending on the nature of the application, e.g. sizes of training and testing database, clutter and variability of the background, noise, occlusion, and finally, speed requirements, some of the subtasks could be very challenging. Assuming that segmentation and normalization haven been done, we focus on the subtask of object recognition and verification, and demonstrate the performance using several sets of images.

Diverse paradigms have been used in the development of algorithms for image recognition, some of them are: artificial neural networks [7, 8], principal component analysis [9, 10], fuzzy models [11, 12], genetic algorithms [13, 14] and Auto-Associative memory [15]. The following paragraphs describe some work done with these paradigms.

Abrishambaf *et al* designed a fingerprint recognition system based in Cellular Neural Networks (CNN). The system includes a preprocessing phase where the input fingerprint image is enhanced and a recognition phase where the enhanced fingerprint image is matched with the fingerprints in the database. Both preprocessing and recognition phases are realized by means of CNN approaches. A novel application of skeletonization method is used to perform ridgeline thinning which improves the quality of the extracted lines for further processing, and hence increases the overall system performance [6].

In [16], Yang and Park developed a fingerprint verification system based on a set of invariant moment features and a nonlinear Back Propagation Neural Network (BPNN) verifier. They used an image-based method with invariant moment features for fingerprint verification to overcome the demerits of traditional minutiae-based methods and other image-based methods. The proposed system contains two stages: an off-line stage for template processing and an on-line stage for testing with input fingerprints. The system preprocesses fingerprints and reliably detects a unique reference point to determine a Region of Interest (ROI). A total of four sets of seven invariant moment features are extracted from four partitioned sub-images of an ROI. Matching between the feature vectors of a test fingerprint and those of a template fingerprint in the database is evaluated by a nonlinear BPNN and its performance is compared with other methods in terms of absolute distance as a similarity measure. The experimental results show that the proposed method with BPNN matching has a higher matching accuracy, while the method with absolute distance has a faster matching speed. Comparison results with other famous methods also show that the proposed method outperforms them in verification accuracy.

In [17] the authors presents a classifier based on Radial Basis Function Network (RBFN) to detect frontal views of faces. The technique is separated into three main steps, namely: preprocessing, feature extraction, classification and recognition. The curvelet transform, Linear Discriminant Analysis (LDA) are used to extract features from facial images first, and RBFN is used to classify the facial images based on features. The use of RBFN also reduces the number of misclassification caused by not-linearly separable classes. 200 images are taken from ORL database and the parameters like recognition rate, acceptance ratio and execution time performance are calculated. It is shown that neural network based face recognition is robust and has better performance of recognition rate 98.6% and acceptance ratio 85 %.

Bhowmik *et al.* designed an efficient fusion technique for automatic face recognition. Fusion of visual and thermal images has been done to take the advantages of thermal images as well as visual images. By employing fusion a new image can be obtained, which provides the most detailed, reliable, and discriminating information. In this method fused images are generated using visual and thermal face images in the first step. At the second step, fused images are projected onto eigenspace and finally classified using a radial basis function neural network. In the experiments Object Tracking and Classification Beyond Visible Spectrum (OTCBVS) database benchmark for thermal and visual face images have been used. Experimental results show that the proposed approach performs well in recognizing unknown individuals with a maximum success rate of 96% [8].

Zeng and Liu described state of the art of important advances of type-2 fuzzy sets for pattern recognition [18]. The success of type-2 fuzzy sets has been largely attributed to their three-dimensional membership functions to handle more uncertainties in real-world problems. In pattern recognition, both feature and hypothesis spaces have uncertainties, which motivate us of integrating type-2 fuzzy sets with conventional classifiers to achieve a better performance in terms of the robustness, generalization ability, or recognition accuracy.

A face recognition system for personal identification and verification using Genetic algorithm (GA) and Back-propagation Neural Network (BPNN) is described in [19]. The system consists of three steps. At the very outset some pre-processing are applied on the input image. Secondly face features are extracted, which will be taken as the input of the Back-propagation Neural Network and Genetic Algorithm in the third step and classification is carried out by using BPNN and GA. The proposed approaches are tested on a number of face images. Experimental results demonstrate the higher degree performance of these algorithms.

In [20], Blahuta *et al.* applied pattern recognition on finite set brainstem ultrasound images to generate neuro solutions in medical problems. For analysis of these images the method of Principal Component Analysis (PSA) was used. This method is the one from a lot of methods for image processing, exactly to pattern recognition where is necessary a feature extraction. Also the used artificial neural networks (ANN) for this problem and compared the results. The method was implemented in NeuroSolutions software that is very sophisticated simulator of ANN with PCA multilayer (ML) NN topology.

Pandit and Gupta proposed a Neural Network model that has been utilized to train the system for image recognition. The NN model uses Auto-Associative memory for training. The model reads the image in the form of a matrix, evaluates the weight matrix associated with the image. After training process is done, whenever the image is provided to the system the model recognizes it appropriately. The evaluated weight matrix is used for image pattern matching. It is noticed that the model developed is accurate enough to recognize the image even if the image is distorted or some portion/ data is missing from the image. This model eliminates the long time consuming process of image recognition [15].

In [21], authors present the design of three types of neural networks with different features for image recognition, including traditional backpropagation networks, radial basis function networks and counterpropagation networks. The design complexity and generalization ability of the three types of neural network architectures are tested and compared based on the applied digit image recognition problem. Traditional backpropagation networks require very complex training process before being applied for classification or approximation. Radial basis function networks simplify the training process by the specially organized 3-layer architecture. Counterpropagation networks do not need training process at all and can be designed directly by extracting all the parameters from input data. The experimental results show the good noise tolerance of both RBF networks and counterpropagation network on the image recognition problem, and somehow point out the poor generalization ability of traditional backpropagation networks. The excellent noise rejection ability makes the RBF networks very proper for image data preprocessing before applied for recognition.

The remaining sections of this Chapter are organized as follows. In next Section, a brief theoretical background of the Learning Algorithm for Multivariate Data Analysis (LAMDA) is given. In Section 3 we describe the proposed search algorithm for image recognition based on LAMDA algorithm. Then, in Section 4 we present the implementation results obtained by the proposed approach. Finally, Section 5 contains the conclusions of this Chapter.

2. Learning Algorithm for Multivariate Data Analysis

The Learning Algorithm for Multivariate Data Analysis (LAMDA) is an incremental conceptual clustering method based on fuzzy logic, which can be applied in the processes of formation and recognition of concepts (classes). LAMDA has the following features [22-24]:

- The previous knowledge of the number of classes is not necessary (unsupervised learning).
- The descriptors can be qualitative, quantitative or a combination of both.
- LAMDA can use a supervised learning stage followed by unsupervised one; for this reason, it is possible to achieve an evolutionary classification.
- Formation and recognition of concepts are based on the maximum adequacy (MA) rule.
- This methodology has the possibility to control the selectivity of the classification (exigency level) through the parameterα.
- LAMDA models the concept of maximum entropy (homogeneity). This concept is represented by a class denominated Non-Informative Class (NIC). The NIC concept plays the role of a threshold of decision, in the concepts formation process.

Traditionally, the concept of similarity between objects has been considered fundamental to determine whether the descriptors are members of a class or not. LAMDA does not uses similarity measures between objects in order to group them, but it calculates a degree of adequacy. This concept is expressed as a membership function between the descriptor and any of the previously established classes [22, 25].

2.1. Operation of LAMDA

The objects X (input vectors) and the classes C are represented by a number of descriptors denoted by$(d_1, \ldots, d_n)$. Then, every d_i has its own value inside the setD_k, the n-ary product of theD_k, written as$D_1 \times, \ldots, \times D_p$, with$\{(d_1, \ldots, d_n) : d_i \in D_k \, for 1 \leq i \leq n, \; 1 \leq k \leq p\}$ and it is denominated Universe (U).

The set of objects can be described by $X = \{\boldsymbol{x}^j : j = 1, 2, \ldots, M\}$ and any object can be represented by a vector $\boldsymbol{x}^j = (x_1, \ldots, x_n)$ where$x_i \in U$, so every component x_i will correspond to the value given by the descriptor $d_i$for the object$\boldsymbol{x}^j$. The set of classes can be described by $C = \{\boldsymbol{c}^l : l = 1, 2, \ldots, N\}$ and any class can be represented by a vector $\boldsymbol{c}^l = (c_1, \ldots, c_n)$ where $c_i \in U$, so every component c_i will corresponds to the value given by the descriptor $d_i$for the class $\boldsymbol{c}^l$[23].

2.1.1. Marginal Adequacy Degree

Given an object $\mathbf{x}^j$ and a class$\mathbf{c}^l$, LAMDA computes for every descriptor the so-called *marginal adequacy degree (MAD)* between the value of component x_i of object $\mathbf{x}^j$ and the value that the component c_i takes in$\mathbf{c}^l$, which is denoted as:

$$MAD(x_i^j / c_i^l) = \mathbf{x}^j \times \mathbf{c}^l \rightarrow [0,1]^n \tag{1}$$

Hence, one MAD vector can be associated with an object $\mathbf{x}^j$ (see Figure 1). To maintain consistency with fuzzy logic, the descriptors must be normalized using (1). This stage generates N MADs, and this process is repeated iteratively for every object with all classes [26].

$$x_i = \frac{\tilde{x}_i - x_{min}}{x_{max} - x_{min}} = \frac{\tilde{x}_i}{2^L \quad 1} \tag{2}$$

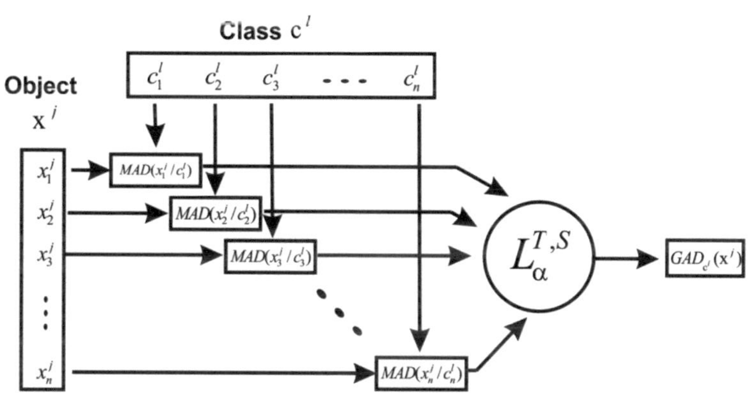

Figure 1. LAMDA basic structure.

Membership functions, denoted as $\mu_X(x)$, are used to associate a degree of membership of each of the elements of the domain to the corresponding fuzzy set. This degree of membership indicates the certainty (or uncertainty) that the element belongs to that set. Membership functions for fuzzy sets can be of any shape or type as determined by experts in the domain over which the sets are defined. Only must satisfy the following constraints [27].

- A membership function must be bounded from below0and from above1.
- The range of a membership function must therefore be [0, 1].
- For each$x \in U$, the membership function must be unique. That is, the same element cannot map to different degrees of membership for the same fuzzy set.

The MAD is a membership function derived from a fuzzy generalization of a binomial probability law [26]. As before, $x^j = (x_1, \ldots, x_n)$, and let E be a non-empty, proper subset of X. We have an experiment where the result is considered a "success" if the outcome x_i is in E. Otherwise, the result is considered a "failure". Let $P(E) = \rho$ be the probability of success so $P(E') = q = 1 - \rho$ is the probability of failure; then intermediate values have a degree of success or failure. The probability mass function of X is defined as [28].

$$f(x) = (\rho)^{(x)} (1 - \rho)^{(1-x)} \tag{3}$$

where $\rho \in [0, 1]$. The following Fuzzy Probability Distributions are typically used by LAMDA methodology to calculate the MADs [25],[29].

- Fuzzy Binomial Distribution.
- Fuzzy Binomial-Center Distribution.
- Fuzzy Binomial-Distance Distribution.
- Gaussian Distribution.

2.1.2. *Global Adequacy Degree*

Global Adequacy degree (GAD) is obtained by aggregating or summarizing of all marginal information previously calculated (see Figure 1), using mathematical aggregation operators (T-norms and S-conorms) given N MADs of an object x^j relative to class c^l, through a linear convex T-S function $L_\alpha^{T,S}$. Some T-norms and their dual S-conorm used in LAMDA methodology are shown in Table 1 [22, 23].

The aggregation operators are mathematical objects that have the function of reducing a set of numbers into a unique representative number. This is simply a function, which assigns a real number y to any n-tuple $(x_1, x_2, \ldots x_n)$ of real numbers, $y = A(x_1, x_2, \ldots x_n)$[30].

The T-norms and S-conorms are two families specialized on the aggregation under uncertainty. They can also be seen as a generalization of the Boolean logic connectives to multi-valued logic. The T-norms generalize the conjunctive 'AND' (intersection) operator and the S-conorms generalize the disjunctive 'OR' (union) operator [30].

Linear convex T-S function is part of the so-called compensatory functions, and is utilized to combine a T-norm and a S-conorm in order to compensate their opposite effects. Zimmermann and Zysno [30] discovered that in a decision making context humans neither follow exactly the behavior of a T-norm nor a S-conorm when aggregating. In order to get closer to the human aggregation process, they proposed an operator on the unit interval based on T-norms and S-conorms.

http://dx.doi.org/10.5772/ 52179

Name	T-Norm (Intersection)	S-Conorm (Union)
Min-Max	$\min(x_1, \ldots, x_n)$	$\max(x_1, \ldots, x_n)$
Product	$\prod_{i=1}^{n} x_i$	$1-\left(\prod_{i=1}^{n} x_i\right)$
Lukasiewicz	$\max\left\{1-n+\sum_{i=1}^{n} x_i,\ 0\right\}$	$\min\left\{\sum_{i=1}^{n} x_i,\ 1\right\}$
Yaguer	$1-\min\left\{\left(\sum_{i=1}^{n}(1-x_i)^{\frac{1}{\lambda}}\right)^{\lambda},\ 1\right\}$	$\min\left\{\left(\sum_{i=1}^{n}(x_i)^{\frac{1}{\lambda}}\right)^{\lambda},\ 1\right\}$
Hammacher	$\frac{1}{1+\sum_{i=1}^{n}\left(\frac{1-x_i}{x_i}\right)}$ 0, if it exist $x_i=0$	$\frac{\sum_{i=1}^{n}\left(\frac{x_i}{1-x_i}\right)}{1+\sum_{i=1}^{n}\left(\frac{x_i}{1-x_i}\right)}$ 1, if it exist $x_i=1$

Table 1. T-norms and S-conorms.

One class of non-associative T-norm and T-conorm-based compensatory operator is the linear convex T-S function [31]:

$$L_{\alpha}^{T,S}(x_1,\ldots,x_n)=(\alpha)\cdot T(x_1,\ldots,x_n)+(1-\alpha)\cdot S(x_1,\ldots,x_n) \tag{4}$$

where $\alpha\in[0,\ 1]$, $T\le L_{\alpha}^{T,S}\le S$, $T=L_{1}^{T,S}$(intersection) and $S=L_{0}^{T,S}$ (union). The parameter α is called exigency level [22, 25].

Finally, once computed the GAD of the object $\boldsymbol{x}^{j}$ related to all classes, and according to the MA rule, $\boldsymbol{x}^{j}$ will be placed in the highest adequation degree class [23]. The MA rule is defined as

$$MA=\max\left(GAD_{c^1}\left(x^j\right),GAD_{c^2}\left(x^j\right),\ldots,GAD_{c^l}\left(x^j\right)\right) \tag{5}$$

LAMDA has been applied to different domains: medical images [32], pattern recognition [33], detection and diagnosis of failures of industrial processes [34], biological processes [35], distribution systems of electrical energy [36], processes for drinking water production [29], monitoring and diagnosis of industrial processes [37], selection of sensors [38], vector quantization [39].

3. Image recognition based on Learning Algorithm for Multivariate Data Analysis

In this section the image recognition algorithm based on LAMDA is described. Our proposal is divided into two phases, training and recognition. At training phase, a codebook is generated based on LAMDA algorithm, let us name it LAMDA codebook. At recognition phase, we propose a search algorithm based on LAMDA and we show its application in image recognition process.

3.1. Training phase

The LAMDA codebook is calculated in two stages, see Figure 2.

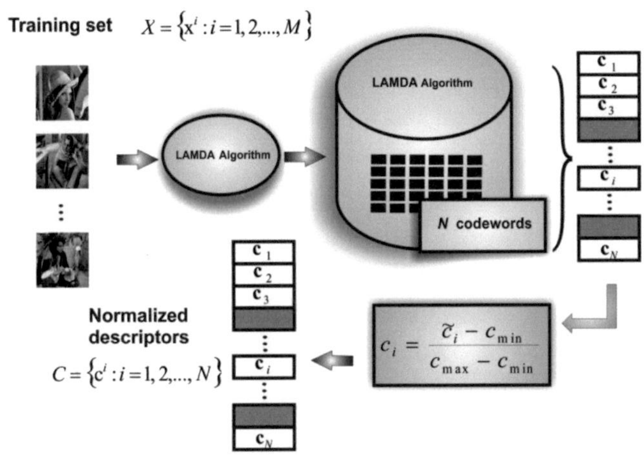

Figure 2. LAMDA codebook generation scheme

Stage 1. *LAMDA codebook generation.* At this stage, a codebook based on LAMDA algorithm is generated. This stage is a supervised process; the training set used in the codebook generation is formed by a set of images.

Let $x=[x_i]_n$ be a vector, which represents an image; the training set is defined as $A=\{x^j : j=1, 2, ..., M\}$. The result of this stage is a codebook denoted as $C=\{c^l : l=1, 2, ..., N\}$, wherec$=[c_i]_n$.

Stage 2. *LAMDA codebook normalization.* Before using the LAMDA codebook, it must be normalized:

$$c_i = \frac{\tilde{c}_i - c_{\min}}{c_{\max} - c_{\min}} = \frac{\tilde{c}_i}{2^L - 1} \tag{6}$$

where $i = 1, 2, \ldots, n$, $\tilde{c}_i$ is the descriptor before normalization, c_i is the normalized descriptor, $0 \le c_i \le 1$, $c_{\min} = 0$ and $c_{\max} = 2^L - 1$; in the context of image processing, L is the number of bits necessary to represent the value of a pixel. The limits (minimum and maximum) of the descriptors values are the limits of the data set.

3.2. Search algorithm for image recognition based on LAMDA

The proposed search algorithm performs the recognition task according to a membership criterion, computed in four stages.

Stage 1. *Image normalization:* Before using the descriptors of the image in the search algorithm LAMDA, it must be normalized:

$$x_i = \frac{\tilde{x}_i - x_{\min}}{x_{\max} - x_{\min}} = \frac{\tilde{x}_i}{2^L - 1} \tag{7}$$

where $i = 1, 2, \ldots, n$, $\tilde{x}_i$ is the descriptor before normalization, x_i is the normalized descriptor, $0 \le x_i \le 1$, $x_{\min} = 0$ and $x_{\max} = 2^L - 1$, L is the number of bits necessary to represent the value of a pixel. The limits (minimum and maximum) of the descriptors values are the limits of the data set.

Stage 2. *Marginal Adequacy Degree (MAD).* MADS are calculated for each descriptor x_i^j of each input vector x^j with each descriptor c_i^l of each class c^l. For this purpose, we can use the following fuzzy probability distributions:

Fuzzy Binomial Distribution:

$$MAD(x_i^j / c_i^l) = \left(\rho_i^l\right)^{\left(x_i^j\right)} \left(1 - \rho_i^l\right)^{\left(1 - x_i^j\right)} \tag{8}$$

where $i = 1, 2, \ldots, n$; $j = 1, 2, \ldots, M$ and $l = 1, 2, \ldots, N$. For all fuzzy probability distributions, $\rho_i^l = c_i^l$.

Fuzzy Binomial-Center Distribution:

$$MAD(x_i^j / c_i^l) = \frac{\left(\rho_i^l\right)^{\left(x_i^j\right)} \left(1 - \rho_i^l\right)^{\left(1 - x_i^j\right)}}{\left(x_i^j\right)^{\left(x_i^j\right)} (1 - x_i^j)^{(1 - x_i^j)}} \tag{9}$$

Fuzzy Binomial-Distance Distribution:

$$MAD(x_i^j / c_i^l) = (a)^{(1-x_{dist})} (1-a)^{(x_{dist})} \tag{10}$$

where $a = \max(\lfloor \rho_i^l \rfloor, \lfloor 1-\rho_i^l \rfloor)$, $\lfloor \cdot \rfloor$ denotes a rounding operation to the largest previous integer value and $x_{dist} = abs(x_i^j - \rho_i^l)$.

Gaussian Function:

$$MAD(x_i^j / c_i^l) = e^{-\frac{1}{2}\left(\frac{x_i^j - \rho_i^j}{\sigma^2}\right)^2} \tag{11}$$

where $\sigma^2 = \frac{1}{n-1}\sum_{i=1}^{n}(x_i^j - \bar{x})^2$ and $\bar{x} = \frac{1}{n}\sum_{i=1}^{n} x_i^j$ are the variance and arithmetic mean of the vector $\mathbf{x}^j$, respectively.

Stage 3. *Global Adequacy Degree (GAD)*. This stage determines the grade of membership of each input vector x^j to each class c^l, by means of a convex linear function (12) and the use of mathematical aggregation operators (T-norms and S-conorms), these are shown in Table 2.

$$GAD_{c^l}(\mathbf{x}^j) = L_\alpha^{T,S} = (\alpha)\cdot T(MAD\ \ (x_i^j / c_i^l)) + (1-\alpha)\cdot S(MAD\ \ (x_i^j / c_i^l)) \tag{12}$$

Operator	T-Norm (Intersection)	S-Conorm (Union)
Min-Max	$\min(MAD(x_i^j / c_i^l))$	$\max(MAD(x_i^j / c_i^l))$
Product	$\prod_{i=1}^{n} MAD(x_i^j / c_i^l)$	$1-\left(\prod_{i=1}^{n} MAD(x_i^j / c_i^l)\right)$

Table 2. Mathematical aggregation operators

Stage 4. *Obtaining the index*. Finally, this stage generates the index of the class to which the input vector belongs. The index is determined by the GAD that presents the maximum value (MA rule):

$$index = \max\left(GAD_{c^1}(\mathrm{x}^j), GAD_{c^2}(\mathrm{x}^j), \ldots, GAD_{c^l}(\mathrm{x}^j)\right) \tag{13}$$

Figure 3 shows the proposed VQ scheme that makes use of the LAMDA algorithm and the codebook generated by LAMDA algorithm.

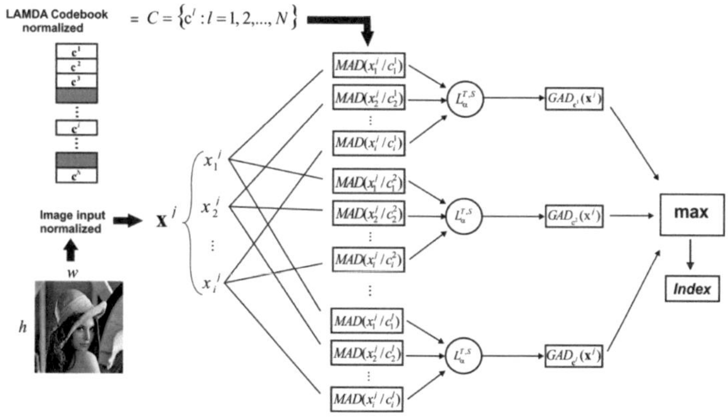

Figure 3. Search algorithm LAMDA

4. Results

In this section, the findings of the implementation of the search algorithm LAMDA, in image recognition of gray-scale are presented. In this implementation the fuzzy probability distributions, binomial and binomial center, and the aggregation operators, product and min-max are only used because only they have a lower computational complexity.

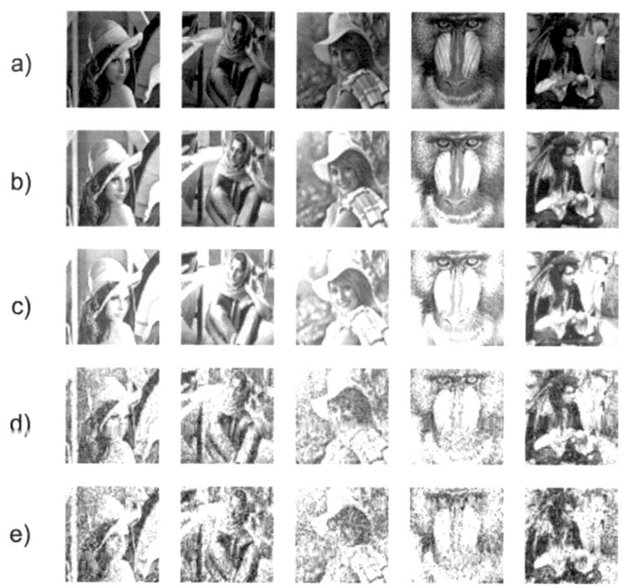

Figure 4. Images of set-1, (a) original image. Altered images, erosive noise (b) 60%, (c) 100%; mixed noise (d) 30 %, (e) 40%

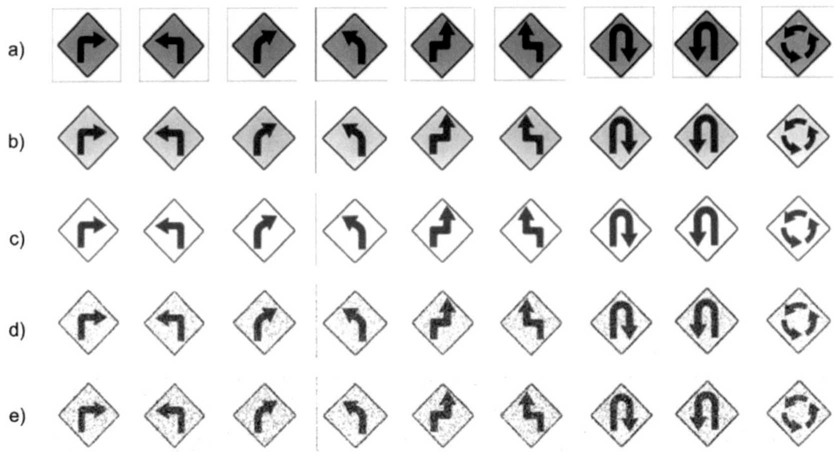

Figure 5. Images of set-2, (a) original image. Altered images, erosive noise (b) 60%, (c) 100%; mixed noise (d) 30 %, (e) 40%

For this experiment we chose two test sets of images, called set-1 and set-2, and their altered versions (see Figures 4, 5). We say that an altered version $\tilde{x}^{\gamma}$ of the image x^{γ} has undergone an *erosive* change whenever$\tilde{x}^{\gamma} \leq x^{\gamma}$, *dilative* change whenever $\tilde{x}^{\gamma} \geq x^{\gamma}$ and *mixed* change when include a mixture of erosive and dilative change. These images were used to training the LAMDA codebook. At this stage, it was determined by means of some tests that if we only used the original images and the altered versions with erosive noise 60%, the best results were obtained for the test images of the set-1. In the case of the test images of the set-2, to obtain the best results we only used the original images and the altered versions with erosive noise 60% and 100%.

To evaluate the proposed search algorithm performance, altered versions of these images distorted by random noise were presented to the classification stage of the search algorithm LAMDA (see Figures 4, 5).

The fact of using two fuzzy probability distributions and two aggregation operators allows four combinations. This way, four versions of the search algorithm LAMDA are obtained: binomial min-max, binomial product, binomial center min-max and binomial center product. Moreover, we proceeded to modify it in the range from 0 to 1 with step 0.1 to determine the value of the level of exigency (α) that provide the best results. Each version of LAMDA was evaluated using two sets of test images. The results of this experiment are shown in Tables 3 and 4.

Table 3 shows the results obtained using the combinations: binomial min-max, binomial product, binomial center min-max y binomial center product and using the set of test images of the set-1.

Image	Fuzzy distribution	Aggregation operator	Exigency level (α)	Distortion percentage added to image				
				original	Erosive noise		Mixed noise	
				0%	60%	100%	30%	40%
				100%	100%	100%	100%	100%
				100%	100%	100%	100%	100%
	Binomial	Min-max	1	100%	100%	100%	100%	100%
				100%	100%	100%	100%	100%
				100%	100%	100%	100%	100%
				100%	100%	100%	100%	100%
				0%	0%	0%	0%	0%
	Binomial	Product	0-1	0%	0%	0%	0%	0%
				0%	0%	0%	0%	0%
				0%	0%	0%	0%	0%
				100%	100%	100%	100%	100%
				100%	100%	100%	100%	100%
	Binomial center	Min-max	1	100%	100%	100%	100%	100%
				100%	100%	100%	100%	100%
				100%	100%	100%	0%	0%
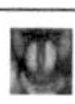				100%	100%	100%	100%	100%
				100%	100%	0%	0%	0%
	Binomial center	Product	1	100%	100%	100%	100%	0%
				100%	100%	0%	0%	0%
				100%	0%	0%	0%	0%

Table 3. Performance results (recognition rate) showed by the proposed search algorithm withaltered versions of the test images of set-1

Image	Fuzzy distribution	Aggregation operator	Exigency level (α)	Distortion percentage added to image				
original		Erosive noise			Mixed noise			
0%	60%	100%		30%		40%		
				100%	100%	100%	100%	100%
				100%	100%	100%	100%	100%
				100%	100%	100%	100%	100%
				100%	100%	100%	100%	100%
	Binomial	Min-max	1	100%	100%	100%	100%	100%
				100%	100%	100%	0%	100%
				100%	100%	100%	100%	100%
				100%	100%	100%	100%	100%
				100%	100%	100%	100%	100%
				100%	100%	100%	100%	100%
				0%	0%	0%	0%	0%
				0%	0%	0%	0%	0%
				0%	0%	0%	0%	0%
	Binomial	Product	1	0%	0%	0%	0%	0%
				0%	0%	0%	0%	0%
				0%	0%	0%	0%	0%
				0%	0%	0%	0%	0%
				0%	0%	0%	0%	0%
				100%	100%	100%	100%	100%
				100%	100%	100%	100%	100%
	Binomial center	Min-max	1	100%	100%	100%	100%	100%
				100%	100%	100%	100%	100%
				100%	100%	100%	100%	100%

http://dx.doi.org/10.5772/ 52179

Image	Fuzzy distribution	Aggregation operator	Exigency level (α)	Distortion percentage added to image				
				100%	100%	100%	0%	100%
				100%	100%	100%	100%	100%
				100%	100%	100%	100%	100%
				100%	100%	100%	100%	100%
				100%	100%	100%	100%	100%
				0%	0%	0%	0%	0%
				0%	0%	0%	0%	0%
				0%	0%	0%	0%	0%
	Binomial center	Product	1	0%	0%	0%	0%	0%
				0%	0%	0%	0%	0%
				0%	0%	0%	0%	0%
				0%	0%	0%	0%	0%
				0%	0%	0%	0%	0%

Table 4. Performance results (recognition rate) showed by the proposed search algorithm withaltered versions of the test images of set-2 .

In the case of the combination of the binomial distribution with the aggregation operator min-max, the best results were obtained with a value of exigency level in the range from 0.8 to 1. We chose the exigency level equal to 1. As a result, the linear convex function is reduced by half, and, consequently, the number of operations is reduced. On the other hand, the combination of the binomial distribution with the aggregation operator product was unable to perform the classification.

In the combination of the binomial center distribution with the aggregation operator min-max, the best results were obtained with a value of exigency level in the range from 0.1 to 1. We chose the exigency level equal to 1. This way, the linear convex function is reduced by half thus reducing the number of operations.

On the other hand, using the combination of the binomial center distribution with the aggregation operator product, the best results were obtained with a value of exigency level equal to 1. Although, as it is shown in Table 3, the classification is not efficient with the images altered with erosive noise of 100% and with mixed noise of 30% and 40%. With this combination, the best results were obtained in comparison to the combination of the binomial distribution with the aggregation operator product.

Table 4 show the results obtained using the combinations: binomial min-max, binomial product, binomial center min-max and binomial center product and using the set of test images of the set-2.

For the combination of the binomial distribution with the aggregation operator min-max, the best results were obtained with a value of exigency level in the range from 0.7 to 1. With the exigency level equal to 1, the linear convex function is reduced by half thus reducing the number of operations. On the other hand, the combination of the binomial distribution with the aggregation operator product was unable to perform classification.

In the combination of the binomial center distribution with the aggregation operator min-max, the best results were obtained with a value of exigency level in the range from 0.1 to 1. Choosing the exigency level equal to 1, the linear convex function is reduced by half and the number of operations is reduced too. On the other hand, the combination of the binomial center distribution with the aggregation operator product was unable to perform classification.

5. Conclusions

In this Chapter, we have proposed the use of LAMDA methodology as a search algorithm for image recognition. It is important to mention that we used LAMDA algorithm both in the training phase and in the recognition phase.

The advantage of the LAMDA algorithm is its versatility which allows obtaining different versions making the combination of fuzzy probability distributions and aggregation operators. Furthermore, it also has the possibility to vary the exigency level, and we can locate the range or the value of the exigency level where the algorithm has better results.

As it was shown in Tables 3 and 4, the search algorithm is competitive, since acceptable results were obtained in the combinations: binomial min-max, binomial center min-max with both sets of images. As you can see the product aggregation operator was not able to perform the recognition. In both combinations the exigency level was equal to 1, this fact allowed to reduce the linear convex function.

Finally, from these two combinations it is better to choose the binomial min-max, because with this combination fewer operations are performed.

Author details

Juan G. Zambrano[1], E. Guzmán-Ramírez[2] and Oleksiy Pogrebnyak[2*]

*Address all correspondence to: mmortari@unb.br

*Address all correspondence to: olek@cic.ipn.mx

1 Universidad Tecnológica de la Mixteca, México

2 Centro de Investigación en Computación, IPN, México

References

[1] Kala, R., Shukla, A., Tiwari, R., Rungta, S., & Janghel, R. R. (2009). Mobile Robot Navigation Control in Moving Obstacle Environment Using Genetic Algorithm, Artificial Neural Networks and A* Algorithm. *World Congress on Computer Science and Information Engineering*, 4, 705-13.

[2] Zhu, Y., Yuan, Q., Wang, Q., Fu, Y., & Wang, H. (2009). Radar HRRP Recognition Based on the Wavelet Transform and Multi-Neural Network Fusion. *Electronics Optics & Control*, 16(1), 34-8.

[3] Esbati, H., & Shirazi, J. (2011). Face Recognition with PCA and KPCA using Elman Neural Network and SVM. World Academy of Science, Engineering and Technology.; , 58, 174-8.

[4] Bowyer, K. W., Hollingsworth, K., & Flynn, P. J. (2008). Image understanding for iris biometrics: A survey. *Computer Vision and Image Understanding*, 110(2), 281-307.

[5] Anagnostopoulos-N, C., Anagnostopoulos, E., Psoroulas, I. E., Loumos, I. D., Kayafas, V., & , E. (2008). License Plate Recognition From Still Images and Video Sequences: A Survey. *IEEE Transactions on Intelligent Transportation Systems*, 9(3), 377-91.

[6] Abrishambaf, R., Demirel, H., & Kale, I. (2008). A Fully CNN Based Fingerprint Recognition System. *11th International Workshop on Cellular Neural Networks and their Applications*, 146-9.

[7] Egmont-Petersen, M., Ridder, D., & Handels, H. (2002). Image processing with neural networks-a review. *Pattern Recognition Letters*, 35(10), 2279-301.

[8] Bhowmik, M. K., Bhattacharjee, D., Nasipuri, M., Basu, D. K., & Kundu, M. (2009). Classification of Fused Images using Radial Basis Function Neural Network for Human Face Recognition. *World Congress on Nature & Biologically Inspired Computing*, 19-24.

[9] Yang, J., Zhang, D., Frangi, A., & Yang, J-y. (2004). Two Dimensional PCA: A New Approach to Appearance-Based Face Representation and Recognition. *IEEE Transactions on Pattern Analysis and Machine Intelligence*, 26(1), 131-7.

[10] Gottumukkal, R., & Asari, V. (2004). An improved face recognition technique based on modular PCA approach. *Pattern Recogn Letters*, 25(4), 429-36.

[11] Bezdek, J. C., Keller, J., Krisnapuram, R., & Pal, N. (2005). Fuzzy Models and Algorithms for Pattern Recognition and Image Processing (The Handbooks of Fuzzy Sets):. *Springer-Verlag New York, Inc.*

[12] Mitchell, H. B. (2005). Pattern recognition using type-II fuzzy sets. *Inf. Sciences*, 170(2-4), 409-18.

[13] Bandyopadhyay, S., & Maulik, U. (2002). Geneticclustering for automaticevolution of clusters and application to imageclassification. *Pattern Recognition*, 35(6), 1197-208.

[14] Bhattacharya, M., & Das, A. (2010). Genetic Algorithm Based Feature Selection In a Recognition Scheme Using Adaptive Neuro Fuzzy Techniques. *Int. Journal of Computers, Communications & Control* [4], 458-468.

[15] Pandit, M., & Gupta, M. (2011). Image Recognition With the Help of Auto-Associative Neural Network. *International Journal of Computer Science and Security*, 5(1), 54-63.

[16] Yang, J. C., & Park, D. S. (2008). Fingerprint Verification Based on Invariant Moment Features and Nonlinear BPNN. *International Journal of Control, Automation, and Systems*, 6(6), 800-8.

[17] Radha, V., & Nallammal, N. (2011). Neural Network Based Face Recognition Using RBFN Classifier. *Proceedings of the World Congress on Engineering and Computer Science*, 1.

[18] Zeng, J., & Liu-Q, Z. (2007). Type-2 Fuzzy Sets for Pattern Recognition: The State-of-the-Art. *Journal of Uncertain Systems*, 11(3), 163-77.

[19] Sarawat Anam., Md, Shohidul, Islam. M. A., Kashem, M. N., Islam, M. R., & Islam, M. S. (2009). Face Recognition Using Genetic Algorithm and Back Propagation Neural Network. *Proceedings of the International MultiConference of Engineers and Computer Scientists*, 1.

[20] Blahuta, J., Soukup, T., & Cermak, P. (2011). The image recognition of brain-stem ultrasound images with using a neural network based on PCA. *IEEE International Workshop on Medical Measurements and Applications Proceedings*, 5(2), 137-42.

[21] Yu, H., Xie, T., Hamilton, M., & Wilamowski, B. (2011). Comparison of different neural network architectures for digit image recognition. *4th International Conference on Human System Interactions*, 98-103.

[22] Piera, N., Desroches, P., & Aguilar-Martin, J. (1989). LAMDA: An Incremental Conceptual Clustering Method. *LAAS Laboratoired'Automatiqueetd'Analyse des Systems.;Report* [89420], 1-21.

[23] Piera, N., & Aguilar-Martin, J. (1991). Controlling Selectivity in Nonstandard Pattern Recognition Algorithms. *IEEE Transactions on Systems, Man and Cybernetics*, 21(1), 71-82.

[24] Aguilar-Martin, J., Sarrate, R., & Waissman, J. (2001). Knowledge-based Signal Analysis and Case-based Condition Monitoring of a Machine Tool. *Joint 9th IFSA World Congress and 20th NAFIPS International Conference Proceedings*, 1, 286-91.

[25] Aguilar-Martin, J., Agell, N., Sánchez, M., & Prats, F. (2002). Analysis of Tensions in a Population Based on the Adequacy Concept. *5th Catalonian Conference on Artificial Intelligence, CCIA*, 2504, 17-28.

[26] Waissman, J., Ben-Youssef, C., & Vázquez, G. (2005). Fuzzy Automata Identification Based on Knowledge Discovery in Datasets for Supervision of a WWT Process. *3rd International Conference on Sciences of Electronic Technologies of Information and Telecommunications.*

[27] Engelbrecht, A. P. (2007). Computational intelligence. *Anintoduction: John Wiley & Sons Ltd.*

[28] Buckley, J. J. (2005). Simulating Fuzzy Systems. *Kacprzyk J, editor: Springer-Verlag Berlin Heidelberg.*

[29] Hernández, H. R. (2006). Supervision et diagnostic des procédés de productiond'eau potable. *PhD thesis.l'Institut National des Sciences Appliquées de Toulouse.*

[30] Detyniecki, M. (2000). Mathematical Aggregation Operators and their Application to Video Querying. *PhD thesis.Université Pierre et Marie Curie.*

[31] Beliakov, G., Pradera, A., & Calvo, T. (2007). Aggregation Functions: A Guide for Practitioners. *Kacprzyk J, editor: Springer-Verlag Berlin Heidelberg.*

[32] Chan, M., Aguilar-Martin, J., Piera, N., Celsis, P., & Vergnes, J. (1989). Classification techniques for feature extraction in low resolution tomographic evolutives images: Application to cerebral blood flow estimation. *In 12th Conf GRESTI Grouped'Etudes du Traitement du Signal et des Images.*

[33] Piera, N., Desroches, P., & Aguilar-Martin, J. (1990). Variation points in pattern recognition. *Pattern Recognition Letters*, 11, 519-24.

[34] Kempowsky, T. (2004). Surveillance de Procedes a Base de Methodes de Classification: Conception d'un Outild'aide Pour la Detection et le Diagnostic des Defaillances. *PhD Thesis. l'Institut National des Sciences Appliquées de Toulouse.*

[35] Atine-C, J., Doncescu, A., & Aguilar-Martin, J. (2005). A Fuzzy Clustering Approach for Supervision of Biological Processes by Image Processing. *EUSFLAT European Society for Fuzzy Logic and Technology*, 1057-63.

[36] Mora, J. J. (2006). Localización de fallas en sistemas de distribución de energía eléctrica usando métodos basados en el modelo y métodos basados en el conocimiento. *PhDThesis. Universidad de Girona.*

[37] Isaza, C. V. (2007). Diagnostic par Techniquesd'apprentissageFloues :Conceptiond'uneMethode De Validation Et d'optimisation des Partitions. *PhD Thesis. l'Université de Toulouse.*

[38] Orantes, A., Kempowsky, T., Lann-V, M., Prat, L., Elgue, L., Gourdon, S., Cabassud, C., & , M. (2007). Selection of sensors by a new methodology coupling a classification technique and entropy criteria. *Chemical engineering research & design Journal*, 825-38.

[39] Guzmán, E., Zambrano, J. G., García, I., & Pogrebnyak, O. (2011). LAMDA Methodology Applied to Image Vector Quantization. *Computer Recognition Systems 4*, 95, 347-56.

Chapter 2

Ant Algorithms for Adaptive Edge Detection

Aleksandar Jevtić and Bo Li

Additional information is available at the end of the chapter

http://dx.doi.org/10.5772/52792

1. Introduction

Edge detection is a pre-processing step in applications of computer and robot vision. It transforms the input image to a binary image that indicates either the presence or the absence of an edge. Therefore, the edge detectors represent a special group of search algorithms with the objective of finding the pixels belonging to true edges. The search is performed following certain criteria, as the edge pixels are found in regions of an image where the distinct intensity changes or discontinuities occur (e.g. in color, gray-intensity level, texture, etc.).

In applications domains such as robotics, vision-based sensors are widely used to provide information about the environment. On mobile robots, images from sensors are processed to detect and track the objects of interest and allow safe navigation. The purpose of edge detection is to segment the image in order to extract the features and objects of interest. No matter what method is applied, the objective remains the same, to change the representation of the original image into something easier to analyze. Digital images may be obtained under different lighting conditions and using different sensors. These may produce noise and deteriorate the segmentation results.

In recent years, algorithms based on swarming behavior of animal colonies in nature have been applied to edge detection. Swarm Intelligence algorithms use the bottom-up approach; the patterns that appear at the system level are the result of local interactions between its lower-level components [2]. The initial purpose of Swarm Intelligence algorithms was to solve optimization problems [7], but recent studies show they can be a useful image-processing tool. The emerging properties inherent to swarm intelligence make these algorithms adaptive to the changing image patterns. This is a useful feature for real-time image processing.

In this work, two edge-detection methods inspired by the foraging behavior of natural ant colonies are presented. Ants use pheromone trails to mark the path to the food source. In digital images, pixels define the discrete space in which the artificial ants move and the edge pixels represent the food. The edge detection operation is performed on a set of grayscale images. The first proposed method extracts the edges from the original grayscale image. The second method finds the missing broken-edge segments and can be used as a complementary tool in order to improve the edge-detection results. Finally, the study on the adaptability of the first edge detector is performed using a set of grayscale images as a dynamically changing environment.

The chapter is organized as follows. Section 2 provides an overview of the state-of-the-art edge detectors. Section 3 introduces the basic Ant System algorithm. In Section 4 the proposed Ant System-based edge detector is described. The discussion of the simulation results is also given in this section. Follows the description of the proposed broken-edge linking algorithm and the simulation results in Section 5. The study on the adaptability of the proposed Ant System-based edge detector is given in Section 6. Finally, in Section 7 the conclusions are made.

2. Related work

Edges represent important contour features in the image since they are the boundaries where distinct intensity changes or discontinuities occur. In practice, it is difficult to design an edge detector capable of finding all the true edges in image. Edge detectors give ambiguous information about the location of object boundaries for which they are usually subjectively evaluated by the observers [30].

Several conventional edge detection methods have been widely cited in literature. The Prewitt operator [25] extracts contour features by fitting a Least Squares Error (LSE) quadratic surface over an image window and differentiate the fitted surface. The edge detectors proposed in [31] and [3] use local gradient operators, sometimes with additional smoothing for noise removal. The Laplacian operator [9] applies a second order differential operator to find edge points based on the zero crossing properties of the processed edge points.

Although conventional edge detectors usually perform linear filtering operations, there are various nonlinear methods proposed. In [23], authors proposed an edge detection method based on the Parameterized Logarithmic Image Processing (PLIP) and a four directional Sobel detector, achieving a higher level of independence from scene illumination. In [10], an edge detector based on bilateral filtering was proposed, which achieves better performance than single Gaussian filtering. In [21], authors proposed using Coordinate Logic Filters (CLF) to extract the edges from images. CLF constitute a class of nonlinear digital filters that are based on the execution of Coordinate Logic Operations (CLO). An alternative method for calculating CLF using Coordinate Logic Transforms (CLT) was introduced in [4]; the authors presented a new threshold-based technique for the detection of edges in grayscale images.

In recent years, Swarm Intelligence algorithms have shown its full potential in terms of flexibility and autonomy, especially when it comes to design and control of complex systems that consist of a large number of agents. Metaheuristics such as Ant Colony Optimization (ACO) [6], Particle Swarm Optimization (PSO) [17] and Bees Algorithm (BA) [24] include sets of algorithms that demonstrate emergent behavior as a result of local interactions between the members of the swarm. They tend to be decentralized, self-organized, autonomous and adaptive to the changes in the environment. The adaptability and the ability to learn are very important for systems that are designed to be autonomous.

ACO is a metaheuristic that exploits the self-organizing nature of real ant colonies and their foraging behavior to solve discrete optimization problems. The learning ability, in natural and artificial ant colonies, consists in storing information about the environment by laying pheromone on the path that leads to a food source. The emerging pheromone structures serve as the swarm's external memory that can be used by any of its members. Although a single ant can only detect the local environment, the designer of a swarm-based system can observe the emergent global patterns that are a result of the cooperative behavior.

ACO algorithms have been applied to image processing. Some of the proposed applications include image retrieval [28] and image segmentation [11, 14, 18]. Several ACO-based edge detection methods have also been proposed in literature. Among others, these include modifications to Ant System (AS) [22] or Ant Colony System (ACS) algorithms [1, 8, 32] for a digital image habitat, combined with local gray-intensity comparison for different pixel's neighborhood matrices. Some studies showed that an improved detection can be obtained using a hybrid approach with an artificial neural network classifier [26].

In order to apply artificial ant colonies to edge detection one needs to set the rules for local interactions between the ants and define the "food" that ants will search for. For the edge detection problem, the food are the edge pixels in digital images.

3. Ant System algorithm

Artificial ants, unlike their biological counterparts, move through a discrete environment defined with nodes, and they have memory. When traversing from one node to another, ants leave pheromone trails on the edges connecting the nodes. The pheromone trails attract other ants that lay more pheromone, which consequently leads to pheromone trail accumulation. Negative feedback is applied through pheromone evaporation that, importantly, restrains the ants from taking the same route and allows continuous search for better solutions.

Ant System (AS) is the first ACO algorithm proposed in literature and it was initially applied to the Travelling Salesman Problem (TSP) [5]. A general definition of the TSP is the following. For a given set of cities with known distances between them, the goal is to find the shortest tour that allows each city to be visited once and only once. In more formal terms, the goal is to find the Hamiltonian tour of minimal length on a fully connected graph.

AS consists of a colony of artificial ants that move between the nodes (cities) in search for the minimal route. The probability of displacing the kth ant from node i to node j is given by:

$$p_{ij}^{k} = \begin{cases} \frac{(\tau_{ij})^{\alpha}(\eta_{ij})^{\beta}}{\sum_{h \notin tabu_k,} (\tau_{ih})^{\alpha}(\eta_{ih})^{\beta}} & \text{if } j \notin \text{tabu}_k \\ 0 & \text{otherwise} \end{cases} \tag{1}$$

where τ_{ij} and η_{ij} are the intensity of the pheromone trail on edge (i,j) and the visibility of the node j from node i, respectively, and α and β are control parameters ($\alpha, \beta > 0$; $\alpha, \beta \in \Re$). The tabu$_k$ list contains nodes that have already been visited by the kth ant. The definition of the node's visibility is application-related, and for the TSP it is set to be inversely proportional to the node's Euclidean distance:

$$\eta_{ij} = \frac{1}{d_{ij}} \tag{2}$$

It can be concluded from the equations (1) and (2) that the ants favor the edges that are shorter and contain a higher concentration of pheromone.

AS is performed in iterations. At the end of each iteration, pheromone values are updated by all the ants that have built a solution in the iteration itself. The pheromone update rule is described with the following equation:

$$\tau_{ij(new)} = (1-\rho)\tau_{ij(old)} + \sum_{k=1}^{m} \Delta\tau_{ij}^{k} \tag{3}$$

where ρ is the pheromone evaporation rate ($0 < \rho < 1$, $\rho \in \Re$), m is the number of ants in the colony, and $\Delta\tau_{ij}^{k}$ is the amount of pheromone laid on the edge (i,j) by the kth ant, and is given by:

$$\Delta\tau_{ij}^{k} = \begin{cases} \frac{Q}{L_k} & \text{if edge } (i,j) \text{ is traversed by the k}th \text{ ant} \\ 0 & \text{otherwise} \end{cases} \tag{4}$$

where L_k is the length of the tour found by the kth ant, and Q is a scaling constant ($Q > 0$, $Q \in \Re$).

The algorithm stops when the satisfactory solution is found or when the maximum number of iterations is reached.

4. Ant System-based edge detector

In this section, the AS-based edge detector proposed by [15] is described. The method generates a set of images from the original grayscale image using a nonlinear image enhancement technique called Multiscale Adaptive Gain [19], and then the modified AS algorithm is applied to detect the edges on each of the extracted images. The resulting set of pheromone-trail matrices is summed to produce the output image. Threshold and edge thinning, which are optional steps, are finally applied to obtain a binary edge image. The block diagram of the proposed method is shown in Figure 1.

4.1. Multiscale Adaptive Gain

Image enhancement techniques emphasize important features in the image while reducing the noise. Multiscale Adaptive Gain is applied to obtain contrast enhancement by suppressing pixels with the grey intensity values of very small amplitude and enhancing only those pixels with values larger than a certain threshold within each level of the transform space. The nonlinear operation is described with the following equation:

$$G(I) = A[sigm(k(I-B)) - sigm(-k(I+B))] \tag{5}$$

where

$$A = \frac{1}{sigm(k(1-B)) - sigm(-k(1+B))} \tag{6}$$

where $I = I(i,j)$ is the grey value of the pixel at (i,j) of the input image and $sigm(x)$ is defined as

$$sigm(x) = \frac{1}{1+e^{-x}} \tag{7}$$

and B and k control the threshold and rate of enhancement, respectively. ($0 < B < 1$, $B \in \Re$; $k \in \aleph$). The transformation function (5) relative to the original image pixel values is shown in Figure 2. It can be observed that $G(I)$ is continuous and monotonically increasing; therefore, the enhancement will not introduce new discontinuities into the reconstructed image.

4.2. Ant System algorithm for edge detection

The generic Ant System algorithm described in Section 3 was used as a base for the proposed edge detector. In digital images, discrete environment in which the ants move is defined by pixels, i.e. their gray-intensity values, $0 \leq I(i,j) \leq I_{max}$, $i = 1,2,\ldots,N$; $j = 1,2,\ldots,M$. Possible ant's moves to the neighboring pixels are shown in Figure 3.

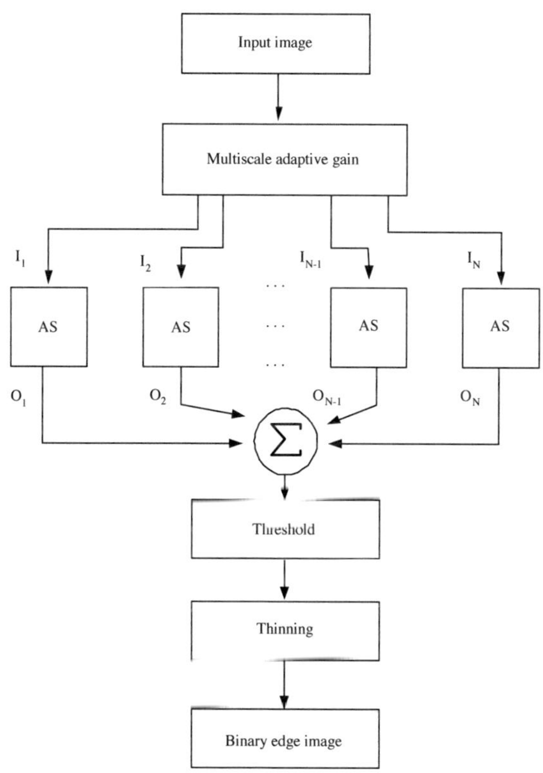

Figure 1. Block diagram of the proposed edge detection method

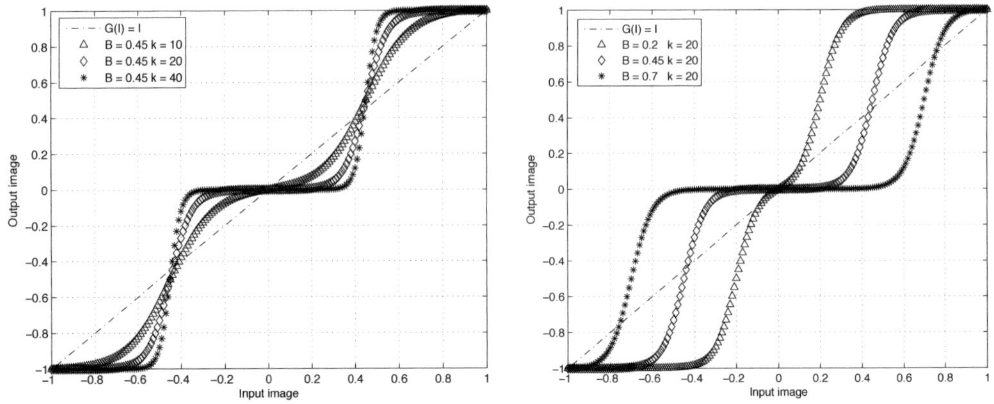

(a) (b)

Figure 2. Transformation function $G(I)$ in respect to the original image pixel values: (a) $B = 0.45$; $k = 10$, 20 and 40; (b) $B = 0.2$, 0.45 and 0.7; $k = 20$.

Unlike the cities' visibility in the TSP, the visibility of the pixel at (i, j) is defined as follows:

$$\eta_{ij} = \frac{1}{I_{max}} \cdot max \begin{bmatrix} |I(i-1, j-1) - I(i+1, j+1)|, \\ |I(i-1, j+1) - I(i+1, j-1)|, \\ |I(i, j-1) - I(i, j+1)|, \\ |I(i-1, j) - I(i+1, j)| \end{bmatrix} \tag{8}$$

where I_{max} is the maximum gray-intensity value in the image ($0 \leq I_{max} \leq 255$). For the pixels in regions of distinct gray-intensity changes the higher visibility values are obtained, which makes those pixels more attractive to ants.

The AS algorithm is an iterative process which includes the following steps:

1. Initialization: the number of ants proportional to $\sqrt{N \cdot M}$ is randomly distributed on the pixels in the image. Only one ant is allowed to reside on a pixel within the same iteration. Initial non-zero pheromone trail value, τ_0, is assigned to each pixel, otherwise the ants would never start the search.

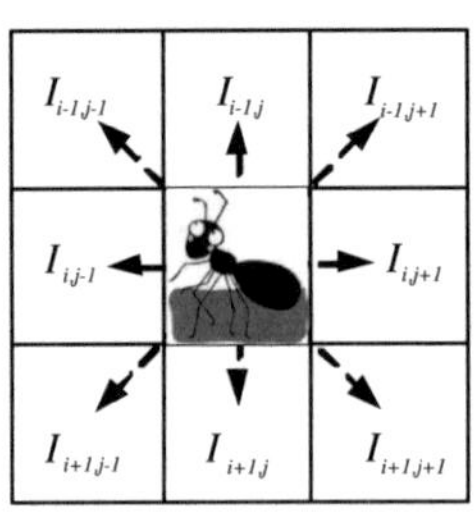

Figure 3. Proposed pixel transition model

2. Pixel transition rule: Unlike their biological counterparts, artificial ants have memory. $Tabu_k$ represents the list of pixels that the kth ant has already visited. If ant is found surrounded by the pixels that are either in the tabu list or occupied by other ants, it is randomly displaced to another unoccupied pixel that is not in the tabu list. Otherwise, the displacement probability of the kth ant to a neighboring pixel (i,j) is given by:

$$p^k_{(i,j)} = \begin{cases} \frac{(\tau_{ij})^\alpha (\eta_{ij})^\beta}{\sum_u \sum_v (\tau_{uv})^\alpha (\eta_{uv})^\beta} & (i,j) \text{ and } (u,v) \text{ are allowed nodes} \\ 0 & \text{otherwise} \end{cases} \tag{9}$$

where τ_{ij} and η_{ij} are the intensity of the pheromone trail and the visibility of the pixel at (i,j), respectively, and α and β are control parameters ($\alpha, \beta > 0$; $\alpha, \beta \in \Re$).

3. Pheromone update rule: Negative feedback is implemented through pheromone evaporation according to:

$$\tau_{ij(new)} = (1-\rho)\tau_{ij(old)} + \Delta\tau_{ij} \tag{10}$$

where

$$\Delta\tau_{ij} = \sum_{k=1}^{m} \Delta\tau^k_{ij} \tag{11}$$

and

$$\Delta\tau^k_{ij} = \begin{cases} \eta_{ij} \text{ if } \eta_{ij} \geq T \text{ and } k\text{th ant displaces to pixel } (i,j) \\ 0 \quad \text{otherwise.} \end{cases} \tag{12}$$

T is a threshold value which prevents ants from staying on the background pixels hence enforcing the search for the true edges. The existence of the pheromone evaporation rate, ρ, prevents the algorithm stagnation. Pheromone trail evaporates exponentially from the repeatedly not-visited pixels.

4. Stopping criterion: The steps 2 and 3 are repeated in a loop and algorithm stops executing when the maximum number of iterations is reached.

4.3. Simulation results and discussion

The proposed method was tested on four different grayscale images of 256×256 pixels resolution: "Cameraman", "Lena", "House" and "Peppers". As seen from the block diagram in Figure 1, first the Multiscale Adaptive Gain defined in (5) is applied to the input image: $0 \leq I(i,j) \leq I_{max}$, $i = 1,2,\ldots,N$; $j = 1,2,\ldots,M$. ($N = M = 256$.) The values of B and k were varied to obtain a set of nine enhanced images: $B = \{0.2, 0.45, 0.7\}$; $k = \{10, 20, 40\}$.

Figure 4. Effects of the transformation function $G(I)$; "Cameraman", 256 × 256 pixels: (a) original image; (b) $B = 0.2$, $k = 10$; (c) $B = 0.45$, $k = 20$; (d) $B = 0.7$, $k = 40$.

The effects of the transformation function on the image "Cameraman" are shown in Figure 4. It can be observed that, by changing the transformation function's parameters, some features in the image become highlighted while others get attenuated.

Afterwards, the AS-based edge detector is applied to each of the nine enhanced images. The algorithm's parameters are set as proposed in [22]: $\tau_0 = 0.01$, $\alpha = 1$, $\beta = 10$, $\rho = 0.05$ and $T = 0.08$. The number of ants equal to $\sqrt{N \cdot M} = 256$ was randomly distributed over the pixels in the image with the condition that no two ants were placed on the same pixel. The memory (tabu list) length for each ant was set to 10. The algorithm was stopped after 100 iterations generating a pheromone-trail matrix of the same resolution as the original image. After each of the nine enhanced images was processed, the sum of the pheromone-trail matrices produced the final pheromone-trail image (Figure 5(e)–(h)). The parameters values such as the number of ants, the memory length, and the number of iterations were obtained as a result of trial and error and their further optimization will be a part of future work.

The effectiveness of the proposed method was compared with the ant-based edge detectors proposed by Tian *et al.* [32] and Nezamabadi-pour *et al.* [22], and the results are shown in Figure 6. To provide a fair comparison, the threshold and morphological edge-thinning operations are neglected. The simulation results show that the proposed method outperforms the other two methods in terms of visual quality of the extracted edge information and sensitivity to weaker edges. The qualitative results of the edge detector proposed in [8] were presented after applying the thinning step, hence a fair comparison with the here-presented results could not be made. It is worth mentioning that the number of iterations used in experiments in [8] was much higher (1000 iterations) than required by our algorithm. The performance evaluation is given in Subsection 4.3.1.

The main contribution of the proposed edge-detection method is the preprocessing step and the parallel execution of the Ant System-based edge detector on a set of images that finally produce the output edge image. The execution time of the proposed method is high for real-time image processing, which would require additional algorithm's code optimization in a different programming environment. The presented experiments were performed in Matlab software that offers an easy high-level implementation but is ineffective in terms of speed.

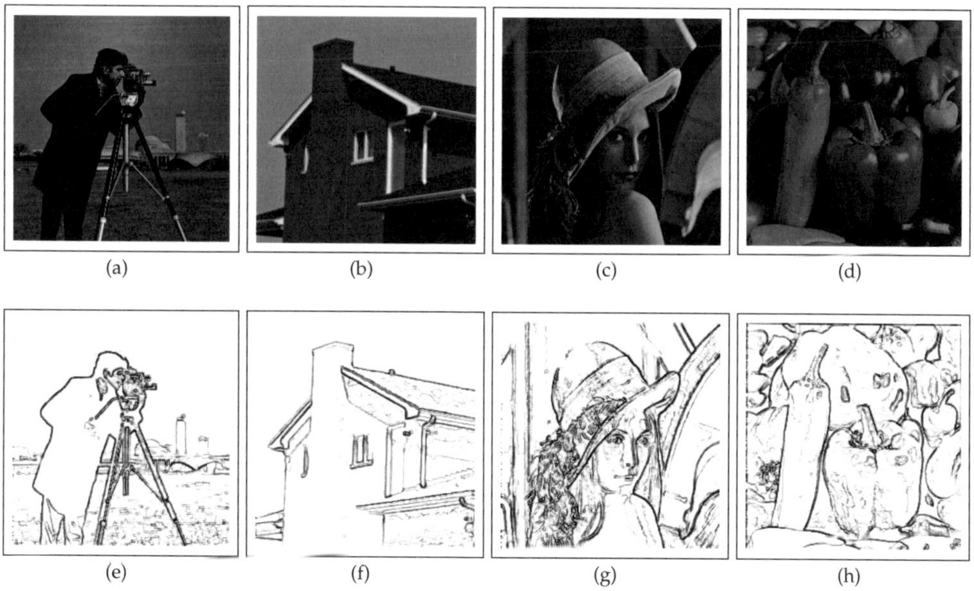

Figure 5. Qualitative results of the proposed method, 256 × 256 pixel images: (a) "Cameraman" original image; (b) "House" original image; (c) "Lena" original image; (d) "Peppers" original image; (e) "Cameraman" pheromone trail image; (f) "House" pheromone trail image; (g) "Lena" pheromone trail image; (h) "Peppers" pheromone trail image.

4.3.1. Performance evaluation

In the complexity-performance trade-off, it was found that varying the values of algorithm's parameters can affect its performance. A set of experiments was performed on a synthetic test image (Figure 7) to show how the number of ants and iterations will be related to the number of detected edge points. The results of this analysis are shown in Figure 8. The number of ants is proportional to the square root of the image resolution $n = \sqrt{N \cdot M}$. The number of edge points was 780.

It can be observed that when the number of ants was increased, the required number of iterations was reduced to achieve a similar performance. Figure 8 shows that the algorithm needs more than 130 iterations to reach good performance when the number of ants was

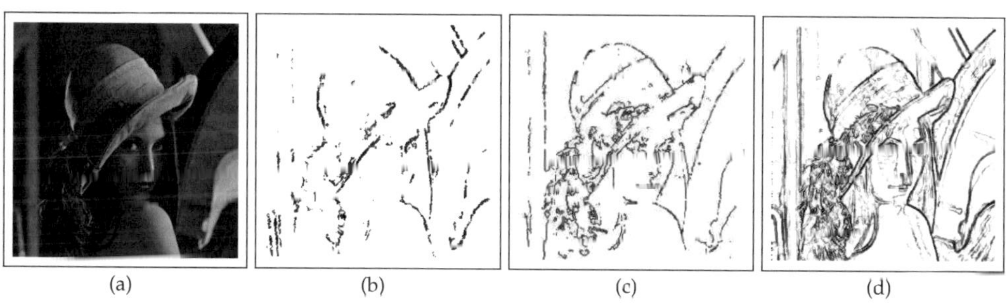

Figure 6. Comparative results with other ant-based edge detectors, "Lena" 256 × 256 pixels: (a) original image; (b) Tian *et al.*; (c) Nezamabadi-pour *et al.*; (d) the proposed method.

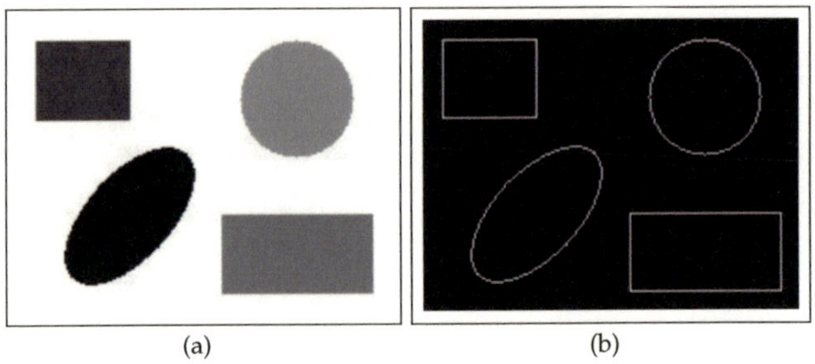

Figure 7. Test image, 256 × 256 pixels: (a) original image; (b) ground-truth edge image.

set to $1 \cdot n$. However, the results not presented here showed that the algorithm was able to detect the maximal number of edge pixels after 400 iterations. Future work may include the optimization of parameters with respect to the computation time.

5. Ant System-based broken-edge linking algorithm

Conventional image edge detection always results in missing edge segments. Broken-edge linking is an improvement technique that is complementary to edge detection. It is used to connect the broken edges in order to form the closed contours that separate the regions of interest. The detection of the missing edge segments is a challenging task. A missing segment is sought between two endpoints where the edge is broken. The noise that is present in the original image may limit the performance of edge-linking algorithms.

Many broken-edge linking techniques have been proposed to compensate the edges that are not fully connected by the conventional edge detectors. [16] applied morphological image enhancement techniques to detect and preserve thin-edge features in the low contrast regions of an image. [33] applied Sequential Edge-Linking (SEL) algorithm that provided full connectivity of the edges but for a rather simplified two-region edge-detection problem.

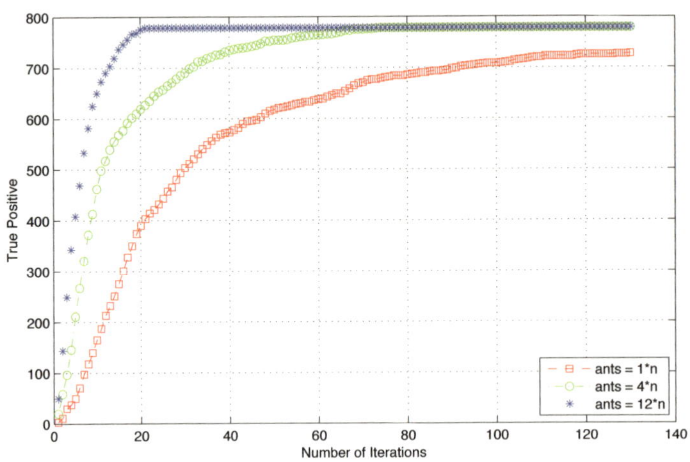

Figure 8. Extracted features vs. number of iterations for different ant-colony size.

Authors proposed this method to extract the contour of a breast as the region of interest in mammogram. [29] applied adaptive structuring elements to dilate the broken edges along their slope directions. [20] proposed improvement to the traditional Ant Colony Optimization (ACO) based method for broken-edge linking to reduce the computational cost.

In this section, the Ant System-based broken-edge linking algorithm proposed by [13] is presented. As inputs are used: the Sobel edge image and the original grayscale image. The Sobel edge image is a binary image obtained after applying the Sobel edge operator [31] to the original grayscale image. From this image the endpoints are extracted that will be used afterwards as the starting pixels for the ants' routes.

The original image is used to produce the *grayscale visibility* matrix, which for the pixel at (i,j) is calculated as follows:

$$\xi_{ij} = \frac{1}{I_{max}} \cdot max \begin{bmatrix} |I(i-1,j-1) - I(i+1,j+1)|, \\ |I(i-1,j+1) - I(i+1,j-1)|, \\ |I(i,j-1) - I(i,j+1)|, \\ |I(i-1,j) - I(i+1,j)| \end{bmatrix} \tag{13}$$

where I_{max} is the maximum gray value in the image, so ξ_{ij} is normalized $(0 \leq \xi_{ij} \leq 1)$. For the pixels in regions of distinct gray intensity changes the higher values are obtained. The matrix of *grayscale visibility* will be the initial pheromone trail matrix. It is also used to calculate the fitness value of a route chosen by ant. The resulting image will contain the routes (connecting edges) with the highest fitness values found as optimal routes between the endpoints. In order to discard non-optimal routes, a fitness threshold is applied. Finally, the output image is the improved image that is a sum of the Sobel edge image and the connecting edges. The block diagram of the proposed method is shown in Figure 9.

The proposed AS-based algorithm for broken-edge linking includes the following steps:

1. Initialization: The number of ants equals the number of endpoints found in the Sobel edge image, and each endpoint will be a starting pixel of a different ant. Initial pheromone trail for each pixel is set to its *grayscale visibility* value.
2. Pixel transition rule: Possible ant's transitions to the neighboring pixels are defined by 8-connection pixel transition model shown in Figure 3. The admissible neighboring pixels for the kth ant to move to are the ones not in the $tabu_k$ list. The probability for the kth ant to move from pixel (r,s) to pixel (i,j) is calculated as follows:

$$p^k_{(r,s)(i,j)} = \begin{cases} \frac{(\tau_{ij})^\alpha (\eta_{ij})^\beta}{\sum_u \sum_v (\tau_{uv})^\alpha (\eta_{uv})^\beta} & \text{if} (i,j) \text{ and } (u,v) \notin \text{tabu}_k \\ & r-1 \leq i,u \leq r+1, \\ & s-1 \leq j,v \leq s+1 \\ \\ 0 & \text{otherwise} \end{cases} \tag{14}$$

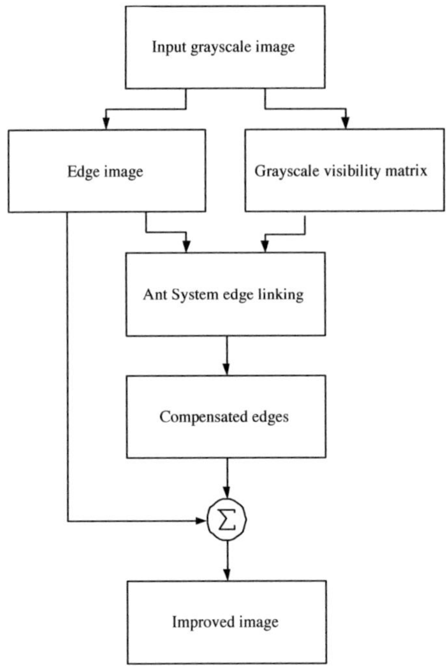

Figure 9. Block diagram of the proposed edge linking method

where τ_{ij} and η_{ij} are the intensity of the pheromone trail and the visibility of the pixel at (i, j), respectively, and α and β are control parameters ($\alpha, \beta > 0$; $\alpha, \beta \in \Re$). The visibility of a pixel should not be misinterpreted as its *grayscale visibility*, and for the pixel at (i, j) it is defined as:

$$\eta_{ij} = \frac{1}{d_{ij}} \tag{15}$$

where d_{ij} is the Euclidean distance of the pixel at (i, j) from the closest endpoint.

3. Pheromone update rule: Negative feedback is demonstrated through the pheromone trails evaporation according to:

$$\tau_{ij(new)} = (1 - \rho)\tau_{ij(old)} + \Delta\tau_{ij} \tag{16}$$

where ρ is the pheromone evaporation rate ($0 < \rho < 1$; $\rho \in \Re$), and

$$\Delta\tau_{ij} = \sum_{k=1}^{m} \Delta\tau_{ij}^{k} \tag{17}$$

where

http://dx.doi.org/10.5772/52792

(a) (b) (c) (d)

Figure 10. Qualitative results of the proposed edge-linking method, "Peppers" 256 × 256 pixels: (a) original image; (b) Sobel edge image; (c) resulting image of the proposed method; (d) improved edge image.

$$\Delta\tau_{ij}^{k} = \begin{cases} \frac{f_k}{Q} & \text{if} \quad kth \quad \text{ant displaces to pixel} \quad (i,j) \\ 0 & \text{otherwise.} \end{cases} \tag{18}$$

The fitness value of a pixel, f_k, is equal to the fitness value of the route it belongs to. The proposed fitness function is given by:

$$f_k = \frac{\bar{\xi}}{\sigma_{\xi} \cdot N_p} \tag{19}$$

where $\bar{\xi}$ and σ_{ξ} are the mean value and the standard deviation of the *grayscale visibility* of the pixels in the route, and N_p is the total number of pixels belonging to that route. Pheromone evaporation prevents algorithm stagnation. From the repeatedly not-visited pixels the pheromone trail evaporates exponentially.

4. Stopping criterion: The steps 2 and 3 are repeated in a loop and algorithm stops executing when the maximum number of iterations is reached. An iteration ends when all the ants finish the search for the endpoints, by either finding one or getting stuck and being unable to advance to any adjacent pixel.

5.1. Simulation results and discussion

The simulation results of the proposed algorithm applied to the "Peppers" image of 256 × 256 pixels are shown in Figure 10. The algorithm detects the missing edge segments (Figure 10(c)) as the optimal routes consisted of the edge pixels. The initial pheromone trail for each pixel was set to its *grayscale visibility* value. In this manner, the pixels belonging to true edges have a higher probability of being chosen by ants on their initial routes, which shortens the time needed to find a satisfactory solution, or improves the solution found for a fixed number of iterations. The results were obtained after 100 iterations; this number was chosen on trial and error basis.

Designated values $\alpha = 10$ and $\beta = 1$ were determined on trial and error basis. A large α/β ratio forces the ants to choose the strongest edges. The existence of the control parameter β is important since it inclines the ant's route towards the closest endpoint. Experimental results showed that, by setting the β value to zero, it took more steps for the ants to find the endpoints which made the computation time longer. In some cases, ants were not even able to find the satisfactory solution for a reasonable number of steps, or they just got stuck between already visited pixels.

The effect of the α/β parameter ratio on the resulting image is best presented in Figure 11. It can be observed that the endpoint in the upper-left corner of the ROI image (Figure 11(c)–(e)) was not connected to any of the closer endpoints, and that the ants successfully found the more remote endpoint which was the correct one. The existence of the β parameter keeps the ants away from the low-contrast regions, such as the region of low gray-intensity pixels between two closer endpoints.

The ant's memory, i.e. the length of the tabu list, was set to 10. Larger ant's memory values would improve the quality of the resulting binary image but would as well lead to the prolonged computation time. The designated value was large enough to keep the ants from being stuck in small pixel circles.

The fitness value of a route is dependent on the mean value and the standard deviation of the grayscale visibility of the pixels in the route, and the total number of pixels belonging to that route, as defined in (19). The routes that have higher grayscale visibility mean value are the stronger edges as the gray level contrast of their adjacent pixels is higher. The smaller standard deviation of the grayscale visibility of the pixels in the route results in a higher fitness value. By this, more importance is given to the routes consisted of pixels belonging to the same edge, thus avoiding the ants crossing between the edges and leaving pheromone trails on non-edge pixels. Finally, the shorter routes are more favorable as a solution, therefore by keeping the total number of pixels in the route smaller, the higher fitness values are obtained.

The number of iterations was set to 100, which gave satisfactory results within an acceptable computational time of execution. The lower resolution images, for example 128×128 pixels, allowed a larger number of iterations to be used, since a smaller number of ants was processed for a smaller number of relatively closer endpoints. The execution time of the algorithm was not optimal, and it was measured in minutes. One of the reasons is that the algorithm code was not written in an optimal manner since Matlab as a programming environment is not intended for a fast code execution, but rather for an easy high-level algorithm implementation.

In order to test the proposed method on different input images, simulations were performed on "House", "Lena" and "Cameraman" images of size 256×256 pixels. The results confirm the effectiveness of the method, as shown in Figure 12. It can be noticed that the found edge segments are often not unidirectional, which indicates that the fitness function was adequately defined and the ants found the true edges. The main contribution of the proposed broken-edge-linking method is in using a bottom-up approach that avoids using a global threshold to find the missing segments.

6. Adaptability of the proposed edge detector

The adaptability and the ability to learn are important features of autonomous systems. In ant colonies, natural and artificial, learning consists in changing the environment by laying the pheromone trails while searching for food. The structures that emerge from the

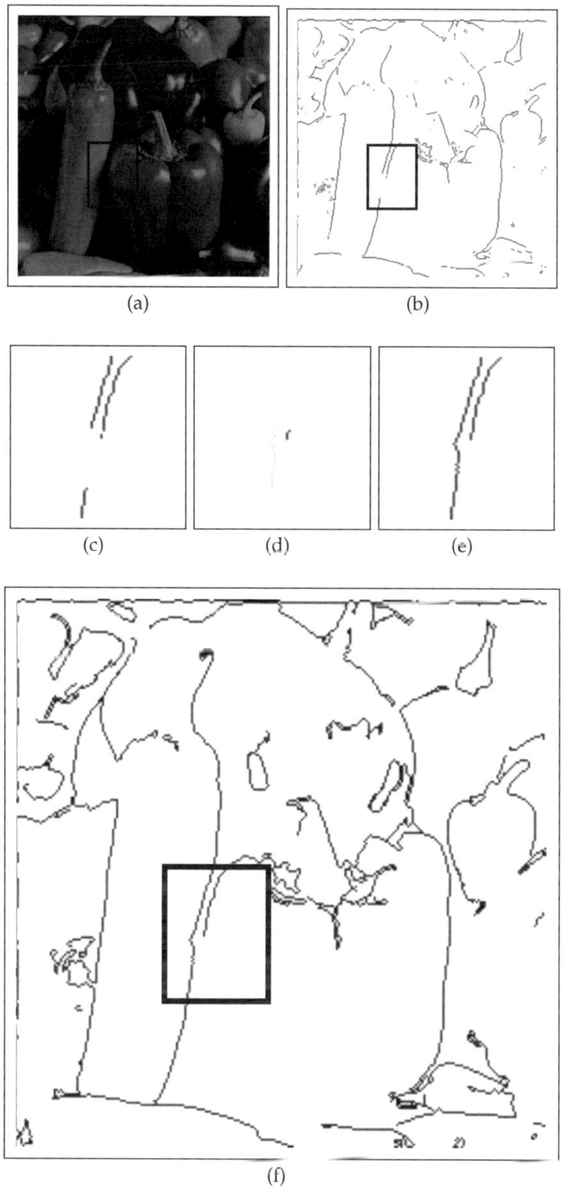

Figure 11. Effect of the control parameters on correct connection of the endpoints: "Peppers" 256 × 256 image: (a) original image with marked region of interest (ROI); (b) Sobel edge image with marked ROI; (c) enlarged ROI: Sobel edge image; (d) enlarged ROI: pheromone trails image; (e) enlarged ROI: improved edge image; (f) improved edge image with marked ROI.

accumulated pheromone represent the stored information about the environment that can be used by any member of the swarm. Although a single ant has no knowledge of the global pattern, the designer of such a swarm-based system is a privileged observer of the emergence that comes as a result of the cooperative behavior.

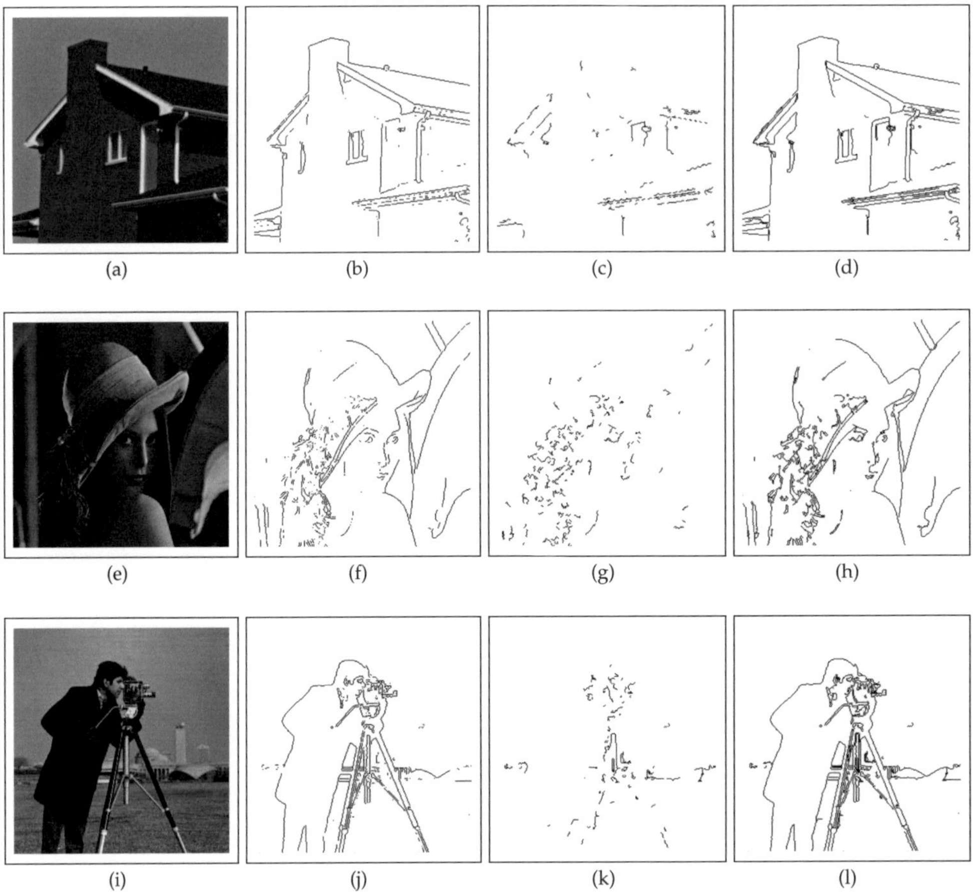

Figure 12. Qualitative results of the proposed method, 256 × 256-pixel images: (a) "House" original image; (b) "House": Sobel edge image; (c) "House": result of the proposed method; (d) "House": improved edge image; (e) "Lena": original image; (f) "Lena": Sobel edge image; (g) "Lena": result of the proposed method; (h) "Lena": improved edge image; (i) "Cameraman": original image; (j) "Cameraman": Sobel edge image; (k) "Cameraman": result of the proposed method; (l) "Cameraman": improved edge image.

The resulting mass behavior in swarms is hard to predict. Although the adaptability can be demonstrated on a variety of applications such as in image segmentation [27], a general theoretical framework on design and control of swarms does not exist. Artificial swarms use bottom-up approach, meaning that the designer of such distributed multi-agent system needs to set the rules for local interactions between the agents themselves and, if required, between the agents and the environment. The indirect communication via environment is referred to as *stigmergy*, and in case of ant colonies, it consists in pheromone-laying and pheromone-following. For each specific application, the food that ants search for must also be defined.

This section presents a study on the adaptability of the algorithm proposed in Section 4 [12]. Experiments with two different sets of grayscale images were performed. In the first experimental setup, a set of three different grayscale images was used to test the adaptability of the proposed AS-based edge detector. The images were obtained by applying

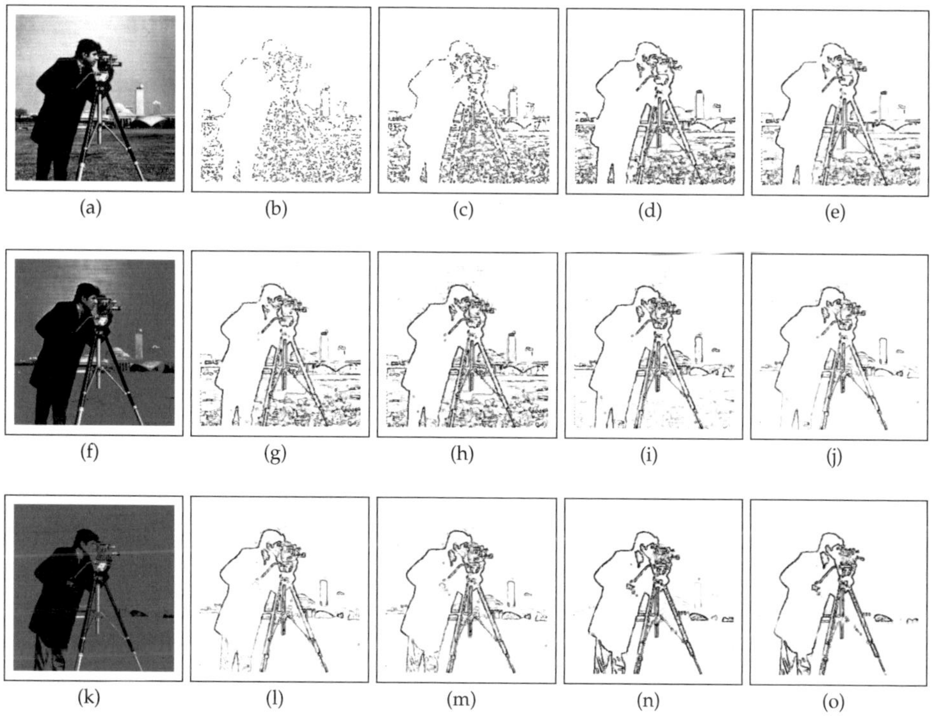

Figure 13. Adaptive edge detection on enhanced "Cameraman" images, 256 × 256 pixels: (a) enhanced image 1; (b) t=5 iterations; (c) t=10 iterations; (d) t=50 iterations; (e) t=100 iterations; (f) enhanced image 2; (g) t=105 iterations; (h) t=110 iterations; (i) t=150 iterations; (j) t=200 iterations; (k) enhanced image 3; (l) t=205 iterations; (m) t=210 iterations; (n) t=250 iterations; (o) t=300 iterations.

a Multiscale Adaptive Gain contrast enhancement to the 256 × 256 pixel "Cameraman" image (see Figure 4). Every $N_i = 100$ iterations one image from the set was replaced by another. The response of the artificial ant colony to the change in the environment was a different distribution of pheromone trails. The number of 100 iterations per image was enough for the new pheromone structure to be established. The algorithm parameters used in the experiments were determined empirically: $\tau_0 = 0.01$, $\rho = 0.5$, $\alpha = 1$, $\beta = 10$, $T = 0.08$ and the tabu list length was set to 10. Parameters could be optimized for a better edge detection, but it is of no importance for this study. It would not affect the adaptability of the algorithm since every image change would result in a change of the pheromone trail structure. Simulation results are shown in Figure 13.

The results show that the Ant System-based edge detector was capable of detecting the changes that occurred as a result of replacing one image from the set with another. The experiments were repeated for a set of four widely used test grayscale images: "Cameraman", "Lena", "House", and "Peppers". The images were used as inputs to the algorithm in that order. Every $N_i = 100$ iterations one image was replaced by the next one from the set. Again, the change in the environment produced by the change of input image resulted in different pheromone patterns, which is shown in Figure 14.

It can be observed that the new pheromone trails accumulated on the pixels belonging to the newly-emerged edges, while the pheromone trails where the edges were no longer present

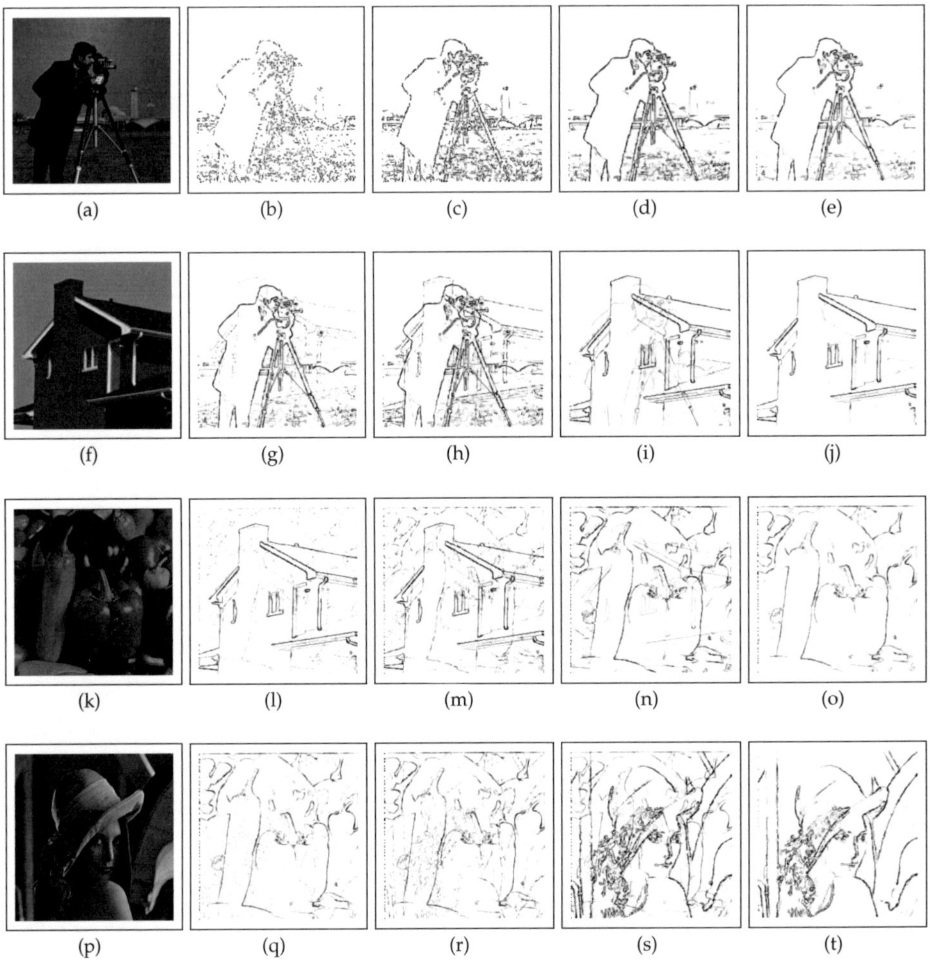

Figure 14. Adaptive edge detection on four test images, 256 × 256 pixels: (a) "Cameraman"; (b) t=5 iterations; (c) t=10 iterations; (d) t=50 iterations; (e) t=100 iterations; (f) "House"; (g) t=105 iterations; (h) t=110 iterations; (i) t=150 iterations; (j) t=200 iterations; (k) "Peppers"; (l) t=205 iterations; (m) t=210 iterations; (n) t=250 iterations; (o) t=300 iterations; (p) "Lena"; (q) t=305 iterations; (r) t=310 iterations; (s) t=350 iterations; (t) t=400 iterations.

gradually disappeared. In order to obtain a quicker transition between different pheromone distributions, the evaporation rate ρ was set to a higher value than for the edge-detection simulations ($\rho = 0.5$). This resulted in disappearing of the "weakest" edges and introduced slightly poorer overall performance of the proposed edge detector. The experimental results show that the algorithm is able to adapt to a dynamically changing environment resulting in different pheromone trail patterns. Even though the images were used in the experiments, the study could be extended to any other type of digital habitat which can lead to a new set of applications for the adaptive artificial ant colonies.

One of the possible applications for the adaptive edge detector could be real-time image processing where online image preprocessing could be used to obtain better image segmentation. By applying various image enhancement techniques, such as contrast enhancement, certain features in the image could be amplified while others could be reduced or even removed. This would enable easier detection of the regions of interest in the image.

7. Conclusions

Two edge-detection methods inspired by the ants foraging behavior were proposed. The first method uses a grayscale image as input and as output produces a pheromone image marking the location of the edge pixels. The second method finds the missing edge segments after the edge detection was applied and can be used as a complementary tool to any edge detector. In our work, the Sobel edge detector was used to produce the binary edge image.

The first method combines a nonlinear contrast enhancement technique, Multiscale Adaptive Gain, and the Ant System algorithm inspired by the ants foraging behavior. The set of enhanced images was obtained after applying the Multiscale Adaptive Gain and the Ant System algorithm generated pheromone patterns where the true edges were found. The experiments showed that our method outperformed other ACO-based edge detectors in terms of visual quality of the extracted edge information and sensitivity in finding weaker edges. The quantitative analysis showed that the performance could further be optimized by varying the number of ants and iterations.

The adaptability of the proposed edge detector was demonstrated in a dynamically changing environment made of a set of digital grayscale images. The algorithm responded to the changes by generating pheromone patterns according to the distribution of the newly-created edges. It also proved to be robust since even an ant colony of a smaller size could detect the edges, even though the number of detected edge pixels was reduced.

The second proposed method uses the ant colony search for the edge segments that connect pairs of endpoints. A novel fitness function was proposed to evaluate the found segments. It depends on two variables: the pixels grayscale visibility and the edge-segment length. The fitness function produces higher values for the segments that consisted of smaller number of pixels, which had grayscale visibility of a higher mean value and a lower variance. Another novelty was to apply the grayscale visibility matrix as the initial pheromone trails matrix so that the pixels belonging to true edges have a higher probability of being chosen by ants on their initial routes, which reduced the computational load. The proposed broken-edge linking method was tested as a complementary tool for the Sobel edge detector, and it significantly improved the output edge image.

Future research will include optimization and automatic detection of the proposed methods' parameters for an improved edge detection results. Until now these parameters were experimentally obtained. An exhaustive analysis of the edge detection method's adaptability will be performed in order to apply it to other digital habitats. Also, the methods optimization for faster execution would make them suitable for real-time image processing.

Acknowledgements

This work has been financed by the EU-funded Initial Training Network (ITN) in the Marie-Curie People Programme (FP7): INTRO (INTeractive RObotics research network), grant agreement no.: 238486.

8. Acronyms

ACO Ant Colony Optimization

ACS Ant Colony System

AS Ant System

BA Bees Algorithm

CLF Coordinate Logic Filters

CLO Coordinate Logic Operations

CLT Coordinate Logic Transforms

LSE Least Square Error

PLIP Parameterized Logarithmic Image Processing

PSO Particle Swarm Optimization

ROI Region Of Interest

SEL Sequential Edge-Linking

TSP Traveling Salesman Problem

Author details

Aleksandar Jevtić[1] and Bo Li[2]

* Address all correspondence to: aleksandar.jevtic@robosoft.fr; bo.li@tfe.umu.se

1 Robosoft, Bidart, France
2 Dept. of Applied Physics and Electronics, Umeå University, Umeå, Sweden

References

[1] Baterina, A. V. & Oppus, C. [2010]. Image edge detection using ant colony optimization, *WSEAS Transactions on Signal Processing* 6(2): 58–67.

[2] Bonabeau, E., Dorigo, M. & Theraulaz, G. [1999]. *Swarm Intelligence: From Natural to Artificial Systems*, Oxford University Press, New York.

[3] Canny, J. [1986]. A computational approach to edge detection, *IEEE Transactions on Pattern Analysis & Machine Intelligence* 8: 679–714.

[4] Danahy, E. E., Panetta, K. A. & Agaian, S. S. [2007]. Coordinate logic transforms and their use in the detection of edges within binary and grayscale images, *IEEE International Conference on Image Processing, 2007. ICIP 2007.*, Vol. 3, pp. III–53–III–56.

[5] Dorigo, M., Maniezzo, V. & Colorni, A. [1996]. Ant system: optimization by a colony of cooperating agents, *IEEE Transactions on Systems, Man, and Cybernetics - Part B* 26(1): 29–41.

[6] Dorigo, M. & Stützle, T. [2004]. *Ant colony optimization*, MIT Press, Cambridge.

[7] Engelbrecht, A. P. [2005]. *Fundamentals of Computational Swarm Intelligence*, John Wiley & Sons, Ltd, Chichester, UK.

[8] Etemad, S. A. & White, T. [2011]. An ant-inspired algorithm for detection of image edge features, *Applied Soft Computing* 11(8): 4883–4893.

[9] Gonzalez, R. C. & Woods, R. E. [2008]. *Digital Image Processing*, 3rd edn, Prentice Hall, New Jersey, USA.

[10] He, X., Jia, W., Hur, N., Wu, Q., Kim, J. & Hintz, T. [2006]. Bilateral edge detection on a virtual hexagonal structure, *in* G. Bebis, R. Boyle, B. Parvin, D. Koracin, P. Remagnino, A. Nefian, G. Meenakshisundaram, V. Pascucci, J. Zara, J. Molineros, H. Theisel & T. Malzbender (eds), *Advances in Visual Computing*, Vol. 4292 of *Lecture Notes in Computer Science*, Springer Berlin / Heidelberg, pp. 176–185.

[11] Huang, P., Cao, H. & Luo, S. [2008]. An artificial ant colonies approach to medical image segmentation, *Computer Methods & Programs in Biomedicine* 92: 267–273.

[12] Jevtić, A. & Andina, D. [2010]. Adaptive artificial ant colonies for edge detection in digital images, *Proceedings of the 36th Annual Conference on IEEE Industrial Electronics Society, IECON 2010*, pp. 2813–2816.

[13] Jevtić, A., Melgar, I. & Andina, D. [2009]. Ant based edge linking algorithm, *Proceedings of 35th Annual Conference of the IEEE Industrial Electronics Society (IECON 2009)*, Porto, Portugal, pp. 3353–3358.

[14] Jevtić, A., Quintanilla-Dominguez, J., Barrón-Adame, J.-M. & Andina, D. [2011]. Image segmentation using ant system-based clustering algorithm, *in* E. Corchado, V. Snasel, J. Sedano, A. E. Hassanien, J. L. Calvo & D. Slezak (eds), *SOCO 2011 - 6th International Conference on Soft Computing Models in Industrial and Environmental Applications*, Vol. 87, Springer Berlin / Heidelberg, pp. 35–45.

[15] Jevtić, A., Quintanilla-Domínguez, J., Cortina-Januchs, M. G. & Andina, D. [2009]. Edge detection using ant colony search algorithm and multiscale contrast enhancement, *Proceedings of 2009 IEEE International Conference on Systems, Man, and Cybernetics (SMC 2009)*, San Antonio, TX, USA, pp. 2193–2198.

[16] Jiang, J. A., Chuang, C. L., Lu, Y. L. & Fahn, C. S. [2007]. Mathematical-morphology-based edge detectors for detection of thin edges in low-contrast regions, *IET Image Processing* 1(3): 269–277.

[17] Kennedy, J. & Eberhart, R. C. [1995]. Particle swarm optimisation, *Proceedings of IEEE International Conference on Neural Networks Vol. IV*, IEEE service center, Piscataway, NJ, USA, pp. 1942–1948.

[18] Khajehpour, P., Lucas, C. & Araabi, B. N. [2005]. Hierarchical image segmentation using ant colony and chemical computing approach, *in* L. Wang, K. Chen & Y. S. Ong (eds), *Advances in Natural Computation*, Vol. 3611 of *Lecture Notes in Computer Science*, Springer Berlin / Heidelberg, pp. 1250–1258.

[19] Laine, A. F., Schuler, S., Fan, J. & Huda, W. [1994]. Mammographic feature enhancement by multiscale analysis, *IEEE Transactions on Medical Imaging* 13(4): 725–740.

[20] Lu, D. S. & Chen, C. C. [2008]. Edge detection improvement by ant colony optimization, *Pattern Recognition Letters* 29(4): 416–425.

[21] Mertzios, B. G. & Tsirikolias, K. [2001]. Applications of coordinate logic filters in image analysis and pattern recognition, *Proceedings of the 2nd International Symposium on Image and Signal Processing and Analysis, 2001. ISPA 2001.*, pp. 125–130.

[22] Nezamabadi-pour, H., Saryazdi, S. & Rashedi, E. [2006]. Edge detection using ant algorithms, *Soft Computing* 10(7): 623–628.

[23] Panetta, K. A., Wharton, E. J. & Agaian, S. S. [2008]. Logarithmic edge detection with applications, *Journal of Computers* 3(9): 11–19.

[24] Pham, D. T., Ghanbarzadeh, A., Koç, E., Otri, S., Rahim, S. & Zaidi, M. [2006]. The bees algorithm - a novel tool for complex optimisation problems, *Proceedings of IPROMS 2006 Conference*, Cardiff, UK, pp. 454–461.

[25] Prewitt, J. M. S. [1970]. *Object Enhancement and Extraction in Picture Processing and Psychopictorics*, Academic Press, New York.

[26] Rahebi, J. & Tajik, H. R. [2011]. Biomedical image edge detection using an ant colony optimization based on artificial neural networks, *International Journal of Engineering Science and Technology (IJEST)* 3(12): 8211–8218.

[27] Ramos, V. & Almeida, F. [2000]. Artificial ant colonies in digital image habitats - a mass behavior effect study on pattern recognition, *Proceedings of ANTS'2000 - 2nd International Workshop on Ant Algorithms (From Ant Colonies to Artificial Ants), M. Dorigo, M. Middendorf, T. Stüzle (Eds.)*, Brussels, Belgium, pp. 113–116.

[28] Ramos, V., Muge, F. & Pina, P. [2002]. Self-organized data and image retrieval as a consequence of inter-dynamic synergistic relationships in artificial ant colonies, *in* J. Ruiz-del Solar, A. Abraham & K. M. (eds), *Frontiers in Artificial Intelligence and Applications, Soft Computing Systems - Design, Management and Applications*, Vol. 87 of *2nd International Conference on Hybrid Intelligent Systems*, Springer Berlin / Heidelberg, Santiago, Chile, pp. 500–509.

[29] Shih, F. Y. & Cheng, S. [2004]. Adaptive mathematical morphology for edge linking, *Information Sciences?Informatics and Computer Science: An International Journal* 167(1-4): 9–21.

[30] Shin, M. C., Goldgof, D. B. & Bowyer, K. W. [2001]. Comparison of edge detector performance through use in an object recognition task, *Computer Vision & Image Understanding* 84(1): 160–178.

[31] Sobel, I. & Feldman, G. [1968]. A 3 × 3 isotropic gradient operator for image processing, *presented at a talk at the Stanford Artificial Project, unpublished but often cited* .

[32] Tian, J., Yu, W. & Xie, S. [2008]. An Ant Colony Optimization algorithm for image edge detection, *Proceedings of the IEEE Congress on Evolutionary Computation, 2008. CEC 2008. (IEEE World Congress on Computational Intelligence)*, Hong Kong, China, pp. 751–756.

[33] Wei, L., Sheng, L., Yi, R. X. & Peng, D. [2008]. A new contour detection in mammogram using sequential edge linking, *2008 Second International Symposium on Intelligent Information Technology Application (IITA '08)*, Vol. 1, Hong Kong, China, pp. 197–200.

Chapter 3

Content-Based Image Feature Description and Retrieving

Nai-Chung Yang, Chung-Ming Kuo and
Wei-Han Chang

Additional information is available at the end of the chapter

http://dx.doi.org/10.5772/45841

1. Introduction

With the growth in the number of color images, developing an efficient image retrieval system has received much attention in recent years. The first step to retrieve relevant information from image and video databases is the selection of appropriate feature representations (e.g. color, texture, shape) so that the feature attributes are both consistent in feature space and perceptually close to the user [1]. There are many CBIR systems, which adopt different low level features and similarity measure, have been proposed in the literature [2-5]. In general, perceptually similar images are not necessarily similar in terms of low-level features [6]. Hence, these content-based systems capture pre-attentive similarity rather than semantic similarity [7]. In order to achieve more efficient CBIR system, active researches are currently focused on the two complemented approaches: region-based approach [4, 8-10] and relevance feedback [6, 11-13].

Typically, the region-based approaches segment each image into several regions with homogenous visual prosperities, and enable users to rate the relevant regions for constructing a new query. In general, an incorrect segmentation may result in inaccurate representation. However, automatically extracting image objects is still a challenging issue, especially for a database containing a collection of heterogeneous images. For example, Jing et al. [8] integrate several effective relevance feedback algorithms into a region-based image retrieval system, which incorporates the properties of all the segmented regions to perform many-to-many relationships of regional similarity measure. However, some semantic information will be disregarded without considering similar regions in the same image. In another study [10], Vu et al. proposed a region-of-interest (ROI) technique which is a sampling-based approach called SamMatch for matching framework. This method can prevent incorrectly detecting the visual features.

On the other hand, the mechanism of relevance feedback is an online-learning technique that can capture the inherent subjectivity of user's perception during a retrieval session. In Power Tool [11], the user is allowed to give the relevance scores to the best matched images, and the system adjusts the weights by putting more emphasis on the specific features. Cox et al. [11] propose an alternative way to achieve CBIR that predicts the possible image targets by Bayes' rule rather than provides with segmented regions of the query image. However, the feedback information in [12] could be ignored if the most likely images and irrelevant images have similar features.

In this Chapter, a novel region-based relevance feedback system is proposed that incorporates several feature vectors. First, unsupervised texture segmentation for natural images is used to partition an image to several homogeneous regions. Then we propose an efficient dominant color descriptor (DCD) to represent the partitioned regions in image. Next, a regional similarity matrix model is introduced to rank the images. In order to attack the possible fails of segmentation and to simplify the user operations, we propose a foreground assumption to separate an image into two parts: foreground and background. The background could be regarded as the irrelevant region that confuses with the query semantics for retrieval. It should be noted that the main objectives of this approach could exclude irrelevant regions (background) from contributing to image-to-image similarity model. Furthermore, the global features extracted from entire image are used to compensate the inaccuracy due to imperfect segmentations. The details will be presented in the following Sections. Experimental results show that our framework improves the accuracy of relevance-feedback retrieval.

The Chapter is organized as follows. Section 2 describes the key observations which explain the basis of our algorithm. In Section 3, we first present a quantization scheme for extracting the representative colors from images, and then introduce a modified similarity measure for DCD. In Section 4, image segmentation and region representation based on our modified dominant color descriptor and local binary pattern are described. Then the image representation and the foreground assumption are explained in Section 5. Our integrated region-based relevance feedback strategies, which consider pseudo query image and relevant images as the relevance information, are introduced in Section 6. Experimental results and discussions of the framework are made in Section 7. Finally, a short conclusion is presented in Section 8.

2. Problem statement

The major goal in region-based relevance feedback for image retrieval is to search perceptually similar images with good accuracy in short response time. For nature image retrieval, conversional region-based relevance feedback systems use multiple features (e.g., color, shape, texture, size) and update weighting scheme. In this context, our algorithm is motivated by the following viewpoints.

1. Computational cost increases as the selected features increased. However, an algorithm with large number of features does not guarantee an improvement of retrieval performance. In theory, the retrieval performance can be enhanced by choosing more compact feature vectors.

2. The CBIR systems retrieve similar images according to the user-defined feature vectors [10]. To improve the accuracy, the region-based approaches [14, 15] segment each image into several regions, and then extract the image features, such as the dominant color, texture or shape. However, the correct detection of semantic objects involves many conditions [16] such as lighting conditions, occlusion and inaccurate segmentation. Since no automatic segmentation algorithm achieves satisfactory performance currently, segmented regions are commonly provided by the user to support the image retrieval. However, semantically correct segmentation is a strict challenge to the user, even some systems provide segmentation tools.

3. The CBIR technique helps the system to learn how to retrieve the results that users are looking for. Therefore, there is an urgent need to develop a convient technique for region-of-interest analysis.

3. A modified dominant color descriptor

Color is one of the most widely used visual features for retrieving images from common semantic categories [12]. MPEG-7 specifies several color descriptors [17], such as dominant colors, scalable color histogram, color structure, color layout and GoF/GoP color. The human visual system captures dominant colors in images and eliminates the fine details in small areas [18]. In MPEG-7, DCD provides a compact color representation, and describes the color distribution in an image[16]. The dominant color descriptor in MPEG-7 is defined as

$$F = \{\{c_i, p_i\},\ i = 1, \ldots N\}, \quad (1)$$

where N is the total number of dominant colors in image, c_i is a 3-D dominant color vector, p_i is the percentage for each dominant color, and $\Sigma pi - 1$.

In order to extract the dominant colors from an image, a color quantization algorithm has to be predetermined. A commonly used approach is the modified generalized Lloyd algorithm (GLA) [19], which is a color quantization algorithm with clusters merging. This method can simplify the large number of colors to a small number of representative colors. However, the GLA has several intrinsic problems associated with the existing algorithm as follows [20].

4. It may give different clustering results when the number of clusters is changed.

5. A correct initialization of the centroid of cluster is a crucial issue because some clusters may be empty if their initial centers lie far from the distribution of data.

6. The criterion of the GLA depends on the cluster "distance"; therefore, different initial parameters of an image may cause different clustering results.

In general, the conventional clustering algorithms are very time consuming [2, 21-24]. On the other hand, the quadratic-like measure [2, 17, 25] for dominant color descriptor in MPEG7 does not matching human perception very well, and it could cause incorrect ranks for images with similar color distribution [3, 20, 26]. In this Chapter, we adopt the linear block algorithm (LBA) [20] to extract the representative colors, and measure the perceptual similar dominant colors by the modified similarity measure.

Considering two dominant color features $F_1=\{\{c_i, p_i\}, i=1, \cdots, N_1\}$ and $F_2=\{\{b_j, q_j\}, j=1, \ldots N_2\}$, the quadratic-like dissimilarity measure between two images F_1 and F_2 is calculated by:

$$D^2(F_1, F_2) = \sum_{i=1}^{N_1} p_i^2 + \sum_{j=1}^{N_2} q_j^2 - \sum_{i=1}^{N_1}\sum_{j=1}^{N_2} 2a_{i,j} p_i q_j \tag{2}$$

where $a_{i,j}$ is the similarity coefficient between color clusters c_i and b_j, and it is given by

$$a_{i,j} = \begin{cases} 1 - d_{i,j} / d_{\max} & d_{i,j} \le T_d \\ 0 & d_{i,j} > T_d \end{cases}. \tag{3}$$

The threshold T_d is the maximum distance used to judge whether two color clusters are similar, and $d_{i,j}$ is Euclidean distance between two color clusters c_i and b_j; $d_{\max}=\alpha T_d$, notation α is a parameter that is set to 2.0 in this work.

The quadratic-like distance measure in Eq. (2) may incorrectly reflect the distance between two images. The improper results are mainly caused by two reasons. 1) If the number of dominant colors N_2 in target image increases, it might cause incorrect results. 2) If one dominant color can be found both in target images and query image, a high percentage q_j of the color in target image might cause improper results. In our earlier work [19], we proposed a modified distance measure that considers not only the similarity of dominant colors but also the difference of color percentages between images. The experimental results show that the measure in [20] provides better match to human perception in judging image similarity than the MPEG-7 DCD. The modified similarity measure between two images F_1 and F_2 is calculated by:

$$D^2(F_1, F_2) = 1 - \sum_{i=1}^{N_1}\sum_{j=1}^{N_2} a_{i,j} S_{i,j}, \tag{4}$$

$$S_{i,j} = [1 - |p_q(i) - p_t(j)|] \times \min(p_q(i), p_t(j)), \tag{5}$$

where $p_q(i)$ and $p_t(j)$ are the percentages of the ith dominant color in query image and the jth dominant color in target image, respectively. The term in bracket, $1 - | p_q(i) - p_t(j) |$ is used to measure the difference between two colors in percentage, and the term $\min(p_q(i), p_t(j))$ is the intersection of $p_q(i)$ and $p_t(j)$ that represents the similarity between two colors in percentage. In Fig. 1, we use two real images selected from Corel as our example, where the color and percentage values are given for comparison.

Query image Q	Target image F1	Target image F2
{(33,31,33),0.794240}	{(66,41,29), 0.108795}	{(60,55,53), 0.378306}
{(184,179,180),0.20576}	{(203,47,71), 0.334035}	{(139,123,115),0.073598}
	{(207,193,59), 0.067861}	{(198,194,188),0.548096}
	{(228,98,161), 0.219045}	
	{(230,162,203),0.270264}	

Figure 1. Example images with the dominant colors and their percentage values. First row: 3-D dominant color vector c_i and the percentage p_i for each dominant color. Middle row: the original images. Bottom row: the corresponding quantized images.

In Fig. 1, we calculate this example by using the modified measure and quadratic-like measure for comparison. In order to properly reflect similarity coefficient between two color clusters, the parameter is set to 2 and $T_d = 25$ in Eq(3). Since the pair-wised distance between Q and F_1 in Fig. 1 is exceed Td, the quadratic-like dissimilarity measure can be determined by

$D^2(Q, F_1) = 0.6732 + 0.249 = 0.9222.$

However, using the quadratic-like dissimilarity measure between the Q and F_2 is:

$D^2(Q, F_2) = 0.6732 + 0.4489 - 2 \times (1 - 22/50) \times 0.20576 \times 0.548096 = 0.9958.$

It can be seen that the comparison result of $D^2(Q, F_2) > D^2(Q, F_1)$ is not consistent with human perception. Whereas, using the dissimilarity measure in [19], we have

$$D^2(Q, F_1)=1-0=1$$

and

$$D^2(Q, F_2)=1-\{(1-22/50)\times(1-|0.20576-0.548096|)\times 0.20576\}=0.9242$$

Query image Q	Target image F_1	Target image F_2
{(47,44,46), 0.626495} {(194,188,184), 0.373505}	{(43,46,45), 0.822144} {(189,23,55),0.177856}	{(40,41,30), 0.641947} {(138,137, 108), 0.072815} {(202,194, 185), 0.28524}

Figure 2. Example images with the dominant colors and their percentage values. First row: 3-D dominant color vector c_i and the percentage p_i for each dominant color. Middle row: the original images. Bottom row: the corresponding quantized images.

In DCD, the quadratic-like measure results incorrect matches due to the existence of high percentage of the same color in target image. For example, consider the quantized images in Fig. 2. We can see that the percentage of dominant colors of F_1 (rose) and F_2 (gorilla) are 82.21% and 92.72%, respectively. In human perception, Q is more similar to F_2 . However, the quadratic-like similarity measure is $D^2(Q, F_2)>D^2(Q, F_1)$. Obviously, the result causes a wrong rank. The robust similarity measure [19] is more accurate to capture human perception than that of MPEG-7 DCD. In our experiments, the modified DCD achieves 16.7% and 3% average retrieval rate (ARR) [27] improvements than Ma [28] and Mojsilovic [29], respectively. In this Chapter, the modified dominant color descriptor is chosen to support the proposed CBIR system.

4. Image segmentation and region representation

4.1. Image segmentation

It has been mentioned that segmentation is necessary for those region-based image retrieval systems. Nevertheless, automatic segmentation is still unpractical for the applications of region-based image retrieval (RBIR) systems [8, 30-32]. Although many systems provide segmentation tools, they usually need complicated user interaction to achieve image retrieval. Therefore, the processing is very inefficient and time consuming to the user. In the following, the new approach will propose to overcome this problem. In our algorithm, the user does not need to provide precisely segmented regions, instead, the boundary checking algorithm are used to support segmented regions.

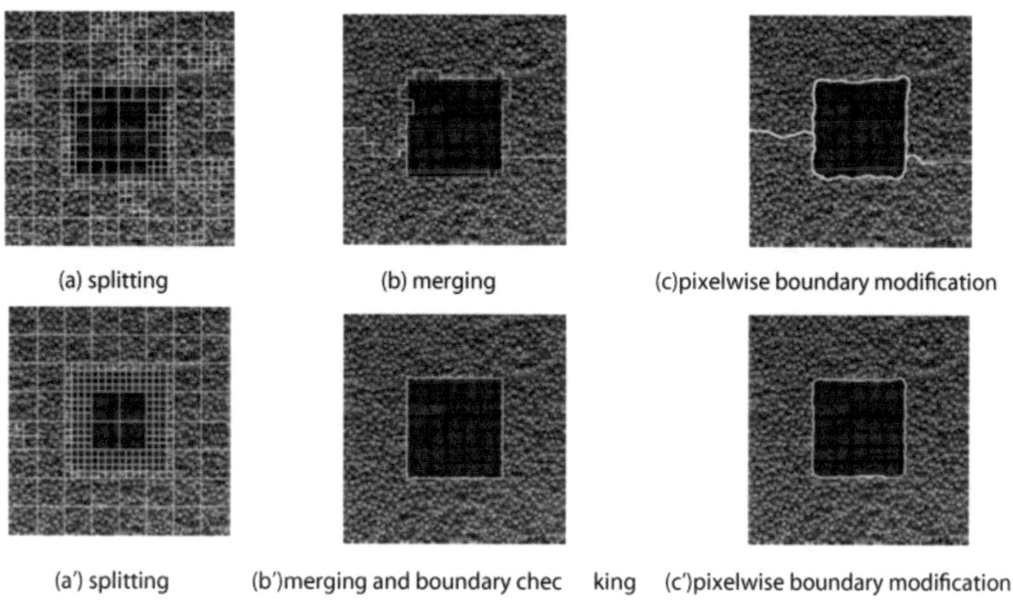

Figure 3. a), (b) and (c) are the results by using the method of T. Ojala et. al. (a'), (b') and (c') are the results by using our earlier segmentation method

For region-based image retrieval, we adopt the unsupervised texture segmentation method [30, 33]. In [30], Ojala et al. use the nonparametric log-likelihood-ratio test and the G statistic to compare the similarity of feature distributions. The method is efficient for finding homogeneously textured image regions. Based on this method, a boundary checking algorithm [34] has been proposed to improve the segmentation accuracy and computational cost. For more details about our segmentation algorithm, we refer the reader to [33]. In this Chapter, the weighted distribution of global information CIH (color index histogram) and local information LBP (local binary pattern) are applied to measure the similarity of two adjacent regions.

An example is shown in Fig. 3. It can be seen that boundary checking algorithm segments the test image correctly, and it costs only about 1/20 processing time of the method in [30]. For color image segmentation, another example is shown in Fig. 4. In Fig. 4(c) Fig. 4(c′), we can see that the boundary checking algorithm achieves robustness segmentation for test image "Akiyo" and another nature image.

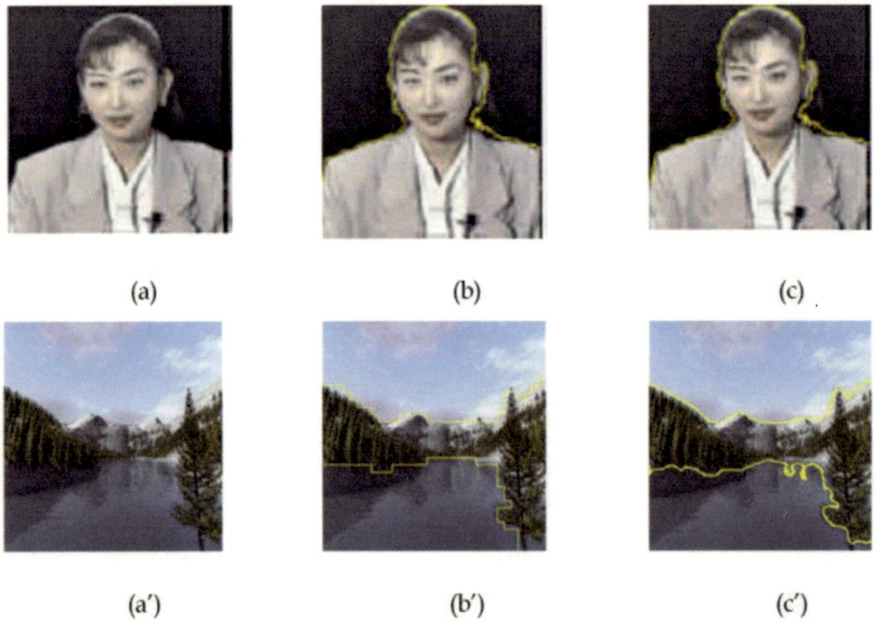

Figure 4. The segmentation processes for test image "Akiyo" and a nature image. (a), (a′) Original image. (b), (b′) Splitting and Merging. (c), (c′) Boundary checking and modification.

4.2. Region representation

To achieve region-based image retrieval, we use two compact and intuitive visual features to describe a segmented region: dominate color descriptor (DCD) and texture. For the first one, we use our modified dominant color descriptor in [19, 26]. The feature representation of a segmented region R is defined as

$$R_{DCD} = \left\{ \left\{ R_{c_i}, R_{p_i} \right\},\ 1 \le i \le 8 \right\}, \tag{6}$$

where R_{c_i} and R_{p_i} are the ith dominant color and its percentage in R, respectively.

For the second one, the texture feature of a region is characterized by the weighted distribution of local binary pattern (LBP) [6, 25, 32]. The advantages of LBP include its invariant property to illumination change and its low computational cost [32]. The value of kth bin in LBP histogram is given by:

$$R_{LBP\_h_K} = \frac{n_K}{P}, \tag{7}$$

where n_K represents the frequency of LBP value at kth bin, and P is the number of pixels in a region. Therefore, the texture feature of region R is defined as

$$R_{texture} = \left\{ \left\{ R_{LBP\_h_k} \right\},\ 1 \le k \le 256 \right\}. \tag{8}$$

In addition, we define a feature R_{poa} to represent the *percentage of area* for region R in the image. Two regions are considered to be visual similar if both of their content (color and texture) and area are similar.

4.3. Image representation and definition of the foreground assumption

For image retrieval, each image in database is described by a set of its non-overlapping regions. For an image I that contains N non-overlaping regions, i.e., $I = \{I_{R^1}, I_{R^2}, \ldots, I_{R^N}\}$, $\cup_{i=1}^{N} I_{R^i}$ and $I_{R^i} \cap I_{R^j} = 0$, where I_{R^i} represents the ith region in I. Although the region-based approaches perform well in [9, 11], their retrieval performances are strongly depends on success of image segmentation because segmentation techniques are still far from reliable for heterogeneous images database. In order to address the possible fails of segmentation, we propose a foreground assumption to "guess" the foreground and background regions in images. For instance, we can readily find a gorilla sitting on the grass as shown in Fig. 5. If Fig. 5 is the query image, the user could be interested in the main subject (gorilla) rather than grass-like features (color, texture, etc). In most case, user would pay more attention to the main subject.

The main goal of foreground assumption is to simply distinguish main objects and irrelevant regions in images. Assume that we can divide an image into two parts: foreground and background. In general, the foreground stands the central region of an image. To emphasize the importance of central region of an image, we define

$$\begin{aligned} R_{foreground} &= \{(x,y) : \tfrac{1}{8}h \le x \le \tfrac{7}{8}h,\ \tfrac{1}{8}w \le y \le \tfrac{7}{8}w\} \\ R_{background} &= \{(x,y) : x < \tfrac{1}{8}h \text{ or } x > \tfrac{7}{8}h, y < \tfrac{1}{8}w \text{ or } y > \tfrac{7}{8}w\}, \end{aligned} \tag{9}$$

where $R_{foreground}$ and $R_{background}$ are the occupied regions of foreground and background, respectively; h and w is height and width of the image.

Figure 5. The definition of foreground and background based on foreground assumption.

In region-based retrieval procedure, segmented regions are required. It can be provided by the users or be generated by the system automatically. However, the criterion for similarity measure is based on the overall distances between feature vectors. If an image in database has background regions that is similar to the foreground object of the query image, this image will be considered as similar image based on the similarity measure. In this case, the accuracy of region-based retrieval system decreases. Therefore, we modify our region representation by adding a Boolean model $BV \in \{0, 1\}$ to determine whether the segmented region R belongs to the background of the query image or not.

$$BV = \begin{cases} 1 & R \in R_{background} \\ 0 & R \notin R_{background} \end{cases} \tag{10}$$

Note that the variable is designed to reduce the segmentation error.

http://dx.doi.org/10.5772/45841

On the other hand, we extract the global features for an image to compensate the inaccuracy of segmentation algorithms. The features F^I includes three feature sets: 1) dominant color $F^I_{R_{DCD}}$ for each region, 2) texture $F^I_{R_{texture}}$ for each region, and 3) dominant color F^I.

$$F^I_{R_{DCD}} = \left\{ \left\{ \left\{ \left\{ R^j_{c_i}, R^j_{p_i} \right\}, 1 \le i \le 8 \right\}, R^j_{poa}, BV_j \right\}, 1 \le j \le N \right\} \tag{11}$$

$$F^I_{R_{texture}} = \left\{ \left\{ \left\{ R^j_{LBP\_h_k} \right\}, 1 \le k \le 256 \right\}, 1 \le j \le N \right\} \tag{12}$$

$$F^I = \left\{ F^I_{global}, F^I_{foreground}, F^I_{background} \right\} \tag{13}$$

where N is the number of partitioned regions in image I; $F^I_{R_{DCD}}$ represents the dominant color vectors; $F^I_{R_{texture}}$ describes the texture distribution for each region; F^I_{global}, $F^I_{foreground}$ and $F^I_{background}$ represent the global, foreground and background color features, respectively. In brief, the images are first segmented using the fast color quantization scheme. Then, the dominant colors, texture distribution and the three color features are extracted in the image.

5. Integrated region-based relevance feedback framework

In region-based image retrieval, an image is considered as relevant if it contains some regions with satisfactory similarity to the query image. The retrieval system can reconstruct a new query that includes only the relevant regions according to user's feedback. In this way, the system can capture the user's query concept automatically. For example, Jing et al. [8] suggest that information in every region could be helpful in retrieval, and group all regions of positive examples by K-means algorithm iteratively to ensure the distance between all the clusters not exceeding a predefined threshold. Then, all regions within a cluster are merged into a new region. However, the computational cost for merging new regions is proportional to the number of positive examples. Moreover, users might be more interested in some specified regions or main objects rather than the positive examples.

To speed up the system, we introduce a similarity matrix model to infer the region-of-interest sets. Inspired by the query-point movement method [8, 31], the proposed system performs similarity comparisons by analyzing the salient region in pseudo query image and relevant images based on user's feedback information.

5.1. The formation of region-of-interest set

5.1.1. Region-based similarity measure

In order to perform region-of-interest (ROI) queries, the relevant regions are obtained by the measurement of region-based color similarity $R\_S(R, R')$ and region based texture similarity $R\_S_T(R, R')$ in Eq. (14) and (15), respectively. This similarity measure allows users to select their relevant regions accurately. Note that the conventional color histogram could not be applied on DCD directly because the images do not have exact numbers of dominant colors [12]. The region-based color similarity between two segmented regions R and R' can be calculated by

$$
\begin{aligned}
&R\_S(R,R') = R\_S_c(R,R') \times R\_S_{poa}(R,R') \\
&R\_S_c(R,R') = \sum_{i=1}^{m}\sum_{j=1}^{n}\min\left(R_{p_i}, R'_{p_j}\right), \quad if \;\; d\left(R_{c_i}, R'_{c_j}\right) < T_d,
\end{aligned}
\tag{14}
$$

where m and n are the number of dominate colors in R and R', respectively; $R\_S_c(R, R')$ is the maximum similarity between two regions in similar color percentage. If the pair-wise Euclidean distance of two dominate color vector c_i and c_j is less than a predefined threshold T_d, it is set to 25 in our work. The notation $R\_S_{poa}(R, R')$ is used to measure the similarity of the area percentage for region pair (R, R'). To measure the texture similarity between two regions, we define

$$
R\_S_T(R,R') = \frac{\sum_{k=1}^{256}\min\left(R_{LBP\_hk}, R'_{LBP\_hk}\right)}{\min\left(R_{Pxl}, R'_{Pxl}\right)},
\tag{15}
$$

where R_{Pxl} and R'_{Pxl} represent the number of pixels in regions R and R', respectively; $\min(R_{LBP\_h\,k}, R'_{LBP\_h\,k})$ is the intersection of LBA histogram for the kth bin.

Theoretically, visual similar is achieved when both color and texture are similar. For example, two regions should be considered as non-similar if they are similar in terms of color but not texture. This can be achieved by imposing

$$
R\_S > 0.8 \;\text{ and }\; R\_S_T > 0.9. \tag{16}
$$

5.1.2. Similarity matrix model

In the following, we introduce a region-based similarity matrix model. The regions of positive examples, which helps the system to find the intention of user's query, are able to exclude the irrelevant regions flexibly. The proposed similarity matrix model is described as follows.

The region similarity measure is performed for all regions. The relevant image set is denoted as $R_s = \{I^i;\ i=1, \ldots, N\}$, where N represents the number of positive images from user's feedback, and each positive image I^i contains several segmented regions. See Fig. 6.

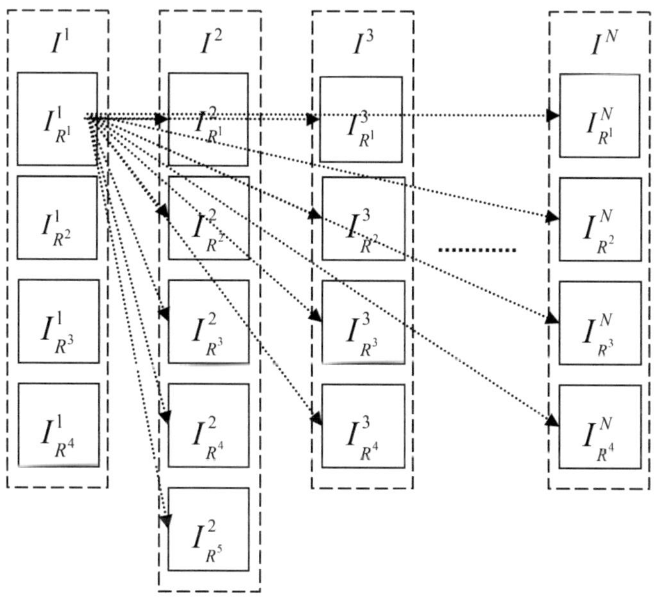

Figure 6. The similarity matching for region pairs.

As an example, let $R_s = \{I^1, I^2, I^3\}$ contains three relevant images, where $I^1 = \{I^1_{R1}, I^1_{R2}, I^1_{R3}\}$, $I^2 = \{I^2_{R1}, I^2_{R2}, I^2_{R3}\}$ and $I^3 = \{I^3_{R1}, I^3_{R2}\}$. Our similarity matrix model to infer the user's query concept is shown in Fig. 7, where the symbol "1" means that two regions are regarded as similar. On the contrary, the symbol "0" represents that two regions are non-similar in content.

To support ROI queries, we perform the one-to-many relationships to find a collection of similar region sets, e.g., $\{I^1_{R1}, I^2_{R1}, I^3_{R2}\}$, $\{I^1_{R2}, I^2_{R2}\}$, $\{I^1_{R3}, I^2_{R2}, I^2_{R3}\}$, $\{I^2_{R1}, I^1_{R1}, I^3_{R2}\}$, $\{I^2_{R2}, I^1_{R2}, I^1_{R3}\}$, $\{I^2_{R3}, I^1_{R3}\}$, $\{I^3_{R1}\}$ and $\{I^3_{R2}, I^1_{R1}, I^2_{R1}\}$, see Fig. 8. After this step, several region-of-interest sets can be obtained by merging all similar region sets. For example, the first set $\{I^1_{R1}, I^2_{R1}, I^3_{R2}\}$ contains three similar regions. Each region will be merged together with the above eight similar region sets. In this example, three region-of-interest sets can be obtained by the merging operation, i.e., $\{I^1_{R1}, I^2_{R1}, I^3_{R2}\}$, $\{I^1_{R2}, I^1_{R3}, I^2_{R2}, I^2_{R3}\}$ and $\{I^3_{R1}\}$. Since user may be interested in some repeated similar regions, the single region set $\{I^3_{R1}\}$ could be assumed to be irrelevant in our approach. Therefore, we have

$ROI^1 = \{I^1_{R^1}, I^2_{R^1}, I^3_{R^2}\}$ and $ROI^2 = \{I^1_{R^2}, I^1_{R^3}, I^2_{R^2}, I^2_{R^3}\}$ as shown in Fig. 8. The two sets are considered as region-of-interests that reflect user's query perception.

	$I^1_{R^1}$	$I^1_{R^2}$	$I^1_{R^3}$	$I^2_{R^1}$	$I^2_{R^2}$	$I^2_{R^3}$	$I^3_{R^1}$	$I^3_{R^2}$
$I^1_{R^1}$	×	×	×	1	0	0	0	1
$I^1_{R^2}$	×	×	×	0	1	0	0	0
$I^1_{R^3}$	×	×	×	0	1	1	0	0
$I^2_{R^1}$	1	0	0	×	×	×	0	1
$I^2_{R^2}$	0	1	1	×	×	×	0	0
$I^2_{R^3}$	0	0	1	×	×	×	0	0
$I^3_{R^1}$	0	0	0	0	0	0	×	×
$I^3_{R^2}$	1	0	0	1	0	0	×	×

Figure 7. Our proposed matrix structure comparison. ×: no comparison for those regions in the same image, 1: similar regions and 0: non-similar regions.

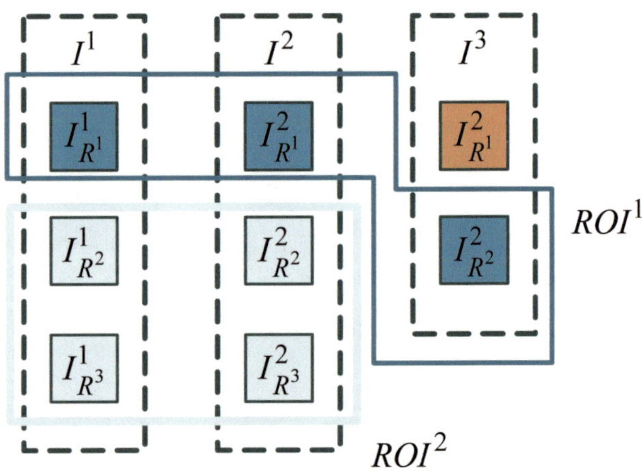

Figure 8. The region-of-interest sets based on the proposed matrix structure comparison.

If users are interested in many regions, the simple merging process can be used to capture the query concept. In Fig. 8, for example, $\{I^1_{R^2}, I^1_{R^3}\}$ and $\{I^2_{R^2}, I^2_{R^3}\}$ are the regions belong to the same relevant image I^1 and I^2, respectively. It can be seen that the similar matrix approach is consistent with human perception and is efficient for region-based comparison.

5.1.3. Salient region model

To improve retrieval performance, all the region-of-interest sets from the relevant image set R_s will be integrated for the next step during relevance feedback. As described in previous subsection, each region-of-interest set could be regarded as a collection of regions, and extracted information can be used to identify the user's query concept. However, correctly capturing the semantic concept from the similar regions is still a difficult task. In this stage, we define salient region as all similar regions within each ROI set. The features of the new region are equal to the weighted average features of individual regions.

In order to emphasize the percentage of area feature, we modified the dominant color descriptor in Eq. (1). The feature representation of the salient region *SR* is described as

$$F_{SR} = \left\{ \left\{ \left\{ \overline{C}_i, \ \overline{P}_i \right\}, \ 1 \le i \le 8 \right\}, \ \overline{R}_{poa} \right\}, \tag{17}$$

where $\overline{C}_i$ is the i th average dominant color of similar region.

All similar regions in ROI can be determined from the eight uniformly divided partitions in RGB color space as shown in Fig. 9.

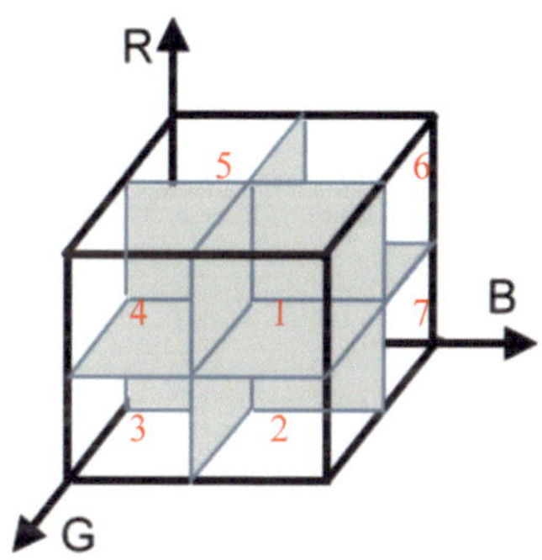

Figure 9. The division of *RGB* color space.

$$\overline{C}_i = \left(\sum_{j=1}^{N_{c_i}} \frac{R_{p_i}^j \times R_{c_i}^j(R)}{\sum_{j=1}^{N_{c_i}} R_{p_i}^j}, \ \sum_{j=1}^{N_{c_i}} \frac{R_{p_i}^j \times R_{c_i}^j(G)}{\sum_{j=1}^{N_{c_i}} R_{p_i}^j}, \ \sum_{j=1}^{N_{c_i}} \frac{R_{p_i}^j \times R_{c_i}^j(B)}{\sum_{j=1}^{N_{c_i}} R_{p_i}^j} \right), \ 1 \le i \le 8 \tag{18}$$

where N_{c_i} is the number of dominant colors in cluster i ; $R_{c_i}^j(R)$, $R_{c_i}^j(G)$ and $R_{c_i}^j(B)$ represent the dominant color components of R, G and B located within partition i for the region j, respectively; $R_{P_i}^{j}$ represents the percentage of its corresponding 3-D dominant color vector in R^{j}; $\overline{P}_i$ is the average percentage of dominant color in the ith coarse partition, i.e., $\overline{P}_i = \frac{\sum_{j=1}^{N_{C_i}} R_{P_i}^{j}}{N_{C_i}}$; $\overline{R}_{poa}$ is the average percentage of area for all similar regions in ROI.

5.2. The pseudo query image and region weighting scheme

To capture the inherent subjectivity of user perception, we define a pseudo image I^+ as the set of salient regions, $I^+ = \{SR^1,\ SR^2,\ ...,\ SR^n\}$. The feature representation of I^+ can be written as

$$F_{SR}^{I^+} = \left\{ \left\{ \left\{ (\overline{C}_i^{1}, \overline{P}_i^{1}),\ 1 \le i \le 8 \right\},\ \overline{R}_{poa}^{1} \right\}, ..., \left\{ \left\{ (\overline{C}_i^{n}, \overline{P}_i^{n}),\ 1 \le i \le 8 \right\},\ \overline{R}_{poa}^{n} \right\} \right\}. \tag{19}$$

During retrieval, the user chooses the best matched regions what he/she is looking for. However, the retrieval system cannot precisely capture the user's query intention at the first or second steps of relevance feedback. With the increasing of the returned positive images, query vectors are then constructed to perform better results. Taking average [8] from all the feedback information could introduce redundant, i.e., information from irrelevant regions. Motivated by this observation, we suggest that each similar region in ROI should be properly weighted according to the amount of similar regions. For example, the ROI^2 in Fig. 8 is more important than in ROI^1. The weights associated with the significance of SR in I^+ can be dynamically updated as

$$w_l = \frac{|ROI^l|}{\sum_{l=1}^{n} |ROI^l|}, \tag{20}$$

where $|ROI^l|$ represents the number of similar regions in region-of-interest set l, and n is the number of region-of-interest sets.

5.3. Region-based relevance feedback

In reality, inaccurate segmentation leads to poor matching result. However, it is difficult to ask for precise segmented regions from users. Based on the foreground assumption, we define three feature vectors, which are extracted from entire image (i.e., global dominant color), foreground and background, respectively. The advantage of this approach is that it provides an estimation that minimizes the influence of inaccurate segmentation. To integrate the two regional approaches, we summarize our relevance feedback as follows.

For the initial query, the similarity measure $S(F^{I}_{entireImage}, F^{I'}_{entireImage})$ for the initial query image I and target image I' in database are compared by using Eq. (4). Therefore, a coarse relevant-image set can be obtained. Then, all regions in the initial query image I and the positive images based on the user's feedback information are merged into relevant image set $R_s = \{I, I^1, I^2, ..., I^N\}$. The proposed region-based similarity matrix model performs Eq. (14) and (15) to find the collection of the similar regions. The similar regions can be determined by Eq. (16), and then be merged into salient region SR. For the next iteration, the feature representation of I^+ in Eq. (19) could be regarded as an optimal pseudo query image that is characterized by salient regions.

It should be noted that I^+ and R_s defined above both contain the relevance information that reflects human semantics. The similarity measure for pseudo query image $F^{I^+}_{SR^l}$ and target image $F^{I'}_{R^j_{DCD}}$ is calculated by

$$S_{region\_based}\left(I^+, I'\right) = \sum_{l=1}^{n}\sum_{j=1}^{m} w_l \times \max R\_S\left(F^{I^+}_{SR^l}, F^{I'}_{R^j_{DCD}}\right), \tag{21}$$

where n is the number of salient region sets in I^+; m is the number of color/texture segmented regions in target image I'; w_l is the weight of salient region SR^l. In Eq. (21), the image-to-image similarity matching maximizes the value of region based color similarity by using Eq. (14). If the Boolean model $BV = 1$ for a partitioned region in target image, then the background of the image will be excluded for matching in Eq. (21).

On the other hand, R_s is a collection of relevant images based on the user's feedback information. Since poor matches arise from inaccurate image segmentations, three global features $F^{I}_{\text{entireImage}}$, $F^{I}_{\text{foreground}}$ and $F^{I}_{\text{backgrounde}}$ in Eq. (13) are extracted to compensate the inaccuracy. The similarity between the relevant image set $R_s = \{I, I^1, I^2, ..., I^N\}$ and target image I' in database is calculated by

$$\begin{aligned} S_{\text{entireImage}}(R_s, I') &= \sum_{i=1}^{N} \max S(F^{R_s}_{\text{entireImage}}, F^{I'}_{\text{entireImage}}) \\ S_{\text{forground}}(R_s, I') &= \sum_{i=1}^{N} \max S(F^{R_s}_{\text{foreground}}, F^{I'}_{\text{forground}}) \\ S_{\text{background}}(R_s, I') &= \sum_{i=1}^{N} \max S(F^{R_s}_{\text{background}}, F^{I'}_{\text{background}}) \end{aligned} \tag{22}$$

where $F^{R_s}_{\text{entireImage}}$, $F^{R_s}_{\text{foreground}}$ and $F^{R_s}_{\text{background}}$ are dominant colors, foreground and background for the ith relevant image in R_s, respectively. In Eq. (22), the similarity measure maximizes the similarity score using Eq. (5). To reflect the difference between R_s and target image I', the average similarity measure is given by

$$S_{\text{avg}}(R_s, I') = \frac{(S_{\text{entireImage}}(R_s, I') + S_{\text{foreground}}(R_s, I') + S_{\text{background}}(R_s, I'))}{3}. \tag{23}$$

It is worth to mention that our region-based relevance feedback approach defined above is able to reflect human semantics. In other words, user might aware some relevant image from the initial query, and then provides some positive image.

Considering the ability to capture the user's perceptions more precisely, the system determines the retrieved rank according to average of region-based image similarity measure in Eq. (21) and foreground-based similarity measure in Eq. (23).

$$S = \frac{S_{\text{region\_based}}\left(I^{+}, I'\right) + S_{\text{avg}}(R_s, I')}{2}. \tag{24}$$

6. Experimental results

We use an image database (31 categories about 3991 images) for general-purpose from Corel's photo to evaluate the performance of the proposed framework. The database has a variety of images including animal, plant, vehicle, architecture, scene, etc. It has the advantages of large size and wide coverage [11]. Table 1 lists the labels for 31 classes. The effectiveness of our proposed region-based relevance feedback approach is evaluated.

In order to make a comparison on the retrieval performance, both average retrieval rate (ARR) and average normalized modified retrieval rank (ANMRR) [26] are applied. An ideal performance will consist of ARR values equal to 1 for all values of recall. A high ARR value represents a good performance for retrieval rate, and a low ANMRR value indicates a good performance for retrieval rank. The brief definitions are given as follows. For a query q, the ARR and ANMRR are defined as:

$$\text{ARR}(q) = \frac{1}{NQ} \sum_{q=1}^{NQ} \frac{NF(\beta, q)}{NG(q)}, \tag{25}$$

$$\text{AVR}(q) = \sum_{k=1}^{NG(q)} \frac{Rank(k)}{NG(q)}, \tag{26}$$

http://dx.doi.org/10.5772/45841

Class 1 (gorilla)	Class 2 (bird)	Class 3 (potted plant)	Class 4 (card)	Class 5 (cloud)	Class 6 (sunset)	Class 7 (pumpkin)
			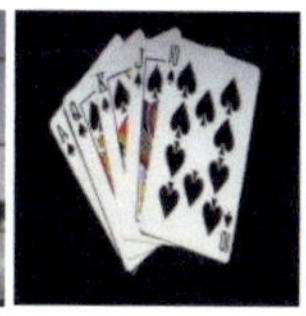	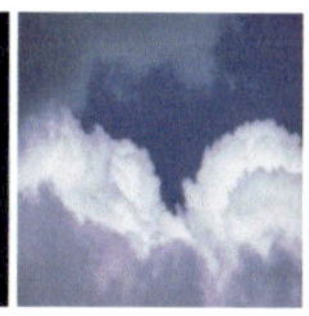		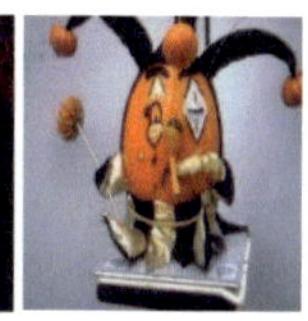
Class 8 (cake)	Class 9 (dinosaur)	Class 10 (dolphin)	Class 11 (elephant)	Class 12 (firework)	Class 13 (flower)	Class 14 (food)

Class 15 (duck)	Class 16 (leopard)	Class 17 (leaf)	Class 18 (car)	Class 19 (cactus)	Class 20 (airplane)	Class 21 (painting)

Class 22 (sea-elephant)	Class 23 (horse)	Class 24 (helicopter)	Class 25 (boat)	Class 26 (snow)	Class 27 (balloon)	Class 28 (waterfall)

Class 29 (building)	Class 30 (stadium)	Class 31 (people)				

Table 1. The labels and examples of the test database.

$$\mathrm{MRR}(q) = \mathrm{AVR}(q) - 0.5 - \frac{NG(q)}{2}, \tag{27}$$

$$\mathrm{NMRR}(q) = \frac{\mathrm{MRR}(q)}{K + 0.5 - 0.5 \times NG(q)}, \tag{28}$$

$$\mathrm{ANMRR}(q) = \frac{1}{NQ}\sum_{q=1}^{NQ} \mathrm{NMRR}(q), \tag{29}$$

where NQ is total number of queries; $NG(q)$ is the number of the ground truth images for a query. The notation is a factor, and $NF(\beta, q)$ is number of ground truth images found within the first $\beta \cdot NG(q)$ retrievals. $Rank(k)$ is the rank of the retrieved signature image in the ground truth. In eq.(28), $K = \min(4 \cdot NG(q); 2 \cdot GTM)$, where GTM is $\max\{NG(q)\}$ for all queries. The NMRR and its average (ANMRR) are normalized to the range of [0 1].

To test the performance of our integrated approach for region-based relevance feedback, we first query an image with a gorilla sits on grass as shown Fig. 10(a).

As mentioned in Section 5.4, the dominant color between query image I and target image I' is used for similarity measure in the initial query. The retrieval results are shown in Fig. 10(b), the top 20 matching images are arranged from left to right and top to bottom in order of decreasing similarity score.

(a)

(b)

Figure 10. The initial query image and positive images. (a) Query image. (b) The 5 positive images in the first row are selected by user.

http://dx.doi.org/10.5772/45841

For better understanding of the retrieval results, the DCD vectors of the query image, rank 6th image and rank 8th image are listed, respectively. See Fig. 11. It can be seen that the query image and the image "lemon" are very similar in the first dominant color (marked by box). If we use the global DCD as the only feature for image retrieval, the system only returns eleven correct matches. Therefore, further investigation on extracting comprehensive image features is needed.

Query image Q	Target image F_1 (rank=6)	Target image F_2 (rank=8)
{(52, 65,18), 0.592}	{(54,64,13), 0.552653}	{(53,59,33), 0.61262}
{(105,139,35), 0.21583}	{(122,130,54), 0.000112}	{(117,124,135), 0.005198}
{(138,117,89), 0.009827}	{(145,130,111), 0.032054}	{(116,137,69), 0.055452}
{(140,125,132), 0.00001}	{(181,165,151), 0.415181}	{(139,112,95), 0.017904}
{(163,174,58), 0.165853}		{(159,120,134), 0.001343}
{(197,181,157), 0.016479}		{(147,153,92), 0.136322}
		{(183,176,169), 0.171163}

Figure 11. Example images with the dominant colors and their percentage values. First row: 3-D dominant color vector c_i and the percentage p_i for each dominant color. Middle row: the original images. Bottom row: the corresponding quantized images.

Assume that the user has selected five best matched images, marked by red box, as shown in Fig. 10(a). In conventional region-based relevance feedback approach, all regions in the initial query image I and the five positive images are merged into relevant image set

$R_s = \{I, I^1, I^2, ..., I^5\}$. The proposed similarity matrix model is able to find the region-of-interest region sets. For the next query, I^+ could be regarded as a new query image which is composed of some salient regions. The retrieval results based on the new query image I^+ are shown in Fig. 12. The following are discussions.

1. The pseudo query image I^+ is capable to reflect user's query perception. Without considering the Boolean model in Eq. (21), the similarity measure by Eq. (21) returns 16 correct matches as shown in Fig. 12.

2. Using the pseudo image I^+ as query image, the initial query image is not ranked first but fifth, as shown in Fig. 12.

Figure 12. The retrieval results based on new pseudo query image I^+ for the first iteration.

3. The retrieval results return three dissimilar images (marked by red rectangle boxes), which ranks are 7th, 8th and 12th, respectively.

4. To analyze the improper result, the dominant color vectors and percentage of area of "cucumber" and "lemon" are listed. See Fig. 13. We can see that each of the images "gorilla", "cucumber" and "lemon" contains three segmented regions. For each region, the number of the dominant colors, percentage of area and BV value are listed and colored red. For similarity matching, the dominant colors (i.e. region#1, region#2 and region#3) of initial image "gorilla" are similar to the dominant color (marked by red rectangle

http://dx.doi.org/10.5772/45841

box) of the image "cucumber". In addition, the percentages of area (0.393911, 0.316813, 0.289276) of initial image "gorilla" are similar to the percentage of area (region#2, 0.264008) of the image "cucumber". The other similarity comparisons between "gorilla" and "cucumber" image are not presented here because the maximum similarity between two regions in Eq. (14) is very small. In brief, without considering the exclusion of irrelevant regions, the region-based image-to-image similarity model in Eq. (21) could cause improper ranks in visualization.

The initial image "gorilla"	The image "cucumber"	The images "lemon"
Region#1 (5, 0.393911, 1) {(54,66,21), 0.358263} {(106,138,38), 0.459055} {(137,118,90), 0.002918} {(162,173,60), 0.175735} {(195,180,156), 0.004029} Region#2 (6, 0.316813, 0) {(54,66,21), 0.901682} {(106,138,38), 0.00366} {(137,118,90), 0.024017} {(142,127,130), 0.000032} {(162,173,60), 0.026554} {(195,180,156), 0.044053} Region#3 (5, 0.289276, 1) {(54,66,21), 0.571122} {(106,138,38), 0.116995} {(137,118,90), 0.003692} {(162,173,60), 0.304955} {(195,180,156), 0.003235}	**Region#1** (4, 0.454976, 1) {(34,41,22), 0.003331} {(121,131,93), 0.000089} {(134,127,118), 0.000022} {(203,199,197), 0.996557} **Region#2** (4, 0.264008, 0) {(43,51,27), 0.985551} {(121,131,93), 0.002004} {(134,127,118), 0.000501} {(203,199,197), 0.011945} **Region#3** (3, 0.281016, 1) {(41,48,26), 0.049412} {(134,127,118), 0.012814} {(203,199,197), 0.937774}	**Region#1** (3, 0.42462, 1) {(5,6,1), 0.035743} {(145,130,111), 0.006157} {(181,165,151), 0.9581} **Region#2** (4, 0.25519, 0) {(52,62,12), 0.933309} {(122,130,54), 0.000438} {(145,130,111), 0.035119} {(181,165,151), 0.031133} **Region#3** (3, 0.32019, 0) {(53,63,12), 0.998539} {(145,130,111), 0.000191} {(181,165,151), 0.001271}

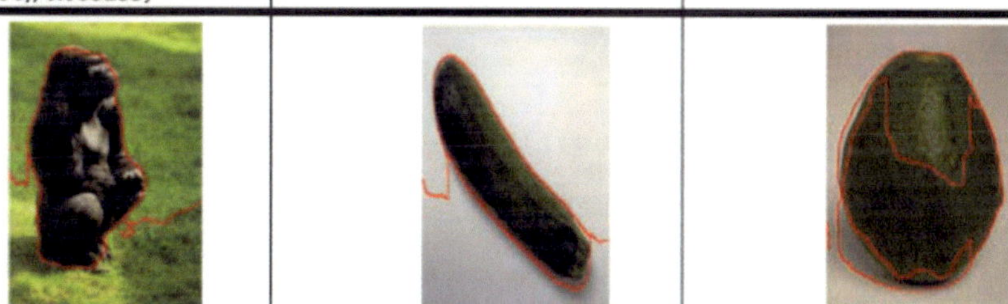

Figure 13. The analysis of retrieval results using the conventional region-based relevance feedback approach. Top row: dominant color distributions and percentage of area *Poa* for each region in initial query image, "cucumber" and "lemon" images. Bottom row: the corresponding segmented images.

The retrieval performance can be improved by automatically determining the user's query perception. In the following, we would like to evaluate the advantages of our proposed relevance feedback approach. For the second query, the integrated region-based relevance feedback contains not only the salient-region information, but also the "specified-region" information based on relevant images set R_s . The retrieval results based on our integrated region-based relevance feedback are shown in Fig. 14. Observations and discussions are described as follows.

1. The system returns 18 correct matches as shown in Fig. 14.

2. In Fig. 13, region#1 and region#3 in query image are two grass-like regions, which are labeled as inner region, i.e., $BV = 1$. On the other hand, the region#2 in image "cucumber" is a green region that is similar to the grass-like regions in query image. In our method, this problem can be solved by examining the BV value in Eq. (21). As we can see, none of the three incorrect images including "cucumber", "lemon" and "carrot" in Fig. 12 appears in the top 20 images in Fig. 14.

3. In contrast, it is possible that the grass-like regions are parts of the user's aspect. In this case, the three feature vectors including entire image, foreground and background can be used to compensate the loss of generality. In Fig. 14 retrieval results indicate that the high performance is achieved by using these features.

4. Our proposed relevance feedback approach can capture the query concept effectively. In Fig. 14, it can be seen that most of the retrieval results are considered to be highly correlated. In this example, 90% of top 20 images are correct images. In general, the features in all retrieval results look similar to gorilla or grass. The results reveal that the proposed method improves the performance of the region-based image retrieval.

Figure 14. The retrieval results based on our integrated region-based relevance feedback.

http://dx.doi.org/10.5772/45841

In Fig. 15-17, further examples are tested to evaluate the performance of the integrated region-based relevance feedback for nature images. In Fig. 15, the contents of the query image include a red car on country road by the side of grasslands. If the user is only interested in the red car, four positive images marked by red boxes will be selected as shown in Fig. 15 (b). In this case, retrieval results (RR=0.25, NMRR=0.7841) are far from satisfactory performance for the initial query.

(a)

(b)

Figure 15. The initial query image and positive images. (a) Query image. (b) The 4 positive images marked by red boxes which are selected by user.

After the submission of pseudo query image I^+ and relevant images set R_s based on user's feedback information, the first feedback retrieval returns 10 images containing "red car" as shown in Fig. 16. For this example, the first feedback retrieval achieves an ARR improvement of 28.6%. More precise results can be achieved by increasing of the number of region-of-interest sets and relevant image set based for the second feedback retrieval as shown in Fig. 17. The retrieval results for the second feedback retrieval returns 11 images containing "red car", and achieve an NMRR improvement of 35% compared to the initial query. Furthermore, the rank order in Fig. 17 is more reasonable than that in Fig. 16.

To show the effectiveness of our proposed region-based relevance feedback approach, the quantitative results for individual class and average performance (ARR, ANMRR) are listed in Table 2 and 3, which show the comparison of the performance for each query. It can be seen that the performance of retrieving precision and rank are relatively poor for the initial query. Through the adding positive examples by user, feedback information could have more potential in finding the user's query concept by means of optimal pseudo query image I^+ and relevant images set R_s as described in Section 5.4. In summary, the first feedback query improves 30.8% of ARR gain and 28% of ANMRR gain, and the second feedback query further improves 10.6% of ARR gain and 11% of ANMRR gain as compared with first feedback query. Although the improvement of retrieval efficiency is decreases progressively after two or three feedback queries, the proposed technique is able to provide satisfactory retrieval results in that few feedback queries.

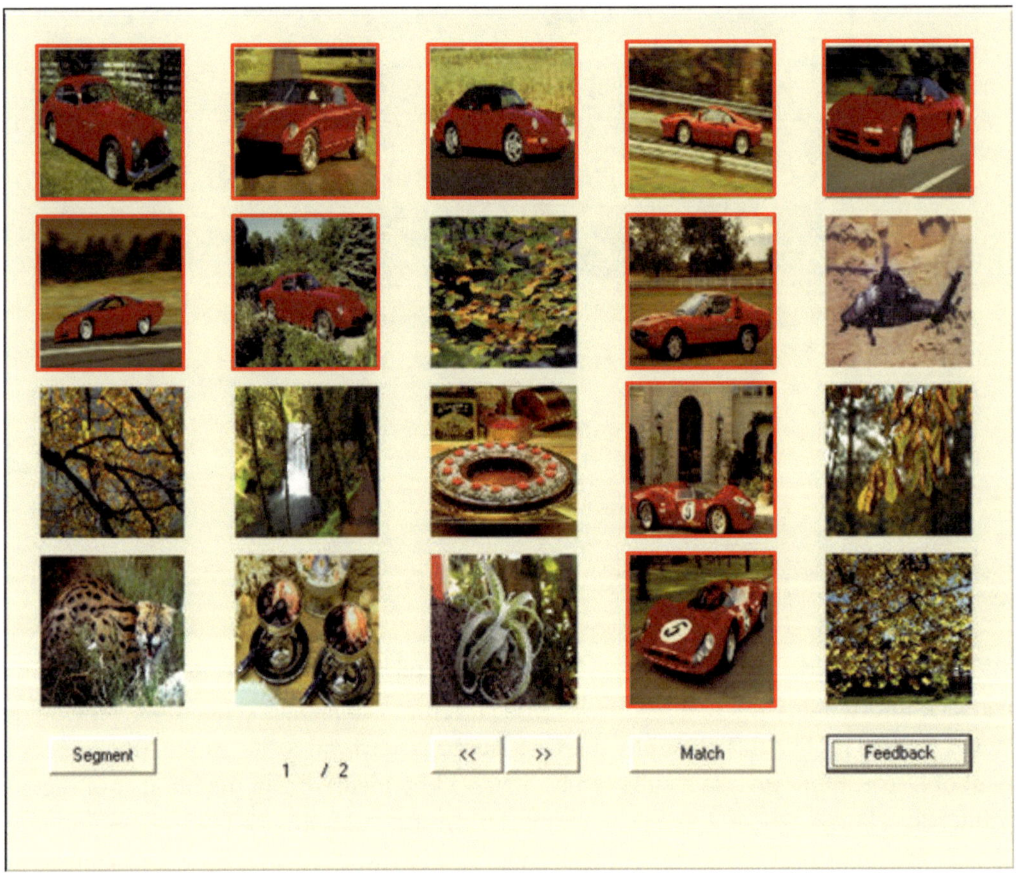

Figure 16. The retrieval results by our integrated region-based relevance feedback for the first iteration.

http://dx.doi.org/10.5772/45841

Figure 17. The retrieval results by our integrated region-based relevance feedback for the second iteration.

7. Conclusion

The conventional existing region-based relevance feedback approaches work well in some specified applications; however, their performances depend on the accuracy of segmentation techniques. To solve this problem, we have introduced a novel region-based relevance feedback for image retrieval with the modified dominant color descriptor. The term "specified area", which combines main objects and irrelevant regions in image, has been defined for compensating the inaccuracy of segmentation algorithm. In order to manipulate the optimal query, we have proposed the similarity matrix model to form the salient region sets. Our integrated region-based relevance feedback approach contains relevance information including pseudo query image I^+ and relevant images set R_s , which are capable to reflect the user's query perception. Experimental results indicate that the proposed technique achieves precise results in general-purpose image database.

Class	Initial query	The 1st feedback query	The 2nd feedback query
1	0.28	0.465	0.635
2	0.56	0.785	0.845
3	0.31	0.53	0.535
4	0.8375	0.85	0.9
5	0.19	0.275	0.32
6	0.255	0.355	0.385
7	0.2	0.29	0.3
8	0.165	0.235	0.245
9	0.73	0.985	1
10	0.345	0.525	0.625
11	0.23	0.345	0.4
12	0.835	1	1
13	0.33	0.52	0.63
14	0.235	0.38	0.4
15	0.655	0.885	0.98
16	0.435	0.625	0.705
17	0.365	0.465	0.515
18	0.235	0.275	0.275
19	0.32	0.505	0.59
20	0.34	0.59	0.635
21	0.37	0.76	0.865
22	0.22	0.355	0.495
23	0.15	0.21	0.225
24	0.31	0.46	0.565
25	0.25	0.43	0.465
26	0.38	0.515	0.61
27	0.245	0.34	0.395
28	0.385	0.415	0.46
29	0.195	0.325	0.41
30	0.4125	0.8	0.8875
31	0.3	0.51	0.61
Avg.	0.357097	0.51629	0.577661

Table 2. Comparisons of ARR performance with different iterations by our proposed integrated region-based relevance feedback approach.

http://dx.doi.org/10.5772/45841

Class	Initial query	The 1st feedback query	The 2nd feedback query
1	0.735	0.399	0.306
2	0.624	0.395	0.326
3	0.741	0.519	0.503
4	0.246	0.135	0.118
5	0.745	0.694	0.643
6	0.744	0.643	0.581
7	0.783	0.721	0.633
8	0.762	0.578	0.537
9	0.215	0.155	0.132
10	0.745	0.571	0.553
11	0.794	0.619	0.557
12	0.331	0.156	0.144
13	0.683	0.591	0.517
14	0.807	0.728	0.709
15	0.514	0.256	0.161
16	0.687	0.559	0.416
17	0.712	0.579	0.554
18	0.836	0.81	0.798
19	0.763	0.512	0.438
20	0.699	0.548	0.488
21	0.716	0.311	0.293
22	0.805	0.664	0.581
23	0.851	0.809	0.797
24	0.725	0.691	0.556
25	0.782	0.645	0.623
26	0.699	0.587	0.503
27	0.791	0.688	0.628
28	0.642	0.613	0.561
29	0.851	0.687	0.649
30	0.662	0.321	0.287
31	0.779	0.587	0.514
Avg.	0.692548	0.541	0.48729

Table 3. Comparisons of ANMRR performance with different iterations by our proposed integrated region-based relevance feedback approach.

Acknowledgement

This work was supported by the National Science Counsel of Republic of China Granted NSC. 97-2221-E-214-053-.

Appendix

BV : Boolean model, which is used to determine whether the segmented region R belongs to the background or foreground.

F : dominant color descriptor

D^2 : similarity measure (dominant color descriptor)

I_{R^i} : the ith non-overlaping region in I

R_{DCD} : dominate color descriptor (DCD) of a segmented region R

$R_{LBP\_h_K}$: the value of kth bin in LBP histogram

R_S : region-based color similarity

R_S_c : the maximum similarity between two regions in similar color percentage

R_S_{poa} : similarity of the area percentage

R_S_T : region based texture similarity

$R_{background}$: defined background based on foreground assumption

$R_{foreground}$: defined foreground based on foreground assumption

R_{poa} : the percentage of area for region R in the image

R_s : relevant image set

$R_{texture}$: texture feature of region R

$a_{i,j}$: similarity coefficient between two color clusters (dominant color descriptor)

c_i : dominant color vector (dominant color descriptor)

$d_{i,j}$: Euclidean distance between two color clusters (dominant color descriptor)

p_i : percentage of each dominant color (dominant color descriptor)

http://dx.doi.org/10.5772/45841

Author details

Nai-Chung Yang[1], Chung-Ming Kuo[1] and Wei-Han Chang[2]

1 Department of Information Engineering, I-Shou University Tahsu, Kaohsiung, Taiwan R.O.C.

2 Department of Information Management, Fortune Institute of Technology, Daliao, Kaohsiung, Taiwan R.O.C.

References

[1] A. Gaurav, T. V. Ashwin and G. Sugata, An image retrieval system with automatic query modification, IEEE Trans. Multimedia, 4(2) (2002) 201-214.

[2] Y. Deng, B. S. Manjunath, C. Kenney, M. S. Moore, and H. Shin, An efficient color representation for image retrieval, IEEE Trans. Image Process., 10(1) (2001) 140–147.

[3] Y. Yang, F. Nie, D. Xu, J. Luo, Y. Zhuang and Y. Pan, A Multimedia Retrieval Framework Based on Semi-Supervised Ranking and Relevance Feedback, IEEE Trans. Pattern Anal. Mach. Intell. , 34(4) (2012) 723-742.

[4] M.Y. Fang, Y.H. Kuan, C.M. Kuo, C.H. Hsieh, "Effective image retrieval techniques based on novel salient region segmentation and relevance feedback," Multimedia Tools and Applications, 57(3), (2012) 501-525.

[5] S. Murala, R. P. Maheshwari and R. Balasubramanian, Local Tetra Patterns: A New Feature Descriptor for Content-Based Image Retrieval, IEEE Trans. Image Process., 21(5) (2012) 2874-2886.

[6] G. Ciocca and R. Schettini, Content-based similarity retrieval of trademarks using relevance feedback, Pattern Recognit., 34(8) (2001) 1639-1655.

[7] X. He, O. King, W. Y. Ma, M. Li, and H. J. Zhang, Learning a Semantic Space from User's Relevance Feedback for Image Retrieval, IEEE Trans. Circ. Syst. Vid. technol., 13(1) (2003) 39-48.

[8] F. Jing, M. J. Li, H. J. Zhang and B. Zhang, Relevance Feedback in Region-Based Image Retrieval, IEEE Trans. Circ. Syst. Vid. technol., 14(5) (2004) 672-681.

[9] T. P. Minka and R. W. Picard, Interactive learning using a society of models, Pattern Recognit., 30(4) (1997) 565–581.

[10] K. Vu, K. A. Hua and W. Tavanapong, Image Retrieval Based on Regions of Interest, IEEE Trans. Knowl. Data Eng., 15(4) 2003 1045-1049.

[11] R. Yong, T. S. Huang, M. Ortega and S. Mehrotra, "Relevance Feedback: A Power Tool for Interactive Content-Based Image Retrieval, IEEE Trans. Circ. Syst. Vid. technol., 8(5) (1998) 644 – 655.

[12] I. J. Cox, M. L. Miller, T. P. Minka, T. V. Papathomas and P. N. Yianilos, The Bayesian Image Retrieval System, PicHunter: Theory, Implementation, and Psychophysical Experiments, IEEE Trans. Image Process., 9(1) (2000) 20–37.

[13] Y. H. Kuo, W. H. Cheng, H. T. Lin and W. H. Hsu, Unsupervised Semantic Feature Discovery for Image Object Retrieval and Tag Refinement, IEEE Trans. Multimedia, 14(9) (2012) 1079-1090.

[14] C. Gao, X. Zhang and H. Wang, A Combined Method for Multi-class Image Semantic Segmentation, IEEE Transactions on Consumer Electronics, 58(2) (2012) 596-604.

[15] J. J. Chen, C. R. Su, W. L. Grimson, J. L. Liu and D. H. Shiue, Object Segmentation of Database Images by Dual Multiscale Morphological Reconstructions and Retrieval Applications, IEEE Trans. Image Process., 21(2) (2012) 828-843.

[16] A. Pardo, Extraction of semantic objects from still images, IEEE International Conference on Image Processing (ICIP '02), vol. 3, 2002, pp. 305 -308.

[17] A. Yamada, M. Pickering, S. Jeannin and L. C. Jens, MPEG-7 Visual Part of Experimentation Model Version 9.0-Part 3 Dominant Color, ISO/IEC JTC1/SC29/WG11/ N3914, Pisa, Jan. 2001.

[18] A. Mojsilovic, J. Hu and E. Soljanin, Extraction of Perceptually Important Colors and Similarity Measurement for Image Matching, Retrieval, and Analysis, IEEE Trans. Image Process., 11 (11) (2002) 1238-1248.

[19] S. P. Lloyd, Least Squares Quantization in PCM, IEEE Trans. Inform. Theory, 28(2) (1982) 129-137.

[20] N. C. Yang, W. H. Chang, C. M. Kuo and T. H. Li, A Fast MPEG-7 Dominant Color Extraction with New Similarity Measure for Image Retrieval, Journal of Visual Communication and Image Representation , 19(2) (2008) 92-105.

[21] Y. W. Lim and S. U. Lee, On the color image segmentation algorithm based on the thresholding and the fuzzy c-means techniques, Pattern Recognit., 23(9) (1990) 935-952.

[22] S. Kiranyaz, M.Birinci and M.Gabbouj, Perceptual color descriptor based on spatial distribution: A top-down approach, Image and Vision Computing 28(8) (2010) 1309-1326.

[23] P. Scheunders, A genetic approach towards optimal color image quantization, IEEE International Conference on Image Processing (ICIP'96), vol. 3, 1996, pp. 1031-1034.

[24] W. Chen, W. C. Liu and M. S. Chen, Adaptive Color Feature Extraction Based on Image Color Distributions, IEEE Trans. Image Process., 19(8) (2010) 2005-2016.

[25] Text of ISO/IEC 15 938-3, "Multimedia Content Description Interface—Part 3: Visual. Final Committee Draft," ISO/IEC/JTC1/SC29/WG11, Doc. N4062, Mar. 2001.

[26] N. C. Yang, C. M. Kuo, W. H. Chang and T. H. Lee, A Fast Method for Dominant Color Descriptor with New Similarity Measure, 2005 International Symposium on Communication (ISCOM2005), Paper ID: 89, Nov. 20-22, 2005.

[27] W. Y. Ma, Y. Deng and B. S. Manjunath, Tools for texture/color based search of images, SPIE Int. Conf. on Human Vision and Electronic Imaging II, 1997, pp. 496- 507.

[28] A. Mojsilovic, J. Kovacevic, J. Hu, R. J. Safranek and S. K. Ganapathy, Matching and Retrieval Based on the Vocabulary and Grammar of Color Patterns, IEEE Trans. Image Process., 9 (1) (2000) 38-54.

[29] T. Ojala and M. Pietikainen, Unsupervised texture segmentation using feature distributions, Pattern Recognit., 32(9) (1999) 447-486.

[30] N. Abbadeni, Computational Perceptual Features for Texture Representation and Retrieval, IEEE Trans. Image Process., 20(1) (2011) 236-246.

[31] M. Broilo, and F. G. B. De Natale, A Stochastic Approach to Image Retrieval Using Relevance Feedback and Particle Swarm Optimization, IEEE Trans. Multimedia, 12(4) (2010) 267-277.

[32] W. C. Kang and C. M. Kuo, Unsupervised Texture Segmentation Using Color Quantization And Color Feature Distributions, IEEE International Conference on Image Processing (ICIP '05), vol. 3, 2005, pp. 1136 - 1139.

[33] S. K. Weng, C. M. Kuo and W. C. Kang, Color Texture Segmentation Using Color Transform and Feature Distributions, IEICE TRANS. INF. & SYST., E90-D(4) (2007) 787-790.

[34] B. S. Manjunath, J. R. Ohm, V. V. Vasudevan and A. Yamada, Color and Texture Descriptors, IEEE Trans. Circ. Syst. Vid. technol., 11(6) (2001) 703-714.

Section 2

Telecommunication Applications

Chapter 4

Multidimensional Optimization-Based Heuristics Applied to Wireless Communication Systems

Fernando Ciriaco, Taufik Abrão and
Paul Jean E. Jeszensky

Additional information is available at the end of the chapter

http://dx.doi.org/10.5772/54152

1. Introduction

In the last two decades, the mobile communications technologies and the Internet have grown almost exponentially, reaching a significant numbers of subscribers around the world. The mobile cellular service got a very large growth of users along with the increase of mobile data services. On the other hand, the Internet provides a great opportunity for users to access the information via fixed and/or wireless networks.

In this scenario, stands out the spread spectrum communication techniques that until the mid-80 were restricted to military applications and is currently in a final technological consolidation phase through the cellular mobile communication systems of third and fourth generations used throughout the world [1].

Such multiple access-based systems use a matched filter bank to detect the interest signal, being however unable to recover the signal in an optimal way, regardless is affected by additive white Gaussian noise (AWGN), flat fading or selective fading channels, since the direct sequence code division multiple access (DS/CDMA) signal is corrupted by multiple access interference (MAI) and severely affected by the near-far effect, resulting in a system whose capacity may remain remarkably below the channel capacity [2] if specific techniques are not introduced to mitigate these effects, such as multiuser detection (MuD) [3], diversity exploration [4, 5] and so forth.

Thus, one of the biggest challenges in the multiuser communication systems development is the interference mitigation. This challenge becomes obvious to the modern and current wireless networks like cellular networks, wireless local area network (WLAN) and wireless metropolitan area network (WMAN), due to the high spectral efficiency need, requiring advanced techniques for frequency reuse and interference mitigation.

The third and fourth generations of cellular mobile systems and wireless networks were designed to support many services through the use of multirate transmission schemes, different quality of service (QoS) requirements and multidimensional diversity (time, frequency and space). Thus, modern systems must accept users transmitting simultaneously in different rates in asymmetric traffic channels (uplink and downlink may be required to work at different rates), and also ensure the minimum specifications of QoS for each offered service.

Hence, current industry standards for wireless networks use a combination of the following techniques to improve the frequency spectrum efficiency: multicarrier, spread spectrum, multiple antennas, spatial multiplexing and coding, reinforcing researches in order to improve the capacity of these systems, considering efficient transmission schemes, multiple diversity combination, multiuser detection methods, among others.

1.1. Multiuser detection

One way to reduce substantially the interference and increase spread spectrum system capacity consists in modifying the detection strategy, using the information of other interfering users signals for detection process of interest user information. This strategy is called multiuser detection (MuD) [3, 6, 7].

In MuD strategy, active user information in the system are used together in order to better detect each individual user, increasing the system performance and/or capacity.

From the 1986 pioneering Verdu's work [3, 6] on optimum multiuser detector (OMuD) to a wide variety of multiuser detectors aiming to improve the performance obtained with the conventional detector in multiple access systems, a remarkable advance in the field has been achieved in the last twenty years. However, given the exponential complexity of the optimum detector, the research efforts has been focused on the development of sub-optimal or near-optimal multiuser detectors with lower complexity.

Alternatives to OMuD include the classical linear multiuser detectors such as Decorrelator [3] and the minimum mean square error (MMSE) [8], the nonlinear MuDs, such as interference cancellation (IC) [9, 10] and zero forcing decision feedback (ZF-DFE) detector [11], and heuristics-based multiuser detectors [12, 13].

However, both classical linear MMSE and Decorrelator multiuser detector algorithms presents two drawbacks; a) for practical system scenarios, both MuD result in performance degradation regarding OMuD; b) they need to perform a correlation matrix inversion, which implies in a high complexity for practical wireless systems with a high number of active users and/or systems with real-time detection in with the active number of users randomly and quickly changes along the time.

The operation principle for the non-linear classical IC and ZF-DF multiuser detectors is the reconstruction of MAI estimates, followed by cancellation (subtraction) for the interest user signal. The operations of MAI reconstruction and cancellation can be repeated in a multistage structure, resulting in more reliable signals canceling each new stage when estimates can be obtained with relative accuracy. The complexity of these detectors increases with the number of necessary stages for demodulation and after a certain number of stages there is no significant performance gain due to the propagation of interference estimation error. This limits the performance of these algorithms. Although the advantage of lower complexity regarding the MMSE and Decorrelator, performance achieved by the non-linear subtractive MuD detectors remain below the MMSE detector for almost all practical interest scenarios.

1.2. Heuristics applied to communication systems

In the last decade, the literature has been collecting sub-optimal solutions proposals based on iterative algorithms and heuristics, particularly evolutionary and local search, applied to inherent multiple access communication systems problems, among which we could cite the following heuristic solutions: optimal multiuser detection [12–19]; spreading sequences selection [20, 21]; parameter estimation, particularly the channel coefficients estimation, delay and power users [19, 22, 23]; power control problem [24, 25]; and resource allocation optimization [26–28].

However, in recent years, the multiuser detection optimization problem in a single DS/CDMA system have been changed for others more complex applications, such as systems with multiple transmit and receive antennas, multirate coded systems with different quality-of-services, and multicarrier CDMA systems.

Differently from the most results reported in the literature, this chapter considers a multidimensional approach which aims to recover optimally (or very close to the optimal point) all the information of all users within the same processing window, considering multipath channels with different power-delay profiles, data rates, time or space-time coding, multiple antenna and multicarrier. In dealing with this sort of system, it will possible to provide various high-rate user's services including voice, data and video under a scenario with growing scarceness of spectrum and energy.

Moreover, to establish quality criteria that meet the acceptance requirements of the scientific community and telecommunications industry standards, this work analyzes convergence and performance aspects of a wide representative heuristic algorithms, considering metrics such as stability, capacity, control and implementation aspects, as well as the algorithm complexity in relation to conventional topologies.

However, aiming at the multidimensional optimization analysis for high performance systems, various techniques based on heuristic algorithms are deployed in this work. Heuristic algorithms have been applied in several optimization problems, showing excellent results in large combination problems for practical cases. Still, there is an inherent difficulty in selecting and setting up the algorithm steps, since correct choices will result in good performances, and contrary, a poorly calibrated parameters may result in a disastrous performance.

Therefore, the manipulation of several variables associated with each heuristic algorithm requires knowledge of the problem to be optimized, experience and keen perception of the algorithm behavior when selecting the parameters. Often, the parameters used are appropriate only for very restrictive settings, and frequently there are no consensus on the (sub-)optimal input parameters to be adopted or even the more conducive internal strategies to adjust those input parameters. Thus, parameters are chosen by past accumulated experience in dealing with other optimization problems or even through non-exhaustive trial tests. This scenario has resulted in a somewhat distrust level of such alternatives application in optimization problems that commonly arise in communications systems.

Thus, the motivation in pursuit of heuristic algorithms to ensure optimal performance is the core of this chapter. To do so, it will be analyzed in a systematic way the main meta-heuristic and hyper-heuristic algorithms deployed in wireless systems, which may be mentioned the genetic algorithm, evolutionary programming, local search (k-optimum), simulated annealing, heuristic algorithm based on Tabu list and a hyper-heuristic-basis selection.

2. System characteristics

Considering the provision of various services with high quality, we opted a transmission/reception scheme that adds several dimensions in order to explore diversity. Figure 1 shows the transmitter and receiver topologies deployed in this work. Hence, lets consider the *k*-th user transmitter and noting that the channel coding stage is necessary to correct the received signal in the presence of errors through the use of redundancy (code diversity). The multirate modulation block aims to ensure the provision of various services to users at different data rates, ensuring the possibility of optimum resource management strategies. The time-spreading code guarantees a rejection level of multiple access interference, and the identification of each DS/CDMA user as well, acting as a kind of time-diversity. The frequency spreading block implements frequency-diversity through the information transmission on different sub-carriers. Finally, the multiple-input-multiple-output (MIMO) antennas block deploys techniques that provide spatial-diversity, either through simple arrangements with various antennas or even by space-time block code (STBC) or trellis code.

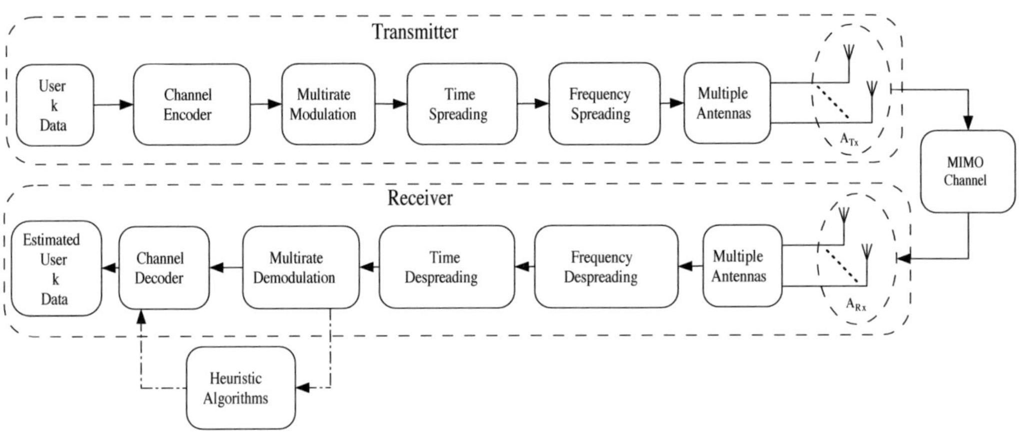

Figure 1. Communication system overview with use of space, time, frequency and coding diversities.

The transmitted signal of the *k*-th user propagates through a channel whose model includes attenuation of small and large scale, i.e., path loss, shadowing and multipath effects.

The *k*-th user signals at the receiver input are demodulated via an antenna array in order to exploit spatial diversity. Structures can be used with several receiving antennas physically separated by a sufficient distance to avoid overlapping signals and block-basis or trellis-basis signal processing techniques. Subsequently, the signals are despread in frequency and time ensuring the channel rejection and multiple access interference rejection, respectively. At this point, it is evident the frequency- and time-diversity exploitation. Thus, the demodulated signals are reassembled considering the *k*-th user transmission rates. This receptor is known as Rake receiver. Finally, the signals are decoded by means of particular techniques, resulting in a type of diversity code.

2.1. Received signal in multipath MIMO channels

Considering the reverse link and assuming a set of bits transmitted (frame) consisting of I bit for each multirate user, the resulting signal propagates through G independent Rayleigh fading paths. Thus, the equivalent baseband received signal (assuming ideal low-pass filter) in one of the antennas is:

$$\begin{aligned} r_{n_{Rx}}(t) = \sum_{i=0}^{I-1}\sum_{k=1}^{K^{(g)}}\sum_{g=1}^{G}\left[\sum_{m=1}^{M}\sum_{j=1}^{m^{(g)}} A'_{k,g}\mathbf{x}_{k^{(g)}}^{(i)}[j]\, s_{Ck}^{(g)}(t-jT)\, s_{k}^{(g)}\left(t-\tau_{k,\ell}^{(g)}-iT\right)\cdot\right. \\ \left.\cdot s_{Fk,m}^{(g)}\cos\left(2\pi f_m t+\phi_{k,m}^{(g)}\right)\right] * h_{k,g}^{(i)}(t)+\eta(t) \end{aligned} \quad (1)$$

where $K^{(g)}$ is the number of physical users belonging to g-th multirate group being $K = K^{(1)} + K^{(2)} + \ldots + K^{(g)} + \ldots + K^{(G)}$ the total number of active users in the physical system, divided into g user groups of same rate, $t \in [0, T]$, M represents the number of subcarriers, the amplitude $A_{k,g'}$ is the amplitude of the received k-th user of g-th multirate group, including the effects of path loss and shadowing channel, and assumed constant over the I bits transmitted base rate, $\mathbf{x}_{k^{(g)}}^{(i)}[j] \in \{\pm 1\}$ is the symbol of coded information passed to the i th symbol interval; s_{Ck}, s_k and $s_{Fk,m}$ represent the sequences of channeling, time and frequency spread, respectively, $\tau_{k,\ell}^{(g)}$ is the random delay, $\phi_{k,m}^{(g)}$ corresponds to the initial k-th user; f_m represents the respective subcarriers frequencies; $h_{k,g}$ is the impulse response of the channel and the term $\eta(t)$ is the AWGN with bilateral power spectral density equal to $N_0/2$.

The k-th user delay of g-th multirate group takes into account the nature of the asynchronous transmission, $d_k^{(g)}$, as well as the propagation delay, $\Delta_{k,\ell}^{(g)}$ for k-th user, ℓ-th path, g-th multirate group, resulting in:

$$\tau_{k,\ell}^{(g)} = \Delta_{k,\ell}^{(g)} + d_k^{(g)} \quad (2)$$

The channel impulse response to the k-th user of g-th multirate group in the range of i-th bit can be written as:

$$h_{k,g}^{(i)}(t) = \sum_{\ell-1}^{L} c_{k,\ell,g}^{(i)}\delta\left(t-\Delta_{k,\ell}^{(g)}-iT\right) \quad (3)$$

where $c_{k,\ell,g}^{(i)} = \beta_{k,\ell,g}^{(i)} e^{j\phi_{k,\ell,g}^{(i)}}$ indicates the complex channel coefficient for the k-th user of g-th multirate group, ℓ-th path and $\delta(t)$ is the unit impulse function. It is assumed that the phase of $c_{k,\ell,g}^{(i)}$ have an uniform distribution $\phi_{k,\ell,g}^{(i)} \in [0, 2\pi)$ and the module channel $\beta_{k,\ell,g}^{(i)}$ represents the small-scale fading envelope with a Rayleigh distribution.

Additionally, we considered normalized channel gain for all users, i.e., $\mathbb{E}\left[\sum_{\ell=1}^{L}|c_{k,\ell,g}|^2\right] = 1$ for $\forall k, g, i$.

Therefore, we can rewrite the received signal in each A_{Rx} antennas replacing the eq.(3) into eq. (1), resulting in:

$$r_{n_{Rx}}(t) = \sum_{i=0}^{I-1}\sum_{k=1}^{K^{(g)}}\sum_{g=1}^{G}\sum_{m=1}^{M}\sum_{j=1}^{m^{(g)}}\sum_{\ell=1}^{L} A'_{k,g,\ell}\mathbf{x}_{k^{(g)}}^{(i)}[j]\, s_{Ck}^{(g)}(t-jT)\, s_{k}^{(g)}\left(t-\tau_{k,\ell}^{(g)}-iT\right)\cdot$$
$$\cdot\, s_{Fk,m}^{(g)}\cos\left(2\pi f_m t+\phi_{k,m}^{(g)}\right)c_{k,\ell,g}^{(i)}\delta(t-\Delta_{k,\ell}^{(g)}-iT)+\eta_{n_{Rx}}(t) \quad (4)$$

For simplicity and without generality loss, we consider ordered random delays, i.e.:

$$0=\tau_{1,1}^{(1)}\leq\tau_{1,2}^{(1)}\leq\cdots\tau_{1,L}^{(1)}\leq\tau_{2,1}^{(1)}\leq\cdots\leq\tau_{K^{(1)},L}^{(1)}\leq\ldots\ \tau_{K^{(G)},L}^{(G)}<T \quad (5)$$

2.2. Conventional SIMO detection systems

Considering the system with only one transmission antenna, one can rewrite eq. (4) considering just the n_{Rx}-th receiving antenna and m-th subcarrier as:

$$r_{n_{Rx},m}(t) = \sum_{i=0}^{I-1}\sum_{k=1}^{K^{(g)}}\sum_{g=1}^{G}\sum_{j=1}^{m^{(g)}}\sum_{\ell=1}^{L} A_{k,g,\ell}\,\mathbf{x}_{k^{(g)}}^{(i)}[j]\, s_{Ck}^{(g)}(t-jT)\; s_{k}^{(g)}(t-\tau_{k,\ell}^{(g)}iT)\, s_{Fk,m}^{(g)}\cdot$$
$$\cdot\,\beta_{k,\ell,g,m}^{(i)}\, e^{\jmath(\omega_m t+\varphi_{k,\ell,g,m}^{(i)})}+\eta_{n_{Rx}}(t) \quad (6)$$

where $\eta_q(t)$ corresponds to the Additive White Gaussian Noise (AWGN) for n_{Rx}-th receiving antenna.

For multipath fading channels, multirate and multicarrier scheme, the receiver for each subcarrier demodulation use the Rake receiver consisting of a bank of KD matched filters to the multirate physical users spread sequences with path diversity order[1] $D \leq L$, followed by the second despreading (channeling) aiming recovering $m^{(g)}$ simultaneously transmitted bits in parallel channels. To be able achieve a perfect synchronism (maximum auto-correlation) of spread sequence at the receiver must use delay accurate estimates for the ℓ-th path of the k-th user of g-th multirate group, $\hat{\tau}_{k,\ell}^{(g)}$. Performance is degraded proportionally when there are errors in the delays estimates.

Thus, the $m^{(g)}$ matched filter outputs for the k-th physical user, g-th multirate group and corresponding to ℓ-th multipath component, m-th subcarrier and n_{Rx}-th antenna, sampled at the end of basic information period T of i-th interval symbol can be expressed as:

$$y_{k,\ell,g,n_{Rx},m}^{(i)}[j] = \frac{1}{\sqrt{N_C}}\int_0^T r_{n_{Rx},m}(t)\, s_k^{(g)}\left(t-\tau_{k,\ell}^{(g)}-iT\right)s_{Ck}^{(g)}(t-jT)\, s_{Fk,m}^{(g)}e^{(-\jmath\omega_m t)}dt$$
$$= \underbrace{A'_{k,g}Tc_{k,\ell,g,m}^{(i)}\mathbf{x}_{k^{(g)}}^{(i)}[j]}_{(I)} + \underbrace{SI_{k,\ell,g,n_{Rx},m}^{(i)}}_{(II)} + \underbrace{MAI_{k,\ell,g,n_{Rx},m}^{(i)}}_{(III)} + \underbrace{n_{k,\ell,g,n_{Rx},m}^{(i)}}_{(IV)} \quad (7)$$

[1] If $D < L$, in each Rake receiver the matched filters to spread sequences are synchronized to D energy major paths.

http://dx.doi.org/10.5772/54152

where $j = 1 : m^{(g)}$. The first term corresponds to the desired signal, the second term to the self-interference (SI), the third to the MAI on the ℓ-th multipath component of the k-th user of g-th multirate group, m-th subcarrier and n_{Rx}-th antenna, as well the last term corresponds to the filtered AWGN.

In this case, the Rake receiver combines the outputs of the matched filters bank available for each user (fingers)[2] and weighted by the respective channel gains [29]. The Maximal Ratio Combiner (MRC) combines the signals from the D correlators in coherent way:

$$\hat{\mathbf{y}}_{k,g}^{(i)} = \sum_{\ell=1}^{D} \sum_{n_{Rx}=1}^{A_{Rx}} \sum_{m=1}^{M} \Re \left\{ \mathbf{y}_{k,\ell,g,n_{Rx},m}^{(i)} \hat{\beta}_{k,\ell,g,n_{Rx},m}^{(i)} e^{-j\hat{\phi}_{k,\ell,g,n_{Rx},m}^{(i)}} \right\} \tag{8}$$

where $\Re\{.\}$ is the real part operator, $\hat{\beta}_{k,\ell,g,n_{Rx},m}^{(i)}$ and $\hat{\phi}_{k,\ell,g,n_{Rx},m}^{(i)}$ are the magnitude and phase estimates of the channel coefficients, respectively, for the i-th processing interval for the k-th user, ℓ-th path, g-th multirate group, n_{Rx}-th antenna and m-th subcarrier. Again, the performance is degraded proportionally when there are errors in the channel estimates.

Finally, the estimates for the $m^{(g)}$ information symbols of k-th user of g-th multirate group are obtained through an abrupt decision rule:

$$\hat{\mathbf{x}}_{k,g}^{(i)} = sgn\left(\hat{\mathbf{y}}_{k,g}^{(i)}\right) \tag{9}$$

Therefore, the estimated symbol frame for all users in the range of i-th bit with $DK_v \times 1$ dimension is given by:

$$\hat{\mathbf{x}}^{(i)} = \left[\hat{\mathbf{x}}_{1,1}^{(i)}\ \hat{\mathbf{x}}_{2,1}^{(i)} \cdots \hat{\mathbf{x}}_{K^{(1)},1}^{(i)} \cdots \hat{\mathbf{x}}_{1,G}^{(i)}\ \hat{\mathbf{x}}_{2,G}^{(i)} \cdots \hat{\mathbf{x}}_{K^{(G)},G}^{(i)}\right]^T \tag{10}$$

The performance obtained with the MRC Rake receiver will be deteriorated considerably when the number of users sharing the same channel grow[3] and/or when the interfering users power increase.

2.3. Optimum multiuser detector

The best performance among the multiuser detectors is achieved with OMuD, where the goal is to maximize the maximum likelihood function [3]. Given the conditional probability:

$$P_r\left(\mathbf{y}^{(i)} | \hat{\mathbf{x}}^{(i)}, i \subset [0, I-1]\right) = e^{\left\{-\frac{1}{2\sigma^2} \int_{t=0}^{I-1} \left[\mathbf{y}^{(i)}\ \ \mathbf{S}_t\left(\hat{\mathbf{x}}^{(i)}\right)\right]^2 dt\right\}} \tag{11}$$

[2] In addition to multipath effects, multiple subcarriers and multiple receiving antennas.

[3] Increasing the MAI, third term of eq. (7).

where the total received signal, reconstructed from the estimated parameters and known at the receiver is:

$$\mathbf{S}_t\left(\hat{\mathbf{b}}^{(i)}\right) = \sum_{k=1}^{K^{(g)}} \sum_{g=1}^{G} \sum_{m=1}^{M} \sum_{j=1}^{m^{(g)}} \sum_{n_{Rx}=1}^{A_{Rx}} \sum_{\ell=1}^{L} A'_{k,g,\ell,n_{Rx}} \hat{\mathbf{x}}_{k,g}^{(i)} \mathbf{s}_{Ck}^{(g)}(t-jT) s_k^{(g)}\left(t-\tau_{k,\ell,n_{Rx}}^{(g)} - iT\right) \cdot \\ \cdot s_{Fk,m}^{(g)} \hat{\beta}_{k,\ell,g,n_{Rx}}^{(i)} e^{j\left(\omega_m t + \hat{\phi}_{k,\ell,g,n_{Rx}}^{(i)}\right)} \tag{12}$$

In this context, the maximum likelihood vector that must be found by OMuD has $DK_v \times 1$ dimension, and by:

$$\hat{\mathbf{x}} = \left[\hat{\mathbf{x}}^{(0)^T} \; \hat{\mathbf{x}}^{(1)^T} \; \hat{\mathbf{x}}^{(2)^T} \ldots \; \hat{\mathbf{x}}^{(I-1)^T}\right]^T \tag{13}$$

Note that the minimum square difference exists in eq. (11) ensures the maximization of the maximum likelihood function. Expanding the quadratic difference in eq. (11), based on the output of the matched filter, vector $\mathbf{y}^{(i)}$, find the maximum likelihood vector $\hat{\mathbf{x}}$ is equivalent to selecting the bit vector $\mathcal{B}$, with same size, which maximizes the Log Likelihood Function (LLF) [3]:

$$\Omega\left(\mathcal{B}\right) = 2 \sum_{n_{Rx}=1}^{A_{Rx}} \Re\left\{\mathcal{B}^T \mathcal{C}^H \mathcal{A} \mathcal{Y}\right\} - \mathcal{B}^T \mathcal{C} \mathcal{A} \mathbf{R} \mathcal{A} \mathcal{C}^H \mathcal{B} \tag{14}$$

where each matrix must be determined for each receiving antenna and $(\cdot)^H$ refers to the Hermitian transpose operator.

The diagonal channel coefficients and amplitudes matrices[4] for n_{Rx}-th receiving antenna of $DK_v I$ dimension are defined, respectively, by:

$$\mathcal{C} = \text{diag}\left[\mathbf{C}^{(0)} \; \mathbf{C}^{(1)} \; \mathbf{C}^{(2)} \ldots \; \mathbf{C}^{(I-1)}\right] \tag{15}$$

$$\mathcal{A} = \text{diag}\left[\mathbf{A}^{(0)} \; \mathbf{A}^{(1)} \; \mathbf{A}^{(2)} \ldots \; \mathbf{A}^{(I-1)}\right] \tag{16}$$

The output vector of the matched filter (MFB), composed by I vectors $\mathbf{y}^{(i)}$ with dimension $DK_v I \times 1$ is given by:

$$\mathcal{Y} = \left[\mathbf{y}^{(0)} \; \mathbf{y}^{(1)} \; \mathbf{y}^{(2)} \ldots \; \mathbf{y}^{(I-1)}\right]^T \tag{17}$$

In general, the MFB output vector $\mathcal{Y}$ is deployed as initial guess in the LLF cost function, eq. (14) when a heuristic multiuser detection is performed. The general rule should be ensure the maximization of the cost function, considering the same $\mathcal{Y}$ for all receiving antennas.

[4] To simplify the notation, hereafter we have omitted the matrix index n_{Rx}.

Finally, the block Toeplitz tridiagonal correlation matrix $\mathbf{R}$, dimension $D\mathcal{K}_vI \times D\mathcal{K}_vI$, is defined as [3]:

$$\mathbf{R} = \begin{bmatrix} \mathbf{R}[0] & \mathbf{R}^T[1] & \mathbf{0} & \cdots & \mathbf{0} & \mathbf{0} \\ \mathbf{R}[1] & \mathbf{R}[0] & \mathbf{R}^T[1] & \cdots & \mathbf{0} & \mathbf{0} \\ \mathbf{0} & \mathbf{R}[1] & \mathbf{R}[0] & \ddots & \mathbf{0} & \mathbf{0} \\ \vdots & \vdots & \ddots & \ddots & \ddots & \vdots \\ \vdots & \vdots & \vdots & \ddots & \mathbf{R}[0] & \mathbf{R}^T[1] \\ \mathbf{0} & \mathbf{0} & \mathbf{0} & \cdots & \mathbf{R}[1] & \mathbf{R}[0] \end{bmatrix} \tag{18}$$

Therefore, the complete frame for the I estimated symbols from all K_v users can be obtained by optimizing eq. (14), resulting in:

$$\widehat{\mathbf{x}} = \arg\left\{ \max_{\mathcal{B}\in\{\pm1\}^{\mathcal{K}_vI}} \left[\Omega\left(\mathcal{B}\right)\right] \right\} \tag{19}$$

The OMuD consists in finding the best data symbols vector in a set with all the possibilities, i.e., it is a NP-complete combination problem [30], which the traditional algorithms are inefficient. Most of these result in exponential complexity growth when one or more of the following factors: number of users, frame, number of receiver antennas, number of paths, number of subcarriers, among others increase.

Therefore, the use of heuristic methods for this class of problems shows up attractive, since it is possible to obtain optimal solutions (or near-optimal) using reduced search spaces. Thus, the proposed strategy in this Chapter aiming for maximize the LLF by testing different candidates symbols vectors at each new iteration/generation of heuristic algorithms. Such attempts seek to maximize the system average performance, approaching or even equaling that obtained by OMuD, but with remarkable reduction in the computational complexity.

3. Heuristic algorithms

This section presents a brief review of heuristic algorithms, specifically local and evolutionary search, describing variants and required parameters. Such variants include encoding (mapping) problem, initialization algorithms step (parameters choice), cost function evaluation, search space scanning step, and replacement candidates step. For the analysis, 1-opt and k-opt local search, simulated annealing, short-term and reactive Tabu search, genetic, as well evolutionary programming algorithms have been considered in this work.

3.1. Encoding problem

The encoding for MuD problem is inherently binary, because the data vector is naturally binary. Therefore, following the Keep it Simple (KIS) principle as much as possible, it is not necessary to perform an encoding (mapping) of candidate solutions differently to binary form. Thus, these candidates vectors will be directly represented by the information bits that will be tested by the cost function, considering only polarized binary encoding, i.e., for each candidate position of the vector $\mathcal{Y}$ is able to assume just only one value in the set $\{\pm1\}$.

3.2. Search space definition

After the encoding step, we must define the problem search space, in which case the MuD problem is characterized by all possible combinations bits that users can transmit. In this case, for K_v virtual users transmiting I bits through a multipath channel with L paths and D processing signal branches at the receiver, the total search universe, considering optimizing the output of matched filter and signals combination from multirate users, is a binary set of dimension:

$$\Theta\left(K_v, I, D\right) = 2^{DK_v I}, \qquad \text{with} \quad 1 \leq D \leq L. \tag{20}$$

It is evident that for the proposed MuD problem, a total search universe should result smaller than $2^{DK_v I}$, since each transmitted bit must be detected in a way that results in a same estimated bit value for all D processing branches, namely:

$$\hat{b}^{(i)}_{k,1,g} = \hat{b}^{(i)}_{k,2,g} = \ldots = \hat{b}^{(i)}_{k,D,g} \in \{+1, -1\} \tag{21}$$

This implies that the search universe covered by the heuristic MuD algorithm should be independent of the number of paths, resulting in:

$$\Theta\left(K_v, I\right) = 2^{K_v I} \tag{22}$$

As a result, the universe of possible solutions is then formed by all vectors candidates that satisfy (21).

The other possibilities belong to the so-called forbidden universe, composing the non-tested candidates set into a heuristic methodology. This guarantees the final solution quality in MuD problem with multipath diversity, because it enable a correct estimate for all paths of the same transmitted bit could be made.

3.3. Evolutionary Programming (EP) algorithms

The evolutionary heuristic algorithms methods are non-deterministic search mechanisms based on natural selection and evolution mechanisms from Darwin's theory [31][5]. This theory explains the life history by the physical processes action and genetic operators in populations or species. These processes are known for reproduction, disturbance, competition and selection.

Considering the computational implementation aspects, the parameters and strategies such as population size, mating pool size, selection strategy, crossover type and rate, mutation type and rate and replacement strategy should be chosen carefully by the user for each class of optimization problem, allowing numerous plausible combinations [33–35].

[5] Because [31] is a rare and difficult access reference, we consider newer editions of the Darwin's work, for instance [32].

http://dx.doi.org/10.5772/54152

3.4. Local Search (LS) algorithms

The Local Search (LS) strategy is based on the better established existing principle for combinatorial optimization methods: trial and error. This is a natural and simple idea, but in fact, surprised by the success degree that this method has the most varied types of combinatorial problems.

The only parameters to be selected corresponds to the search starting point and the neighborhood size. The choice of starting point is usually done by intuition, because very few problems have a guide or direction.

For neighborhood definition, it should be pointed out that small neighborhood set leads in a low complexity algorithm, since the search space consists of few alternatives. On the other hand, the reduced size of the neighborhood set may not provide good solutions due to local minimum or maximum in this reduced region. Large neighborhood sets, on the other hand, provide good solutions but bring much greater complexity, since these sets may result in search space as large as brute force methods [36].

3.5. Simulated Annealing (SA) algorithms

The simulated annealing (SA) algorithm proposed by Kirkpatrick [37] was inspired by the annealing process of physical systems. This was based on the algorithm originally proposed by Metropolis [38] as a strategy for determining equilibrium states (or configurations) of a collection of atoms at a given temperature. The basic idea comes from the statistical thermodynamics, which is a physics branch responsible for theoretical predictions about the behavior of macroscopic systems based on the laws that govern their atoms. Using analogies, the SA algorithm was proposed based on the similarity between the annealing procedure implemented by the Metropolis algorithm and combinatorial optimization processes.

Thus, the concept of the SA algorithm is associated with the principle of thermodynamics, in which a solid heated to a very high temperature and then cooled gradually tends to solidify to form a homogeneous structure with lowest energy [37, 39].

This way, the SA algorithm must be started with one strategy and three parameters: initial temperature, $T(0)$, cooling step, ϵ, size range (plateau) L_{SA}, and acceptance probability equation.

3.6. Tabu search

Tabu search algorithm was originally proposed in 1977 with the pioneering Glover's work [40] and later described in its current form in 1986 [41], being used in various knowledge areas and fields.

The short-term Tabu search (STTS) algorithm is based on the deterministic mode of memory operation. The memory is implemented by recording characteristics of displacement of previously visited solutions. This is described by the Tabu list, which is formed by the recent past search, being called the short-term memory effect. These displacement characteristics are prohibited by Tabu list for a number of iterations. This helps prevent returns to local solutions, promoting diversification in the search.

The reactive Tabu search (RTS) version combines the short-term memory effect with another memory effect to avoid the local maximum returns and ensure efficient search. This effect is known as long-term memory, which alternates between intensification and diversification phases, adapting the prohibition period during the search, provide that the prohibition period takes different values for each iteration [42].

3.7. Hyperheuristic strategies

Over the past 50 years, the well-known meta-heuristic algorithms have been used as optimization tool for a wide range of optimization problems. The ability of the meta-heuristic algorithms to avoid local optimum-solutions offer us the ability to adapt this class of optimization strategy to solve various problems with the robustness and easiness of implementation, contributing to various optimization fields, mainly in those problems where deterministic or traditional optimization methods become inefficient or highly complex.

However, it is not easy or even possible to predict which of the many existing heuristic algorithms is the best choice for a specific optimization problem and be able to produce the same result given the same input parameters. The difficulty of choosing is associated to the performance unpredictability, which constitutes the major factor limiting their use by the scientific community and industry.

Furthermore, for a large optimization problems variety, the input parameter values should be controlled as the search evolves. In this context, recently, the idea of working with a higher level of automation in a heuristic design have been resulted in the development of the so-called hyper-heuristic (HH) strategy [43, 44].

Thus, the HH algorithms consist in applying a high-level methodology to control the selection or generation of generic search strategies using a specific number of different low-level heuristics.

It is worth noting that meta-heuristics are quality techniques to solve complex optimization problems, but efficient implementations of these methods usually require many specific knowledge about the problem being treated. Thus, the HH methodologies have been proposed aiming to build robust optimization algorithms, allowing the use of meta-heuristics methods with minor adaptations.

For HHs based on heuristics selection, one should choose the suitable number of iterations for the HH, the selection strategy, as well as the acceptance strategy.

4. Performance metrics for heuristic algorithms evaluation

Aiming to quantify the performance of the heuristic algorithms in terms of stability and convergence guarantee, it is necessary to know the specified limits for the tolerances calculation, the so-called upper specification limit (USL) and lower specification limit (LSL) [45, 46].

The USL and LSL are simply an upper and lower bounds to measure the algorithm's performance. Thus, as in the case of control charts, it is desired that the algorithm behaves within these two limits. These parameters are often set by the need for quality of solutions

http://dx.doi.org/10.5772/54152

found by the heuristic algorithm and may take values milder or stricter[6]. But this quality analysis should consider the acceptance limits of the solution, i.e., what are the thresholds of deviation from the desired value that can still be accepted as a solution for the optimization problem.

4.1. Algorithm stability and capacity indexes

One of the metrics widely used to evaluate the stability and capacity of the search algorithms is the so-called algorithm stability index (ASI), which corresponds to an ability measure of the algorithm to produce consistent results, described by the ratio between the dispersion of allowed solutions and the dispersion of current solution. The other metric, namely algorithm capability index (ACI) is a measure of how far from the specified limits the solution propitiates by the algorithm is, in terms of quality of the solutions obtained. The ASI and ACI can be calculated as:

$$ASI = \frac{(USL - LSL)}{6\sigma_{\bar{X}}} \tag{23}$$

$$ACI = \frac{(USL - \bar{X})}{3\sigma_{\bar{X}}} \quad \text{or} \quad ACI = \frac{(\bar{X} - LSL)}{3\sigma_{\bar{X}}} \tag{24}$$

In the literature, this methodology is also known as *"Six Sigma"* methodology [47]. The ACI metric measures how close the algorithm's solution is from its purpose, as well as the consistency around their average performance. An algorithm may have a minimal variation, but if it is away from the objective value for one of the specification limits, it will result in a lower ACI value, whereas the ASI metric may still be high. On the other hand, an algorithm could result, on average, in solutions exactly equal to the purpose, but presents a large variation in performance. In this case, ACI is still small and ASI can still be large. Thus, the ACI metric just will be large if and only if it reaches the vicinity of the desirable objective value consistently and with minimal variation.

Note that for practical reasons, it has been considered a good criterion to ensure $ASI > 2$ and $ACI > 1.33$ for most of engineering applications with practical interest [46, 48].

In the next section, the stability and capacity indexes have been considered in the input parameters optimization step of the heuristic algorithms, since the adopted benchmark functions have well-defined values for the global minimum, as described in the following.

4.2. Input parameter optimization

Considering the quality metrics discussed previously, it was decided to hold a *Decathlon* marathon type [43, 49] in order to evaluate the efficiency, stability an convergence capacity of the proposed heuristic MuD algorithms. For this purpose, ten benchmark functions, which correspond to the ten races of marathon, have been deployed aiming to define performance thresholds, as well as parameters determination that provide good solutions for all heuristic algorithms considered.

[6] A usual way is to take $USL = \bar{X} + 3\sigma_X$ and $LSL = \bar{X} - 3\sigma_X$, where σ_X is the standard deviation of the process X.

Hence, in order to optimize the input parameters of each heuristic algorithm, ten benchmarks (test) functions described in Table 1 have been deployed, considering functions commonly used in the literature [34, 44, 50–52], but with different characteristics in terms of local optima and dimensionality. In this study the first three functions of De Jong's work [50] have been considered, and in order to guarantee diversity in the characteristics, a set of seven additional test functions have been chosen.

Name	Definition
De Jong [50]	$F1(\mathbf{x}) = \sum_{i=1}^{n} i \cdot x_i^2$
De Jong [50]	$F2(\mathbf{x}) = \sum_{i=1}^{n-1} 100\left(x_{i-1} - x_i^2\right)^2 + (1 - x_i)^2$
De Jong [50]	$F3(\mathbf{x}) = \sum_{i=1}^{n} \lfloor x_i \rfloor$
Michalewicz [52]	$F4(\mathbf{x}) = -\sum_{i=1}^{n} \sin(x_i)\left(\sin\left(\frac{ix_i^2}{\pi}\right)\right)^{2m}$
Schaffer [53]	$F5(\mathbf{x}) = \left(x_1^2 + x_2^2\right)^{0{,}25}\left(\sin^2\left(50\left(x_1^2 + x_2^2\right)^{0{,}1}\right) + 1\right)$
Ackley [54]	$F6(\mathbf{x}) = x_1^2 + 2x_2^2 - 0{,}3\cos(3\pi x_1) - 0{,}4\cos(4\pi x_2) + 0{,}7$
Rastrigin [55]	$F7(\mathbf{x}) = An + \sum_{i=1}^{n}\left[x_i^2 - A\cos(2\pi x_i)\right]$
Schwefel [56]	$F8(\mathbf{x}) = An + \sum_{i=1}^{n}\left[-x_i \sin\left(\sqrt{\lvert x_i \rvert}\right)\right]$
6-Hump Camelback [57]	$F9(\mathbf{x}) = \left(4 - 2{,}1x_1^2 + \frac{x_1^4}{3}\right)x_1^2 + x_1x_2 + \left(-4 + 4x_2^2\right)x_2^2$
Shubert [58]	$F10(\mathbf{x}) = \prod_{i=1}^{n}\left(\sum_{j=1}^{m} j\cos\left[(j+1)x_i + j\right]\right)$

Table 1. Benchmark functions deployed for the heuristic input parameters optimization.

In order to eliminate eventual bias in the analysis, a large number of simulations have been considered. Hence, in all numerical results presented in this work an average over at least 1000 realization for each numerical parameter determination of each algorithm and for each function have been carried out, aiming to determine means and respective standard deviations, as well as for the calculation of the ASI and ACI quality measures. Thus, the numerical results show confidence intervals that provide consistent analyzes.

As a result of these analyzes, Table 2 presents a summary for the input parameters optimization and adopted strategies (in order to guarantee diversity on the search space) for each (hyper)-heuristic algorithm analyzed in this work. Q_{indiv} indicates the length of each individual-candidate solution which of course is a function of the problem dimensionality.

4.3. Computational complexity

Table 3 presents the generic complexity of the heuristic algorithms for subsequent determination of the heuristic multiuser detectors (MuD) complexity operating under different telecommunications systems scenarios in addition to the presentation of quantitative computational complexity of algorithms for application in Decathlon proof considered in this work (F1 to F10 functions). Notation: $\mathcal{O}_{FC}$ represents the number

GA	Population Size:	$p = 10 \cdot \left\lceil 0.345 \left(\sqrt{\pi (l-1)} + 2 \right) \right\rceil$
	Mating Pool Size:	$T = 0.7p$
	Selection Strategy:	p-sort
	Crossover Type/Rate:	Uniform / $p_c = 50\%$
	Mutation Type/Rate:	Gaussian / $p_m = 10\%$
	Replacement Strategy:	$\mu + \lambda$ (with $\mu = p$)
EP-C	Population Size:	$p = 10 \cdot \left\lceil 0.345 \left(\sqrt{\pi (l-1)} + 2 \right) \right\rceil$
	Cloning Rate:	$I_c = 20\%$
	Selection Strategy:	p-sort
	Mutation Type/Rate:	Gaussian / $p_m = 15\%$
	Replacement Strategy:	$\mu + \lambda$ (with $\mu = p$)
k-opt	Neighborhood Search	Choose neighborhood size (k)
SA	Initial temperature:	$T(0) = \ln(I_t)$
	Step Size (Plateau):	$L_{sa} = 2$
	Cooling Step:	$\varepsilon = \sqrt{\frac{2}{\ln(i)}}$
	Acceptance Probability:	$x(i) = \exp\left[\frac{\lvert\Delta c\rvert}{T(i)}\right] - 1$
STTS	Prohibition Period:	$P = Q_{indiv}/2$
RTS	Initial Prohibition Period:	$P(0) = Q_{indiv}/2$
	Reduction/Increase Rate:	$x = 50\%$
HH	Number of HH iteration:	$I_t(HH) = 10$
	Selection Strategy:	Simply random
	Acceptance Strategy:	Naive

Table 2. A summary for the optimized input parameters and strategies adopted in all considered heuristic algorithms

of operations relevant to the cost function calculation, and G_t is the number of iteration/generation necessary for convergence.

	Flops	Foremost Term
GA	$pG_t\left(\mathcal{O}_{FC} + 11,7Q_{indiv} + 3\log\left(Q_{indiv}\right)\right)$	$pG_t\mathcal{O}_{FC}$
EP-C	$pG_t\left(\mathcal{O}_{FC} + 6,1Q_{indiv} + 3\log\left(Q_{indiv}\right)\right)$	$pG_t\mathcal{O}_{FC}$
1-opt	$Q_{indiv}G_t\left(\mathcal{O}_{FC} + 2Q_{indiv} + 2\right)$	$Q_{indiv}G_t\mathcal{O}_{FC}$
k-opt	$\sum_{i-1}^{k}\binom{Q_{indiv}}{i}G_t\left(\mathcal{O}_{FC} + 2Q_{indiv} + 2\right)$	$\sum_{i=1}^{k}\binom{Q_{indiv}}{i}G_t\mathcal{O}_{FC}$
SA	$Q_{indiv}G_t\left(\mathcal{O}_{FC} + 3Q_{indiv} + 5\right)$	$Q_{indiv}G_t\mathcal{O}_{FC}$
STTS	$Q_{indiv}G_t\left(\mathcal{O}_{FC} + 3,5Q_{indiv} + \log\left(Q_{indiv}\right) + 6\right)$	$Q_{indiv}G_t\mathcal{O}_{FC}$
RTS	$Q_{indiv}G_t\left(\mathcal{O}_{FC} + 3,5Q_{indiv} + \log\left(Q_{indiv}\right) + 7\right)$	$Q_{indiv}G_t\mathcal{O}_{FC}$
HH	$\left(0,6Q_{indiv} + 0,4p\right)G_t\mathcal{O}_{FC} +$ $+0,2Q_{indiv}G_t\left(10Q_{indiv} + 18p + 28\right)$	$\left(0,6Q_{indiv} + 0,4p\right)G_t\mathcal{O}_{FC}$

Table 3. Average complexity in terms of number of operations (Flops) and predominant term, for all considered heuristic algorithms.

5. Numerical results for DS/CDMA systems with multidimensional diversity

This section discuss representative numerical results for multiuser detection obtained under various types of diversity scenarios. First, we present results for a SIMO MC-CDMA systems, i.e. systems in which frequency and space diversities have been deployed jointly, due to the use of multicarrier and multiple receiving antennas. In a second step, a scenario with code and spatial diversity using multiple receive and/or transmission antennas have been analyzed.

The main parameters of the system and channel coefficients are presented in Table 4. In all simulations results, random spreading sequences and slow Rayleigh channel model have been considered. Furthermore, it was assumed that the channel parameters are perfectly known at the receiver side, as well each subcarrier of the MC-CDMA system is subjected to flat frequency fading. Besides, low (BPSK) and high order modulation (M-QAM) format, LDPC and Turbo coding, as well as different spreading codes length, ranging from $N = 8$ to $N = 64$ have been deployed in this section.

Parameters	**Fig.2**	**Fig.3**	**Fig.4**	**Fig.5**	**Fig. 6**
# Users	20	20	1 and 32	1 and 32	64
# Antenas Tx	1	1	4	4	1 and 2
# Antenas Rx	1 to 4	1 to 5	1	4	1 and 2
Modulation	BPSK	BPSK	M-QAM	M-QAM	BPSK
Spread Sequence	$N = 8$	$N = 8$	$N = 32$	$N = 32$	$N = 64$
Subcarriers	$M = 4$	$M = 4$	-	-	$M = 64$
SNR (γ)	9dB	0 to 18dB	-2 to 32dB	-10 to 24dB	0.5 to 5dB
Max. Doppler Freq.	100Hz	100Hz	20Hz	20Hz	30Hz
Channel Coding	- -	- -	short LDPC (204,102) [59]	short LDPC (204,102) [59]	Turbo ($R = 1/2$)
Channel Decoding	- -	- -	Belief Propag.	Belief Propag.	Turbo (MAP)
Space-Time Coding	- -	- -	Rate 1 $R_{STBC} = 1$ [4]	Rate 1 $R_{STBC} = 1$ [4]	Rate 1 $R_{STBC} = 1$ [60]

Table 4. Adopted channel and multicarrier multiple-antenna system parameters.

Figures 2 and 3 consider systems with space and frequency diversity, in SIMO MC-CDMA scenarios. Monte Carlo simulation results indicate that the GA, SA, RTS, STTS, EP-C and HH multiuser detectors result in the same near-optimal performance in terms of solution quality after convergence, but with different complexities. However, the local search algorithm (1-LS and 3-LS) presented ACI and ASI measures below desirable thresholds and are not suitable for applications in scenarios of multiuser detection with multidimensional diversity.

Figure 2 shows the convergence behavior as a function of the number of iterations for all heuristic MuD algorithms considered. Note the equality of BER performance achieved after total convergence for all heuristic techniques. Specifically for convergence evaluation, different initial solutions were considered, while all achieving performances significantly superior to the Conventional detector. However, the number of generations/iterations for convergence proved to be different, which will be analyzed in details on Section 5.1 and 5.2. Note that increasing the number of receiving antennas implies in a significant performance improvement (due to the spatial diversity) for a medium loading system and low signal-to-noise ratio (SNR).

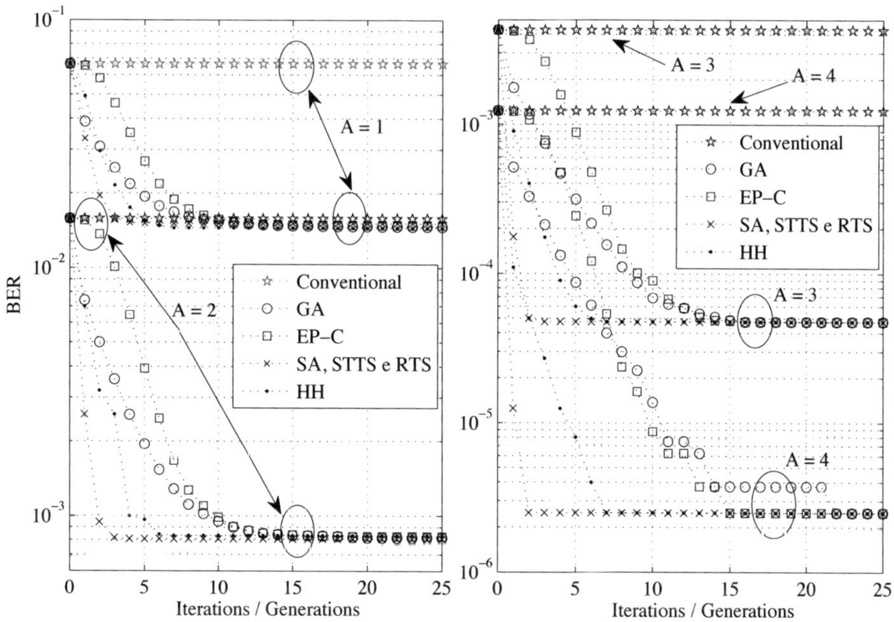

Figure 2. Convergence speed for the conventional and various heuristic algorithm detectors under SIMO MC-CDMA scenarios with $E_b/N_0 = 9$dB and $K = 20$ users.

Figure 3 shows the BER performance for SIMO MC-CDMA with different number of receiving antennas ($A_{Rx} = 1, 2, \ldots 5$) and medium loading system ($K = 20$). Accordingly, a $BER = 10^{-5}$ for a moderate number of A_{Rx} antennas and SNR has been achieved. Thus, there is an expressive performance gain with heuristic MuD strategies regarding the Conventional detector when the number of antennas is increased for signal-to-noise ratio in the range [0; 18] dB.

Systems with multiple-input-multiple-output (MIMO) and space-time block code (STBC) represent a promising solution often incorporated in commercial standards such as Wimax. Furthermore, a better performance $\times$ complexity trade-off can be obtained through the use of low density parity check codes (LDPC). The choice of STBC topology should take into account performance criteria, such as coding gain, diversity gain, multiplexing gain, and obviously the decoder complexity. However, these topics are not the focus of the this work and, therefore, more information can be found in the references [4, 60].

The considered MIMO system is formed by $A_{Tx} = 4$ transmit antennas and $A_{Rx} \geq 1$ receiving antennas, with 4 symbols transmitted simultaneously. Furthermore, the following parameters have been adopted (see Table 4): $A_{Tx}A_{Rx}$ flat fading statistically independent channels, $M-$QAM modulation, quasi-orthogonal STBC (QO-STBC) scheme with rate 1 [4], short LDPC(204,102) code, perfect channel state information knowledge at receiver, random sequences of length $N = 32$ and two scenarios with loading system $\mathcal{L} = \frac{K}{N} = \frac{1}{32}$ and 1.

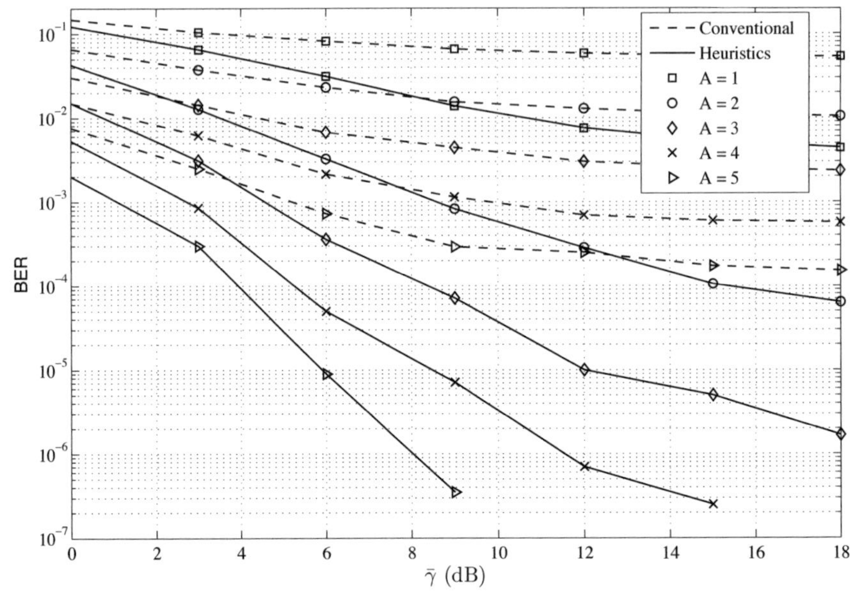

Figure 3. BER performance for heuristic algorithms. SIMO MC-CDMA system with $K = 20$ users.

Figure 4 depicts the BER performance *versus* SNR at the receiver input for different modulation constellations with $A_{Tx} = 4$ and $A_{Rx} = 1$ antennas. As expected, the single-user performance achieves very low BER[7] under smaller SNR than that necessary with high loading system (with $K = 32$ users). However, it is observed that with an increment of 2–3dB in SNR for systems with high loading it is possible to obtain very lower BER, especially for 4-QAM modulation. However, higher order modulations such as 256-QAM enable the transmission of more bits per symbol period, which result in higher throughput systems if more power/energy is available at transmitter. Furthermore, the performance loss by increasing the loading proves be small, enabling the deployment of heuristic MuD algorithms in coded CDMA systems.

Figure 5 compares the heuristic MuD performance in terms of BER *versus* SNR for high order modulation constellations and considering $A_{Tx} = 4$ and $A_{Rx} = 4$ antennas. We observe the same behavior shown in Figure 4. But in this case the number of receiving antennas has been increased to $A_{Rx} = 4$, resulting in significant performance improvement with reduction of $\approx$ 10dB SNR requirement, in order to obtain similar BERs. Again, the performance loss under total loading system is marginal; as a result, all heuristic algorithms discussed herein can be considered suitable for multiuser detection in coded CDMA systems. It is noteworthy that all heuristics algorithms showed the same level of BER performance after 40 generations for GA and EP-C, and 45 iterations for the SA, STTS, RTS and HH algorithms.

Figure 6 shows the BER performance of a multicarrier DS/CDMA (MC-CDMA) system with various types of diversity, considering multiple-input-multiple-output (MIMO) STBC coding, encoding and decoding turbo. This topology[8] was adopted in order to represent a transmission-reception topology with a great diversity order, making possible to obtain

[7] Beyond a certain SNR value, the performance improves sharply.

[8] Several other topologies can be considered for multidimensional analysis. For details, please see [61].

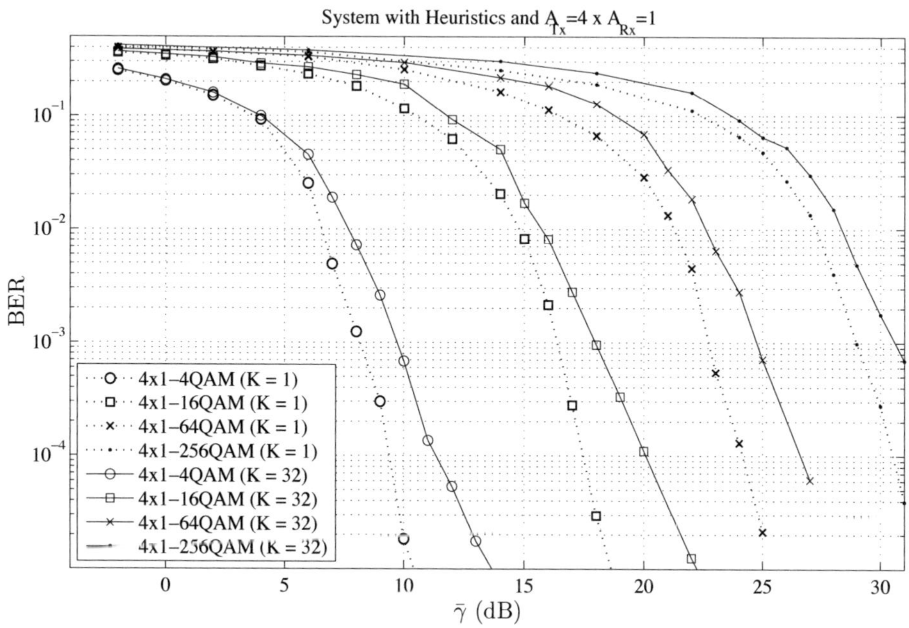

Figure 4. BER performance of the heuristic decoders for QO-STBC MIMO systems with short LDPC(204,102), $A_{Tx} = 4$, $A_{Rx} = 1$ antennas and $K = 1$ or $K = 32$ users.

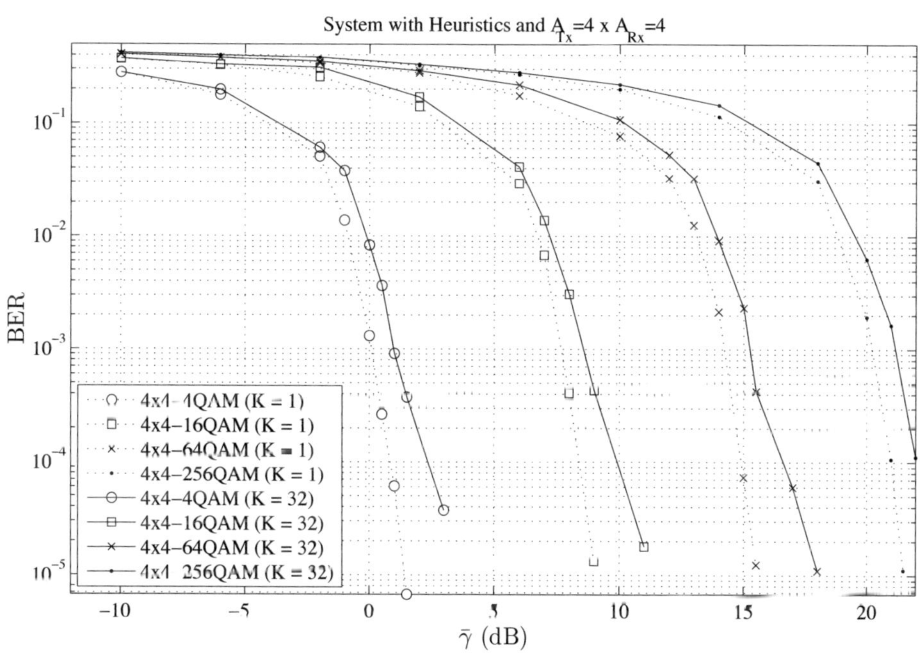

Figure 5. BER performance of the heuristic decoders for QO-STBC MIMO systems with short coding LDPC(204,102), $A_{Tx} = 4$, $A_{Rx} = 4$ antennas and $K = 1$ or $K = 32$ users.

excellent BER performance even for low SNR region. Of course, this topology is promising for adoption as a commercial standard.

Specifically, in the performance of Figure 6, a Turbo encoding and decoding of rate 1/2, which result in a spectral efficiency of 0.5 bps/Hz have been adopted. Again, there is a remarkable performance gain increasing when the transmit-receive antenna array becomes larger for both MMSE and heuristic multiuser detectors. Another important aspect to be mentioned is the extremely low BERs for $\bar{\gamma} \leq 3$ dB achieved with heuristics MuD topologies even under 1×1 antenna array configuration. Increasing the antenna array to 2×2 and adopting Alamouti code [60] it was possible to obtain a performance of $BER \leq 10^{-4}$ even for $\bar{\gamma} \leq 2$ dB. Thus, the performance achieved by MuD heuristics topology approaches the single-user bound (SuB) demonstrating the huge potential applicability in commercial communication systems and standards, specially those ones with high-performance and reliability requirements.

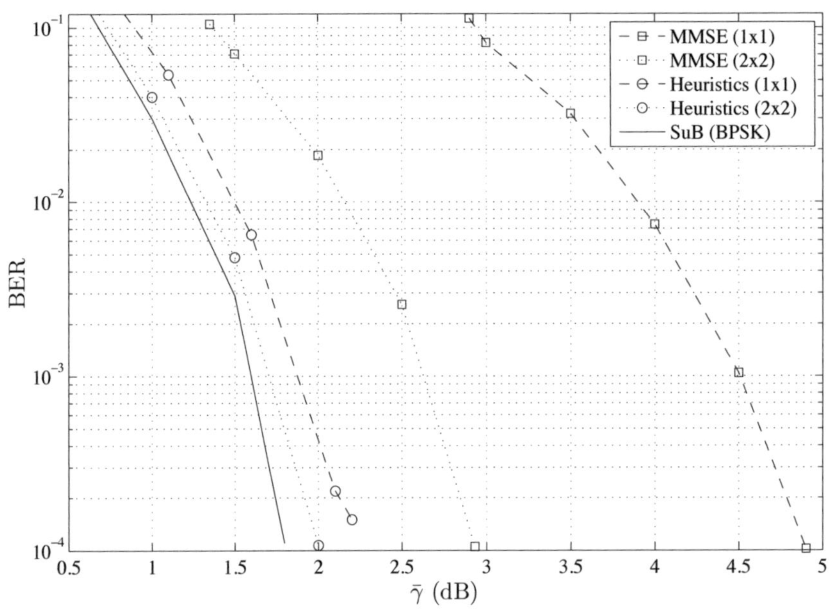

Figure 6. BER performance against SNR for a system with turbo channel coding.

Moreover, different topologies of the chosen one can be analyzed by considering, for example, channel coding (Convolutional or LDPC codes) and spatial diversity with other settings. However, the purpose of this section is to validate the potential of application of heuristics in MuD scenarios with multidimensional diversity and not compare topologies and system settings.

It is noteworthy that all heuristic MuD algorithms showed the same BER performance level after 100 generations for GA and EP-C algorithms, and 120 iterations for the SA, STTS, RTS and HH algorithms.

The presented results considering multiuser detection with different level of diversity exploitation in (non)coded telecommunication systems have demonstrated the effective applicability of the proposed heuristics MuD techniques, due to significant improvement in BER with reduced complexity regarding the optimal multiuser detector (OMuD). Complexity aspects of the proposed heuristic MuDs are discussed in the next section.

5.1. Overall systems complexity

As the complexity of the algorithms in terms of number of operations has been determined, one can determine the complexity of each telecommunication scenario considering the complexity for the cost function calculation.

Both terms of the cost function, defined by $\mathcal{F}_1 = \mathcal{C}^H \mathcal{A}\mathcal{Y}$ and $\mathcal{F}_2 = \mathcal{C}\mathcal{A}\mathbf{R}\mathcal{A}\mathcal{C}^H$ in eq. (14) can be obtained before the loop optimization in each MuD heuristic algorithm. Thus, for each candidate-solution evaluation, $\mathcal{B}^T\mathcal{F}_1$ and $\mathcal{B}^T\mathcal{F}_2\mathcal{B}$ are computed, which in terms of operations is equivalent to $(KID)^2 + 4KID$ operations. For OMuD detector, the number of operations grows exponentially with the number of users, i.e., $\mathcal{O}\left(2^{KI}(KID)^2\right)$. It takes 2^{KI} bit generations of order KID, as well as 2^{KI} cost function calculations for the simultaneous detection of a frame consisting of I bits of K users on a system with multiuser detection operating on fading channels.

Therefore, in order to calculate the cost function, $\mathcal{O}_{FC} = (KID)^2 + 4KID$ operations are needed. It is noteworthy that 42 different scenarios have been analyzed[9]. Besides, scores to define the best strategy have been considered this metric in different scenarios.

Thus, the computational complexity of each proposed heuristic MuD algorithm, in terms of number of operations, has been obtained under different operation scenarios. Table 5 presents such complexities. Strategy with the lowest complexity using the adopted scoring system, for each analyzed scenario has been indicated with bold numbers. It is noteworthy that the scores were normalized considering the higher value with score of 100. For sake of comparison, the optimal MuD complexity is presented in the last column.

Scenario		GA	EP-C	SA	STTS	RTS	HH	OMuD
1 – Fig.2	$(\times 10^6)$	4,597	**4,328**	5,572	5,606	5,607	5,144	$8,22.10^{21}$
2 – Fig.3	$(\times 10^8)$	2,322	**2,303**	3,333	3,336	3,336	2,926	$1,52.10^{45}$
3 – Fig.4	$(\times 10^9)$	4,390	**4,382**	5,794	5,795	5,795	5,451	$1,03.10^{57}$
4 – Fig.5	$(\times 10^{13})$	**2,569**	**2,569**	18,877	18,878	18,878	12,483	$> 10^{300}$
5 – Fig.6	$(\times 10^{13})$	1,028	**1,027**	22,277	22,278	22,278	13,860	$> 10^{300}$
Score		87	100	83	72	56	85	–
Position		2nd	1st	4th	5th	6th	3rd	–

Table 5. Necessary number of operations in all optimized Scenarios.

5.1.1. Computational Time

Furthermore, the complexity in terms of computational time has been determined for each telecommunication scenario. As a result, the computational time for calculating one cost function according each specific scenario has been quantified. We have deployed a personal

[9] But not shown herein due to the lack of space. Additional results can be checked in [61].

computer with the following configuration: Motherboard ASUS P8H67-M EVO, Intel I7-2600 with 3.4 GHz clock and 8MB cache; Memory 8GB Corsair Dominator DDR3 1333MHz and a Video board Radeon HD6950 2GB DDR5.

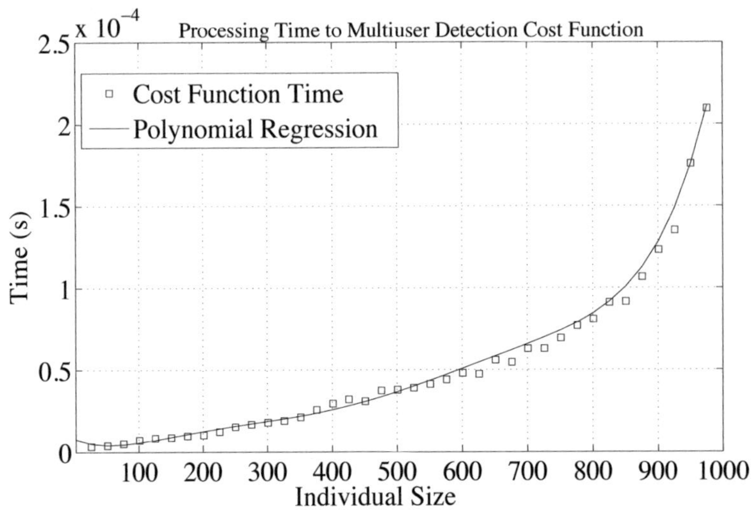

Figure 7. Average time for cost function calculation and respective polynomial approximation.

Figure 7 depicts the time to calculate a cost function as the size of the individual increases, while Table 6 shows the average time required for optimization algorithms in each scenario considered, as well as the respective scores and classification.

Scenario	GA	EP-C	SA	STTS	RTS	HH
1 – Fig.2	0,039	**0,0378**	0,0474	0,0477	0,0477	0,0436
2 – Fig.3	0,297	**0,2892**	0,7509	0,7532	0,7538	0,5948
3 – Fig.4	0,4638	**0,4515**	1,1704	1,1741	1,1749	0,9414
4 – Fig.5	128,126	**128,001**	914,891	914,876	914,906	609,049
5 – Fig.6	5,2576	**5,1283**	24,0933	24,1408	24,1608	18,2711
Score	88	100	73	66	56	80
Position	2nd	1st	4th	5th	6th	3rd

Table 6. Average time (in seconds) spend by the optimized heuristic algorithms under Scenarios 1 – 5.

In conclusion, again the algorithm EP-C presents the lowest complexity in terms of computational time. Note that all heuristic algorithms result in a computational time, as well as number of operation very close each other. Thus, we can adopt any topologies without significant loss in terms of performance $\times$ complexity trade-off metric, validating the deployment of heuristic MuD approach with optimized input parameters as shown in Table 2.

5.2. Quality, stability and topology choice

For all scenarios considered, convergence curves in terms of BER have been resulted in the same level achieved for all algorithms, but after a different number of generations or iterations[10]. In order to evaluate the stability of the algorithms after convergence, some tests have been conducted for a large number of generations/iterations. For all algorithms, the average performance level and number of generations/iterations for total convergence hold under the same boundaries.

Therefore, for the multiuser detection problems presented in this work it is concluded that the proposed heuristic algorithms are able to find the optimum solution or very near-optimum solution with a reduced number of cost function tests, approaching the MuD performance since an adequate and sufficient number of generations/iterations has been available. In this context we are interest in analyze the variation of the solutions when decreasing the number of generations/iterations; as a result, the heuristic MuD algorithms were also classified following the criterion of the smaller deviation values of solutions.

In addition, for the complexity analysis, we chose to average between scores (number of operations and computational time), thus avoiding an unfair analysis among optimization strategies or possible bias.

Table 7 presents the classification for the six heuristic MuD with the adoption of equal weight features for the final score calculation. In conclusion, under equal weight metric for computational complexity and quality/stability of the solutions found, the following choice criteria for MuD problem can be established:

Criterion	Algorithm Choice
Best BER performance:	HH
Lowest Complexity:	EP-C
Performance-Complexity trade-off:	HH (score: 91)

Score	GA	EP-C	SA	STTS	RTS	HH
Number of Operation	87	100	83	72	56	85
Computational Time	88	100	73	66	56	80
Complexity (Average)	88	100	78	69	56	83
Quality and Stability	80	67	78	64	62	100
Final Score (Average)	84	83	78	66	59	91
Final Position	2nd	3rd	4th	5th	6th	1st

Table 7. Scores for all considered heuristic algorithms applicable to wireless communication Scenarios 1 – 5.

6. Main conclusions

Several heuristic techniques applied to multiuser detection problems under different channel and system configuration scenarios, as well as diversity dimensionality, such as time, frequency, space and coding have been analyzed in this work. The main purpose in

[10] This result was obtained when the initial inputs guess for all algorithms considered are the same, indicating that all algorithms were able to achieve convergence after a certain number of generations/iterations, except the LS algorithm, which did not show enough stability for adoption.

combining different types of diversity with heuristic detection is to provide system capacity increasing and/or reliability improvement.

Near single-user bound performance has been achieved by all MuD heuristic algorithms analyzed in this work, considering different system and channel configurations, while offer a dramatic complexity reduction regarding the OMuD with marginal performance loss, even in aggressive fading channels and high loading systems conditions.

Among the analyzed detectors, the best MuD heuristic algorithm choice must take into account that one which offer either smallest computational complexity or the best BER performance, i.e. EP-C or HH multiuser detectors, respectively. Hence, the criteria for topology ranking established in this work allow us to quantify the parameter optimization level, reflecting on the quality and stability of the solutions obtained.

The heuristic input parameter optimization, as well as the proposed methodology for the heuristic MuD topology choice represent the main contribution of this work. Under optimized input parameters condition of all heuristic MuD algorithms, the quality and stability analyses have been carried out deploying ten benchmark functions. The numerical results for the MuD problem confirmed the near-optimal performance achieved by the heuristic algorithms for a wide channel and system configurations, corroborating the methodology adopted for the ranking topology.

Acknowledgements

This work was supported in part by the National Council for Scientific and Technological Development (CNPq) of Brazil under Grants 202340/2011-2, 303426/2009-8 and in part by Londrina State University - Paraná State Government (UEL).

Author details

Fernando Ciriaco[1,*],
Taufik Abrão[1] and Paul Jean E. Jeszensky[2]

* Address all correspondence to: fciriaco@uel.br; abrao@ieee.org; pjj@lcs.poli.usp.br

1 Electrical Engineering Department, State University of Londrina (DEEL-UEL), Londrina, Paraná, Brazil
2 Polytechnic School of the University of Sao Paulo (EPUSP), Sao Paulo, Brazil

References

[1] R. L. Peterson, R. E. Ziemer, and D. E. Borth. *Introduction to Spread Spectrum Communications.* Prentice Hall, New Jersey, 1 edition, 1995.

[2] C. E. Shannon. A mathematical theory of communication. *Bell System Technical Journal,* 27:379–423 and 623–656, July and October 1948.

[3] S. Verdú. *Multiuser Detection.* Cambridge University Press, USA, 1998.

[4] H. Jafarkhani. *Space-Time Coding: Theory and Practice.* Cambridge University Press, 2005.

http://dx.doi.org/10.5772/54152

[5] D. Tse and P. Viswanath. *Fundamentals of Wireless Communication*. Cambridge University Press, Cambridge, UK, 2005.

[6] S. Verdú. Minimum probability of error for asynchronous gaussian multiple-access channels. *IEEE Transactions on Information Theory*, 32:85–96, January 1986.

[7] A. Duel-Hallen, J. Holtzman, and Z. Zvonar. Multiuser detection for cdma systems. *IEEE Personal Communications*, pages 46–58, April 1995.

[8] H. V. Poor and S. Verdú. Probability of error in mmse multi-user detection. *IEEE Transactions on Information Theory*, 43(3):858–881, May 1997.

[9] P. Patel and J. M. Holtzman. Analysis of a single sucessive interference cancellation scheme in a ds/cdma system. *IEEE Journal on Selected Areas in Communications*, 12:796–807, June 1994.

[10] R. Agarwal, B. V. R. Reddy, and K. K. Aggarwal. A reduced complexity hybrid switched mode detector using interference cancellation methods for ds-cdma systems. In *Annual IEEE India Conference*, pages 1 – 4, September 2006.

[11] A. Duel-Hallen. Decorrelating decision-feedback multiuser detector for synchronous cdma channel. *IEEE Transactions on Communications*, 41(2):285–290, February 1993.

[12] P. H. Tan. Multiuser detection in cdma-combinatorial optimization methods. Master's thesis, Chalmers University of Technology, Göteborg, 2001.

[13] F. Ciriaco, T. Abrão, and P. J. E. Jeszensky. Ds/cdma multiuser detection with evolutionary algorithms. *Journal of Universal Computer Science*, 12(4):450–480, 2006.

[14] S. Abedi and R. Tafazolli. Genetically modified multiuser detection for code division multiple access systems. *IEEE Journal on Selected Areas in Communications*, 20(2):463–473, February 2001.

[15] T. Abrão, F. Ciriaco, and P. J. E. Jeszensky. Evolutionary programming with cloning and adaptive cost funciton applied to multi-user ds-cdma systems. In *2004 IEEE International Symposium on Spread Spectrum Techniques and Applications*, pages 160–163, Sydney, August 2004. Australia.

[16] K. Yen and L. Hanzo. Genetic-algorithm-assisted multiuser detection in asynchronous cdma communications. *IEEE Transactions on Vehicular Technology*, 53(5):1413–1422, September 2004.

[17] F. Ciriaco, T. Abrão, and P. J. E. Jeszensky. Multirate multiuser ds/cdma with genetic algorithm detection in multipath channels. In *2006 IEEE International Symposium on Spread Spectrum Techniques and Applications*, Manaus, August 2006. Brazil.

[18] M. S. Arifianto, A. Chekima, L. Barukang, and M.Y. Hamid. Binary genetic algorithm assisted multiuser detector for stbc mc-cdma. In *International Conference on Wireless and Optical Communications Networks*, pages 1 – 5, July 2007.

[19] R. Zhang and L. Hanzo. Harmony search aided iterative channel estimation, multiuser detection and channel decoding for ds-cdma. In *Vehicular Technology Conference Fall*, pages 1 – 5, September 2010.

[20] P. J. E. Jeszensky and G. Stolfi. Cdma systems sequences optimization by simulated annealing. In *IEEE 5th International Symposium on Spread Spectrum Techniques and Applications*, volume 3, pages 706–708, Sun City, 1998. South Africa.

[21] T. M. Chan, S. Kwong, K. F. Man, and K. S. Tang. Sequences optimization in ds/cdma systems using genetic algorithms. In *IEEE Region 10 International Conference on Electrical and Electronic Technology*, volume 2, pages 728–731, Phuket Island, August 2001. Singapore.

[22] F. Ciriaco, T. Abrão, and P. J. E. Jeszensky. Genetic algorithm applied to multipath multiuser channel estimation in ds/cdma systems. In *2006 IEEE International Symposium on Spread Spectrum Techniques and Applications*, Manaus, August 2006. Brazil.

[23] T. Yano, K. Nakamura, T. Tanaka, and S. Honda. Channel parameter estimation in the cdma multiuser detection problem. In *Fourth International Conference on Networked Sensing Systems*, June 2007.

[24] D. Zhao, M. Elmusrati, and R. Jantti. On downlink throughput maximization in ds-cdma systems. In *IEEE 61st Vehicular Technology Conference*, volume 3, pages 1889 – 1893, May 2005.

[25] F. R. Durand, L. Melo, L. R. Garcia, A. J. dos Santos, and Taufik Abrão. Optical network optimization based on particle swarm intelligence. In Taufik Abrão, editor, *Search Algorithms*, volume 1, pages 1–21. InTech Open, 2012.

[26] M. Moustafa, I. Habib, and M. N. Naghshineh. Efficient radio resource control in wireless networks. *IEEE Transactions on Wireless Communications*, 3:2385–2395, November 2004.

[27] T. Abrão, L. D. H. Sampaio, M. L. Proença Jr, and P. J. E. Angélico, B. A.; Jeszensky. Multiple access network optimization aspects via swarm search algorithms. In Nashat Mansour, editor, *Search Algorithms and Applications*, volume 1, pages 261–298. InTech Open, 2011.

[28] M. P. Marques, M. H. Adaniya, T. Abrão, L. H. D. Sampaio, and Paul Jean E. Jeszensky. Ant colony optimization for resource allocation and anomaly detection in communication networks. In Taufik Abrão, editor, *Search Algorithms*, volume 1, pages 1–34. InTech Open, 2012.

[29] J. G. Proakis. *Digital Communications*. Electrical Engineering. Communications and Signal Processing. McGraw-Hill, New York, 2 edition, 1995.

[30] S. Verdú. Computational complexity of optimum multiuser detection. *Algorithmica*, 4:303–312, 1989.

[31] C. R. Darwin. *On The Origin of Species by Means of Natural Selection.* John Murray, London, 1859.

[32] C. R. Darwin. *Origin of Species.* Wilder Publications, 2008.

[33] J. H. Holland. *Adaptation in Natural and Artificial Systems.* University of Michigan Press, Ann Arbor, 1975.

[34] D. E. Goldberg. *Genetic Algorithms in Search Optimization and Machine Learning.* AddisonWesley, Nova York, 1989.

[35] M. Mitchell. *An Introduction to Genetic Algorithms.* MIT Press, Cambridge, 1998.

[36] C. H. Papadimitriou and K. Steiglitz. *Combinatorial Optimization - Algorithms and Complexity.* Dover, 2 edition, 1998.

[37] S. Kirkpatrick, C. D. Gellat, and M. P. Vecchi. Optimization by simulated annealing. *Science*, 220:671–680, 1983.

[38] N. Metropolis, A. W. Rosenbluth, M. N. Rosenbluth, A. H. Teller, and E. Teller. Equation of state calculations by fast computing machine. *Journal Chemical and Physical*, 21:1087–1092, 1953.

[39] V. Cerny. Minimization of continuous functions by simulated annealing. *Research Institute for Theoretical Physics*, 1985.

[40] F. Glover. Heuristic for integer programming using surrogate contraints. *Decision Sciences*, 8:156–166, 1977.

[41] F. Glover. Future paths for integer programming and links to artificial intelligence. *Computers and Operations Research*, 13:533–549, 1986.

[42] R. Battiti and G. Tecchioli. The reactive tabu search. *ORSA, Journal of Computing*, 6:126–140, 1994.

[43] E. Burke, T. Curtois, M. Hyde, G. Kendall, G. Ochoa, S. Petrovic, J. A. V. Rodriguez, and M. Gendreau. Iterated local search vs. hyper-heuristics: Towards general-purpose search algorithms. In *IEEE Congress on Evolutionary Computation (CEC)*, pages 1–8, July 2010.

[44] J. Grobler, A. P. Engelbrecht, G. Kendall, and V. S. S Yadavalli. Alternative hyper-heuristic strategies for multi-method global optimization. In *IEEE Congress on Evolutionary Computation (CEC)*, pages 1–8, July 2010.

[45] W. A. Shewart. *Economic control of Quality of Manufactured Product.* Van Nostrand Reinhold Co., New York, 1931.

[46] E. L. Grant and R. S. Leavenworth. *Statistical Quality Control.* Tsinghua University Press, Beijing: China, seventh edition, 1999.

[47] F. W. Breyfogle. *Implementing Six Sigma: Smarter Solutions Using Statistical Methods*. John Wiley & Sons Ltda, New York, 1999.

[48] Q. L. Zhang. Metrics for meta-heuristic algorithm evaluation. In *International Conference on Machine Learning and Cybernetics*, volume 2, pages 1241–1244, November 2003.

[49] E. K. Burke, T. Curtois, M. Hyde, G. Kendall, G. Ochoa, S. Petrovic, and J. A. Vazquez-Rodriguez. Towards the decathlon challenge of search heuristics. In *Genetic and Evolutionary Computation Conference (GECCO)*, pages 2205–2208, July 2009.

[50] K. A. De Jong. *An analysis of the behavior of a class of genetic adaptive systems*. PhD thesis, Department of computer and Communication Sciences, University of Michigan, Ann Arbor - Michigan, 1975.

[51] D. Whitley, K. Mathias, S. Rana, and J. Dzubera. Building better test functions. In *Proceedings of the Sixth International Conference on Genetic Algorithms*, pages 239–246, 1995.

[52] Z. Michalewicz. A hierarchy of evolution programs: an experimental study. *In Evolutionary Computation*, 1:51–76, 1993.

[53] J. D. Schaffer, R. Caruana, L. J. Eshelman, and R. Das. A study of control parameters affecting online performance of genetic algorithms for function optimization. In *Proceedings of the 3rd International Conference on Genetic Algorithms*, pages 51 – 60, 1989.

[54] D. H. Ackley. *A connectionist machine for genetic hillclimbing*. Kluwer Academic Publishers, Boston, 1987.

[55] L. A. Rastrigin. The convergence of the random search method in the extremal control of a many parameter system. *Automation and Remote Control*, 24(10):1337–1342, 1963.

[56] H. P. Schwefel. *Numerical Optimization of Computer Models*. PhD thesis, Birkhäuser Verlag, traduction by John Wiley and Sons in 1980, Basel, 1977.

[57] L. C. Dixon and G. P. Szego. *The optimization problem: An introduction, Towards Global Optimization II*. North Holland, New York, 1978.

[58] Z. Michalewicz. *Genetic Algorithms + Data Structures = Evolution Programs*. Springer-Verlag Berlin Heideberg, Berlin, 1996.

[59] D. Mackay. Good error-correcting codes based on very sparse matrices. *IEEE Transactions on Information Theory*, 45(2):399–431, March 1999.

[60] S. M. Alamouti. A simple transmit diversity technique for wireless communications. *IEEE Journal on Selected Areas in Communications*, 16(8):1451–1458, October 1998.

[61] Fernando Ciriaco. *Otimização Multidimensional Baseada em Heurísticas Aplicada aos Sistemas de Comunicação Sem Fio*. PhD thesis, Escola Politécnica da Universidade de São Paulo, March 2012. (in portuguese).

## Chapter 5

Ant Colony Optimization for Resource Allocation and Anomaly Detection in Communication Networks

Lucas Hiera Dias Sampaio,
Mateus de Paula Marques, Mário H. A. C. Adaniya,
Taufik Abrão and Paul Jean E. Jeszensky

Additional information is available at the end of the chapter

http://dx.doi.org/10.5772/53338

1. Introduction

Due to the paramount importance of (wireless) communication systems and computer networks, in the last decade both resource allocation (RA) and anomaly detection (AD) problems addressed herein have been intensively studied and numerous solutions have been proposed in the specialized literature [1]. The RA specially in wireless multiple access networks and the AD in computer networks are problems of complex nature and an engineering compromise solution has much appeal in terms of practical and effective deployment.

Resource allocation problems in wireless communication networks include power consumption minimization, information rate and network capacity maximization, battery lifetime maximization, energy-efficient and bandwidth-efficient optimal design among others. In computer networks, anomaly detection system (ADS) consists of a set of techniques aiming to detect anomalies in network operation, helping the administrator to decide which action need to be performed in each situation. Anomaly detection is not an easy task and brings together a range of techniques in several areas, such as machine learning, signal processing techniques based on specification techniques, and data mining among others. Generally, for most scenarios of practical interest, these optimization formulations result in non-convex problems, which is hard to solve or even impossible using conventional convex optimization techniques, even after imposing relaxations to deal with RA problems.

In this sense, heuristics have been widely used to solve problems that deterministic optimization methods result in high computational complexity and therefore have no application in real systems. In this context, the ant colony optimization (ACO) algorithm [2] developed in recent years has attracted a lot of interest from so many professionals due to its robustness and great performance in deal with discrete (combinatorial) and continuous optimization problems.

An important challenge for the future wireless communication systems has been how to acquire higher throughput with lower power consumption. Hence, in order to transfer the exponentially rising amount of available data to the user in an acceptable time, following the "Moore's Law", according to which both the processing power of CPUs and the capacity of mass storage devices doubles approximately every 18 months, the transmission rate in cellular network has been risen at the speed of nearly 10 times every 5 years. Meanwhile, the price paid for this enormous growth in data rates and market penetration is a rising power requirement of information and communication technologies (ICT) – although at a substantially lower speed than "Moore's Law" – the energy consumption doubles every 4-5 years [3].

In order to avoid the collapsing of communication systems and networks resources, an increasing interest and intensive researches in both energy and bandwidth efficient designs have mobilized enormous efforts of research's groups around the globe in the last decade. Against this background, the conventional efficient design of wireless networks mainly focuses on system capacity and spectral efficiency (SE). However, energy-efficient design in wireless networks is of paramount importance and is becoming an inevitable trend, since the deployment of multimedia wireless services and requirement of ubiquitous access have increased rapidly, and as a consequence, energy consumption at both the base station and mobile terminals side have experienced enormous increasing.

In order to achieve high throughput with low total transmit power, system resources such as spectrum (subcarriers, bandwidth), transmit power (energy, battery lifetime) and information rate (QoS requirements) in different multiple access wireless communication systems should be efficiently and appropriately allocated to the different active users. The first part of this chapter is dedicated to deal with the energy-efficient and spectral-efficient designs in DS/CDMA wireless communication systems through the appropriate heuristic optimization of energy and information rate resources.

The Internet has brought to our daily life easy and new ways to execute tasks as searching and gathering information, to communicate and spread ideas and others small gestures that are changing our lives. In order to prevent possible failures and loss of performance, the infrastructure providing theses services must be monitored, which unavoidably increases the responsibility and charge of the network administrator. The administrator is assisted by tools such as firewall, proxy, among others, including the anomaly detection system to help prevent abnormal network operation. Usually the anomaly behavior is a sudden increase or decrease into the network traffic. It can be caused by a simple programming error in some software to hardware failure, among many other causes that affect directly the network operation.

In the next sections of this Chapter the ACO methodology will be applied and analyzed regarding two distinct application of communication scenarios: the resource allocation in a direct sequence code division multiple access (DS/CDMA) systems, which is developed in Section 2 and the anomaly detection in computer networks is discussed in the Section 3. The conclusion remarks for both ACO-communication application problems are offered in Section 4.

http://dx.doi.org/10.5772/53338

2. Resource Allocation in Wireless Multiple Access Networks

The optimized resource allocation in wireless multiple access networks, specially the power rate allocation, is a problem of great interest for telecommunications enterprises and users. It is well known that spectrum and power are valuable resources due to their scarcity, the first one is a natural non renewable resource and the second one is limited by the battery and device size. Therefore, proposing new techniques and algorithms that can allocate this resources in a simple[1] and optimized manner is pretty important.

In the last few decades many researchers have been working on this subject aiming to find a simple yet sturdy algorithm for resource allocation in wireless systems. Among many works recently done we enumerate some notorious in the next section.

2.1. Related Work

Among numerous solutions proposed to resource allocation in wireless multiple access networks we enumerate herein some of great importance works: Foschini and Miljanic [4] distributed power control algorithm (DPCA) stands as the main one. When it comes to metaheuristics, in [5] and [6] a genetic algorithm approach was used to propose the genetic algorithm for mobiles equilibrium, providing the joint power-rate control in CDMA multiple access networks. In [7], the particle swarm optimization (PSO) metaheuristic was used in order to establish a low-complexity power control algorithm. Finally, in [8] a power allocation approach was proposed to solve the parallel interference cancelation in multi-user detectors.

Beyond the metaheuristic approaches, the work developed in [9] exploits an algorithm based on the dynamic cost assignment for downlink power allocation in CDMA networks. Besides, [10] addressed the uplink fairness maximization in a CDMA system with multiple processing gains. In [11], the Verhulst population model, firstly developed to describe the biological species growth with restrictions of space and food, was adapted to the distributed power control problem in a DS/CDMA network. It is noteworthy that this work was the first one to propose a Verhulst model adaptation to resource allocation problems in multiple access networks.

Furthermore, in [12] an analytical approach was proposed for the weighted throughput maximization (WTM) problem, namely MAPEL. The algorithm performs the power control in the interference limited wireless networks, i.e., CDMA and MC/CDMA networks, through a monotonically increasing objective function that is not necessarily convex. This function was formulated as a multiplicative linear fractional programming (MLFP) problem, which is a special case of generalized linear fractional programming (GLFP). So, the GLFP problem presented in [12] was used in [13] in order to formulate a non-decreasing objective function as a weighted SNIR's productory.

Finally, this section presents a heuristic approach through ant colony optimization in the continuous domains ($ACO_{\mathbb{R}}$) applicable to the power and rate allocation problems [13], and is organized as follows. subsection 2.2 describes aspects of the DS/CDMA networks and the power control problem on subsection 2.3 the power control problem and the cost function used with the ACO algorithm are presented; subsection 2.4 deals with the throughput maximization problem and how the ACO algorithm can be applied to solve this optimization

[1] Here, simple is used as a synonym for low computational complexity.

problem, while in subsection 2.5 the ACO algorithm itself is described. Finally, subsection 2.6 introduces the simulations scenarios, the numerical results and conclusions for the first part of this chapter.

2.2. Resource Allocation in DS/CDMA Networks

In DS/CDMA multirate networks, the Bit Error Rate (BER) is usually used as a QoS metric, since it is directly related to the Signal to Noise plus Interference Ratio (SNIR). Thus, the SNIR is associated to the Carrier to Interference Ratio as follows:

$$\gamma_i = \frac{r_c}{r_i} \times \Gamma_i, \qquad i = 1, \ldots, U \tag{1}$$

where γ_i is the i-th user's SNIR, r_c is the chip rate, r_i is the i-th user's information rate, U is the system load and Γ_i is the i-th user's CIR defined as [11], [14]:

$$\Gamma_i = \frac{p_i |g_{ii}|^2}{\sum_{j=1, i \neq j}^{U} p_j |g_{ij}|^2 + \sigma^2}, i = 1, ..., U \tag{2}$$

where p_i is the i-th user's power bounded by p_{max}, U the number of active users on the system, $|g_{ii}|$ the channel gain of the i-th user, $|g_{ij}|$ is the interfering signals gain and σ^2 the Additive White Gaussian Noise (AWGN) at the i-th receiver's input.

$$\mathbf{G}_{\text{upl}} = \begin{bmatrix} g_{11} & g_{12} & \cdots & g_{1U} \\ g_{21} & g_{22} & \cdots & g_{2U} \\ \vdots & \vdots & \ddots & \vdots \\ g_{U1} & g_{U2} & \cdots & g_{UU} \end{bmatrix} \tag{3}$$

where the main diagonal (g_{ii}) shows the i-th user's channel attenuation, while the other values shows the interfering signals gain.

The path loss is inversely proportional to the distance between the mobile unit and the base station; the shadowing is obtained by a log-normal probability distribution random variable, and the multipath fading obtained assuming a Rayleigh probability distribution for cases without line of sight (LOS), and Rice distribution for cases with LOS.

In the DS/CDMA networks with multiple processing gains (MPG), where each user has a different processing gain $F_i > 1$, it is defined as a function of the chip rate:

$$F_i = \frac{r_c}{r_i}, \qquad i = 1, 2, \ldots, U \tag{4}$$

Therefore, from Eq. (1) and (4) follows:

$$\gamma_i = F_i \times \Gamma_i \tag{5}$$

Additionally, the theoretical Shannon channel capacity is defined as [15]:

$$C = W \log(1 + \gamma) \tag{6}$$

where C is the channel capacity in $bits/s$ and γ is the SNIR. It is worthy to note that since this is a theoretical bound a gap can be included, thus, the Shannon equation can be rewritten as [1]:

$$C = W \log(1 + \theta\gamma) \tag{7}$$

where θ is the gap between the theoretical bound and the real information rate. Usually, θ can be defined as [1]:

$$\theta = -\frac{1.5}{\log(5BER)} \tag{8}$$

where BER is the desired bit error rate (BER).

2.3. Power Allocation Problem

The power control objective is to find the minimal transmission power for each user that satisfy its QoS requirement, usually a minimum transmission rate. Since user rate is related to the user SNIR one may use it as a QoS measure. Thus, the power allocation problem may be mathematically stated as:

$$\begin{aligned} \min \quad & \mathbf{p} = [p_1, p_2, \ldots, p_U] \\ \text{s.t.} \quad & \gamma_i \geq \gamma_i^* \\ & 0 \leq p_i \leq p_{\max} \end{aligned} \tag{9}$$

where p_i and γ_i is the ith user power and SNIR, respectively, and γ_i^* is the desired SNIR level.

In order to enable the users to have minimum QoS warranty, the minimum CIR to SNIR relation must be calculated as [16]:

$$\Gamma_{i,\min} = \frac{r_{i,\min} \gamma_i^*}{r_c}, \quad i = 1, ..., U \tag{10}$$

where $\Gamma_{i,\min}$ and $R_{i,\min}$ are the minimum CIR and minimum information rate of each user, respectively, and γ_i^* is the minimum SNIR needed in order to obtain a minimum BER (or QoS) that is acceptable to each user.

This way, the minimum information rate can be mapped in the SNIR through the Shannon's capacity model using the gap introduced in Eq (8):

$$2^{\frac{r_i}{r_c}} = \max\left[1+\theta_i\gamma_i\right] = \max\left[1+\frac{\theta_i F_i \cdot p_i|g_{ii}|^2}{\sum_{i\neq j}^{U} p_j|g_{ij}|^2+\sigma^2}\right] \tag{11}$$

where $2^{\frac{r_i}{r_c}}$ is the normalized information rate for the i-th user, θ_i is the inverse of the gap between the channel's theoretical capacity and the real information rate. Note that for the minimum SNIR γ_i^*, Eq. 11 uses the minimum information rate established by the system, in order to guarantee QoS. Such that one obtain the condition needed for the minimum SNIR to be satisfied, given a minimum information rate:

$$\gamma_i^* = \frac{2^{r_{i,\min}}-1}{\theta_i} \tag{12}$$

Consider the QoS normalized interference matrix **B** [1]:

$$B_{ij} = \begin{cases} 0, & i=j; \\ \frac{\Gamma_{i,\min} g_{ji}}{g_{ii}}, & \text{otherwise.} \end{cases} \tag{13}$$

which $\Gamma_{i,\min}$ can be obtained as follows:

$$\Gamma_{i,\min} = \frac{r_{i,\min}\gamma_i^*}{r_c}, \quad i=1,...,U \tag{14}$$

Now consider the information rate requirements for each user and the QoS normalized noise vector $u = [u_1, u_2, \ldots, u_k]^T$, with elements:

$$u_i = \frac{\Gamma_{i,\min}\sigma_i^2}{g_{ii}} \tag{15}$$

The solution to the power control problem may be analytically obtained solving the following linear system:

$$\mathbf{p}^* = (\mathbf{I}-\mathbf{B})^{-1}\mathbf{u} \tag{16}$$

where $I_{U\times U}$ is the identity matrix. Note that that $(\mathbf{I}-\mathbf{B})$ is invertible only, and only if the maximum eigenvalue of B is smaller than one [17]. Only in this case, the power control problem will present a feasible solution. Nevertheless, due to the limited resources of mobile terminals, the use of this method is not feasible since its computational cost grows prohibitively when the number of users goes beyond some dozens due to a matrix inversion operation. Besides a totally distributed allocation scheme cannot be deployed using this analytical solution. To overcome this issues, this work proposes a metaheuristic approach for the optimum power-rate allocation problems.

In order to use the ACO algorithm to solve the power allocation problem one must map the problem objective into a mathematical function so-called cost function. In [5, 6] a new cost function for power control problem in CDMA systems using genetic algorithms has been proposed. This function was later modified and used with swarm intelligence in [1] to solve the power control problem. Due to the good results obtained in [1] that cost function was used with the ACO algorithm and reproduced hereafter for convenience:

$$
\begin{aligned}
J_1(\mathbf{p}) &= \max \ \frac{1}{U}\sum_{i=1}^{U} \mathbb{F}_i^{\text{th}} \cdot \left(1 - \frac{p_i}{p_{\max}}\right), \qquad \forall i = 1,2,\ldots,U \\
\text{s.t.} \quad & \gamma_i \geq \gamma_i^* \\
& 0 \leq p_i \leq p_{\max} \\
& r_i = r_{i,\min}
\end{aligned} \tag{17}
$$

where the threshold function is defined as $\mathbb{F}_i^{\text{th}} = \begin{cases} 1, & \gamma_i \geq \gamma_i^* \\ 0, & \text{otherwise} \end{cases}$

2.4. Weighted Throughput Maximization (WTM) Problem

The increasing information traffic demand due to multimedia services on third generation networks (3G) and beyond, along with the need of telecommunications companies to improve their profits have motivated development on weighted throughput maximization (WTM) problem, which aims to maximize the system throughput, been formulated as:

$$
\begin{aligned}
\max_{r} \quad & f(\mathbf{p}) \\
\text{s.t.} \quad & r_i \geq r_{i,\min} \\
& 0 \leq p_i \leq p_{\max}
\end{aligned} \tag{18}
$$

where $f(\mathbf{p})$ is a cost function that describes the behaviour of information rate of each user regarding the allocated transmit power vector $\mathbf{p}$; r_i is the i-th user's information rate, $r_{i,\min}$ the minimum rate needed to ensure QoS for user i, $\mathbf{p}$ is the power vector such that $\mathbf{p} = [p_1, p_2, \ldots, p_U]$, and $p_{\max}$ is the maximum transmission power allowed in the system.

Therefore, we must incorporate the multirate criterion to the WTM problem subject to maximum power allowed per user. From this, the optimization problem is formulated as a special case of *generalized linear fractional programming* (GLFP) [18]. This way, the second RA problem can be described as follows:

$$
\begin{aligned}
J_2(\mathbf{p}) = \max \quad & \prod_{i=1}^{U} \left[\frac{f_i(\mathbf{p})}{h_i(\mathbf{p})}\right]^{v_i} \\
\text{s.t.} \quad & 0 < p_i \leq p_{i,\max}, \\
& \frac{f_i(\mathbf{p})}{h_i(\mathbf{p})} \geq 2^{r_{i,\min}}, \qquad \forall i = 1,\ldots,U
\end{aligned} \tag{19}
$$

where $2^{r_{i,\min}}$ is the minimum information rate normalized by the bandwidth of the system (r_c) of the i-th link, including null rate restrictions; $v_i > 0$ is the priority of the i-th user to transmit with satisfied QoS requirements, assumed normalized, such that $\sum_{i=1}^{U} v_i = 1$. It is noteworthy that the second restriction in Eq. (19) is easily obtained from Eqs. (11) and (12), where the minimum information rate given can be transformed in the minimum SNIR through the Shannon capacity equation, considering a maximum tolerable BER for each user or service class. Hence, functions $f_i(\mathbf{p})$ and $h_i(\mathbf{p})$ can be readily defined as:

$$h_i(\mathbf{p}) = \sum_{j \neq i}^{U} p_j |g_{ij}|^2 + \sigma^2 \qquad \text{and} \qquad f_i(\mathbf{p}) = \theta F_i \cdot p_i |g_{ii}|^2 + h_i(\mathbf{p}), \qquad \forall i = 1, \ldots, U. \tag{20}$$

Note that the Eq. (19) is the productory of linear fractional exponentiated functions, and the function $\prod_{i=1}^{U} (z_i)^{v_i}$ is an increasing function in a nonnegative real domain [12]. Based on these properties, problem (19) can be properly rewritten as:

$$\begin{aligned} J_2(\mathbf{p}) = \max \sum_{i=1}^{U} v_i [\log_2 f_i(\mathbf{p}) - \log_2 h_i(\mathbf{p})] \quad &= \quad \max \sum_{i=1}^{U} v_i [\bar{f}_i(\mathbf{p}) - \bar{h}_i(\mathbf{p})] \\ \text{s.t.} \quad 0 < p_i \leq p_{i,\max}, & \\ \bar{f}_i(\mathbf{p}) - \bar{h}_i(\mathbf{p}) \geq r_{i,\min}, \quad \forall i = 1, \ldots, U & \end{aligned} \tag{21}$$

This way, the cost function turns into a sum of logarithms, which results in a monotonic nondecreasing function. With no loss of generality, in this work $v_i = U^{-1}, \forall i$, has been adopted.

2.5. The ACO$_{\mathbb{R}}$ Metaheuristic

The ACO$_{\mathbb{R}}$ is a continuous-valued metaheuristic based on the ants behavior when looking for food. Note that it was first proposed for combinatorial optimization problems. In its discrete version, each ant walks through the points of the input set and deposits pheromone on its edges. The next point selection is done probabilistically, considering the amount of pheromone on each edge, jointly with the heuristic information available in the current algorithm iteration.

Given a set of points next to an ant, the probability of each of this points to be chosen forms a probability mass function (PMF). The main idea of the continuous version ACO$_{\mathbb{R}}$ is the adaptation of this PMF to a Probability Density Function (PDF), allowing each ant to sample a continuous PDF instead of dealing with discrete sampling points. This is due to the fact that the continuous domain has infinite points to be chosen.

The PDF used in this work is Gaussian given its soft capacity in generating random numbers, and due to the fact that it has only one maximum point located at the mean of the process. Nevertheless, this last feature is not useful when the search space has more than one feasible region. To overcome this problem, the ACO$_{\mathbb{R}}$ uses a Gaussian kernel pdf (a weighted sum of Gaussians) to sample each dimension of the problem. Each Gaussian kernel is defined as follows [19]:

$$G^i(x) = \sum_{l=1}^{Fs} \omega_l g_l^i(x) = \sum_{l=1}^{Fs} \omega_l \frac{1}{\sigma_l^i \sqrt{2\pi}} \exp\left[-\frac{(x-\mu_l^i)^2}{2\sigma_l^{i2}}\right], \qquad i = 1, \ldots, U \tag{22}$$

where i is the Gaussian kernel indexer, with U being the number of dimensions of the problem; $\boldsymbol{\omega} = [\omega_1, \omega_2, \ldots, \omega_{Fs}]$ is the weight vector associated to each Gaussian in the kernel; $\boldsymbol{\mu}^i = [\mu_1^i, \mu_2^i, \ldots, \mu_{Fs}^i]$ is the vector of means and $\boldsymbol{\sigma}^i = [\sigma_1^i, \sigma_2^i, \ldots, \sigma_{Fs}^i]$ is the vector of standard deviations. Hence, the cardinality of both vectors is equal to the number of Gaussians in the set, $|\boldsymbol{\omega}| = |\boldsymbol{\mu}^i| = |\boldsymbol{\sigma}^i| = Fs$.

For discrete combinatorial optimization problems, the pheromone informations are kept in a table. This is not possible when we need to deal with continuous problems, since there are an infinite number of points to keep, and as a consequence, an infinite ways to evolve. Thus, a solution file is deployed, where the lth solution s_l, $\forall l = 1, 2, \ldots, Fs$, in the ith dimension, $\forall i = 1, \ldots, U$, is kept on the memory file at the nth iteration, as well as the respective cost function values $f(s_l)$. A schematic grid to understand the file format and the associate ACO input parameters is sketched in Fig. 1.

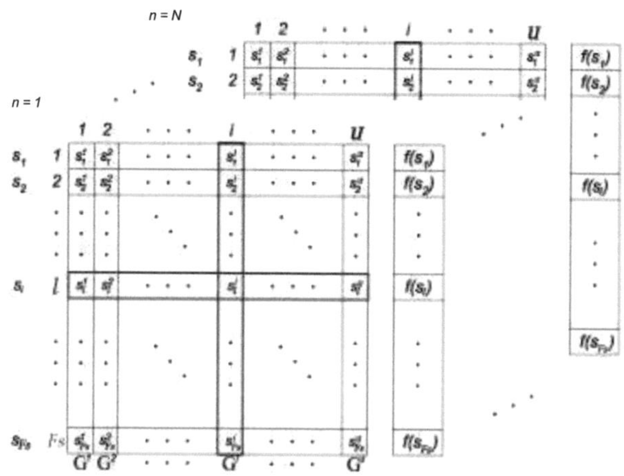

Figure 1. File structure for the ACO algorithm's solutions. Each line of the $Fs \times U$ matrix represents one solution for the problem of dimension U; each column represents one dimension, which, in turn, is sampled by each Gaussian kernel. Cost function vector $\mathbf{J} = [J(s_1), \ldots, J(s_l), \ldots, J(s_{Fs})]$, dimension $Fs \times 1$, represents the solution for the n-th iteration. Finally, each layer (in depth) shows the solution file in each iteration from $n = 1$ to N.

The found solutions are used to generate PDFs dynamically, through a method based on the stored solutions. The vectors $\boldsymbol{\mu}^i$ and $\boldsymbol{\sigma}^i$ at ith dimension and $\boldsymbol{\omega}$ common for all dimensions are calculated through the solutions of the file at each algorithm iteration; so, the Gaussian kernel can be built, and guide the ants throughout the dimensions of the problem.

The solutions file must store Fs solutions. Note that this number is equal to the number of Gaussian PDFs ($g_l^i(x)$) in the i-th kernel (G^i). Thus, one can conclude that the i-th Gaussian kernel will have one Gaussian PDF sampling each i-th variable of each solution. Herein, the greater the value of Fs, the greater will be the number of Gaussian PDFs on the algorithm. Therefore, the parameter Fs leads to the complexity of the set.

A detailed description for the evolution of the ACO solution structure is given in the following. From Fig. 1, note that for an U-dimensional problem, the file solution stores Fs different solutions, the values of its U variables and the results of each solution applied to the cost function. So, s_l^i is the value of the i-th variable of the l-th solution of the file, and $J(s_l)$ is the result of the l-th solution applied to the cost function. For each dimension $i = 1, \ldots, U$ of the problem (in this case, each column of the table), there is a different Gaussian Kernel PDF (G^i) defined. So, for each G^i, the values of the i-th variable of all solutions becomes the elements of the mean vector, $\boldsymbol{\mu}^i = [\mu_1^i, \ldots, \mu_{Fs}^i] = [s_1^i, \ldots, s_{Fs}^i]$, i.e., the l-th value of dimension i is the mean of the l-th gaussian of G^i.

Furthermore, the number of ants (or particles) m is another important input parameter of the ACO algorithm to be adjusted. The ants are responsible for the sampling of G^i, and thus, for the algorithm evolution as well. In that way, on each iteration, each ant chooses one solution of the file probabilistically, through a method based on the weight vector $\boldsymbol{\omega}$. Since the ant has chosen one solution of the file, the next step consists of sampling through the Gaussian kernel. After that, a new solution is generated and attached to the end of the file. As the last ant finishes its sampling, the solution file is sorted based on the value entries in the cost function matrix $\mathbf{J} = [J(s_1), \ldots, J(s_l), \ldots, J(s_{Fs})]$. Hence, for both problems treated in Eq. (17) and (21), the matrix $\mathbf{J}$ must be sorted decreasingly, i.e., $J(s_1) \geq J(s_2) \geq \ldots \geq J(s_{Fs})$.

When the sorting process is completed, a number of worst solutions ordered at the end of the file is discarded, in which is done equal to the number of solutions added on the sampling process. Note that since each ant samples only one solution on each iteration, the number of solutions to be discarded is equal to the number of ants.

At this point, a Gaussian kernel (G^i) is defined for each dimension of the problem, which the l-th variable becomes an element of the $\boldsymbol{\mu}^i$ vector. Thus, considering the G^i defined on the i-th dimension, The weight w_l of each solution is calculated as follows:

$$\omega_l = \frac{1}{qFs\sqrt{2\pi}} \exp\left[-\frac{(l-1)^2}{2q^2Fs^2}\right], \qquad l = 1, \ldots, Fs \tag{23}$$

The weight of the l-th solution can be seen as the probability of a solution to be chosen and sampled by an ant. Hence, the l-th solution's rank in the file, also is the input parameter in Eq. (23), which is a Gaussian PDF with mean 1 and standard deviation $q \cdot Fs$, where q is an input parameter of the ACO algorithm. The q parameter can be interpreted as a diversification parameter, where low values of q enhances the convergence speed of the algorithm; on the other hand, high values entries for q enhances the process robustness. This is due to the fact that, on the normal function for the l-th solution's weight calculation in Eq. (23), the higher the standard deviation values are, more chances to select solutions that are not so near to the mean of the process, which, in turn, is the first solution of the file. So, when the standard deviation $q \cdot Fs$ assumes small values, only the best and a few solutions of the file will be sampled, enabling the algorithm to converge faster. On the other hand, when high-valued standard deviations are admitted, the probability of the file solutions to be chosen becomes more uniform, which makes the algorithm search in a larger space

http://dx.doi.org/10.5772/53338

(more diversity), at the cost of lower convergence speed. Thus, the ACO$_{\mathbb{R}}$'s q parameter corresponds to the best-so-far solution and iteration-best solution concepts. So, Eq. (23) gives rise to an important equilibrium between q and Fs parameters, making their individual calibration sensitive to each one, in order to achieve a good tradeoff among robustness and convergence velocity for each specific optimization problem.

Furthermore, the suitable choice for the population size m plays an important role in order to improve the robustness-speed tradeoff in conjunction with the best attainable $q \cdot Fs$ calibration. Note that m might be overloaded in order to increase the algorithm capacities, at the undesirable cost of greater computational complexity. Finally, the algorithm robustness $\mathcal{R}$ can be thought as the ratio between the number of convergence success $\mathcal{S}$ to the total number of process realizations $\mathcal{T}$:

$$\mathcal{R} = \frac{\mathcal{S}}{\mathcal{T}} \cdot 100 \qquad [\%] \qquad @N \ \text{ iterations}$$

and the speed as the average number of iterations needed to the algorithm achieves convergence in $\mathcal{T}$ trials for a given problem.

Next, important steps of the ACO algorithm are briefly discussed, such as general ACO algorithm structure, the $\boldsymbol{\sigma}^i$ vector computation, as well as the sampling process; the last one will be presented directly on the algorithm structure. The algorithm organization common to various implemented versions of continuous ACO is described in Algorithm 1, in which the functions performed inside are briefly described in following.

Algorithm 1 Overview of ACO

```
while The end conditions aren't met do
    AntBasedSolutionConstruction()
    PheromoneUpdate()
    DaemonActions()
end while
```

- *AntBasedSolutionConstruction()*: Through the decision variables of each solution $s_l^i, i = 1, \ldots, U$, each ant builds the solution by U steps. Since ACO$_{\mathbb{R}}$ uses a Gaussian mixture in each problem dimension (Eq. (22)), and that the number of Gaussians on the mixture is equal to the size Fs of the solutions file, we conclude that at each step i we will have a different sample of G^i

  In order to sample the Gaussian kernel, the vectors $\boldsymbol{\mu}^i$, $\boldsymbol{\sigma}^i$ and $\boldsymbol{\omega}$ must be updated. In this work, the vector $\boldsymbol{\omega}$ will not be used, and the explanation for that is given in the next paragraph.

  In practice, the sample process is made on three stages: First, the elements of $\boldsymbol{\omega}$ vector must be calculated, where should be noted that the solutions ranking parameter l will never change, independently of the change of the solutions order on the file. On the second stage, each ant must choose a solution of the file aiming to sample it, and the probability of this choose must be relative to the normalization of each solution weight for the sum of all weights:

$$p_l = \omega_l \cdot \left(\sum_{r=1}^{k} \omega_r \right)^{-1} \quad (24)$$

that is, the probability of each solution being chosen can be thought as a random number generator of normal distribution, with mean 1 and standard deviation $q \times Fs$, since the choose probability of each rank will never change. Adopting this strategy, the ω vector as well as the first stage of the sampling process will no longer be needed.

Thus, since the ant chosen its solution, it must be sampled stepwise using a normal random number generator. The chosen solution must be sampled dimensionally ($g_l^i, i = 1, \ldots, U$), causing each Gaussian mixture's parameters to be seen only in one dimension a time, smoothing the calculation of the pattern deviation and allowing linear transformations on the problem without result changes.

Therefore, let s_l (Fig. 1) to be the solution chosen by an ant during the ACO's evolution process. It is known that the ant will sample s_l dimensionally, as well as that the sampling is done through a Gaussian function parametrized by the μ_l^i and σ_l^i values. Thus, the σ_l^i is calculated for dimension i as follows:

$$\sigma_l^i = \xi \sum_{e=1}^{k} \frac{|s_e^i - s_l^i|}{k-1} \quad (25)$$

Herein, the σ_l^i value for s_l^i is the mean distance from s_l^i to the other values of dimension i in the other solutions of the file. The process is repeated until the last dimension of the file is reached. This way, the higher the variability of the different solutions, higher will be the standard deviation value σ_l^i. Note that the $\xi \in [0,1]$ parameter aims to reduce the standard deviation, working as a learning factor. Since the σ_l^i value is calculated, the ant will sample the Gaussian PDF $g_l^i(\mu_l^i, \sigma_l^i)$.

The parameter ξ is the same for all dimensions of all solutions, and corresponds to the pheromone evaporation rate, or to the inverse of the learning rate. This way, when ξ is low valued the algorithm speed is enhanced, and when it is high-valued, its robustness is enhanced. It is noteworthy that the algorithm converges when $\sigma \rightarrow 0$ throughout all dimensions of the file.

- *PheromoneUpdate()*: The $ACO_{\mathbb{R}}$ algorithm updates its pheromone informations as follows: At the beginning of the algorithm, the file is initialized with Fs solutions uniformly random distributed. From this, the pheromone updating is done adding the new solutions generated by the ants, as well as removing the same number of worst solutions.

  Finally, the size of the solutions file is a parameter of the algorithm, and must not be smaller than the number of dimensions of the problem if it is enabled to handle variable correlation, and to support linear transformations on the problem being optimized. Nevertheless, these techniques are not used in this work. Furthermore, the size of the file leads to the algorithm diversity, since a big file will cover a greater region of the search space than the small one, enabling the algorithm to overcome local optima, but on the other hand, a small file will make the algorithm to be faster than the big one.

http://dx.doi.org/10.5772/53338

- *DaemonActions()*: This is the optional component of $ACO_{\mathbb{R}}$ that can be used to implement centralized actions of the algorithm that aren't accessible for the ants. In this stage, the found solution must be updated and returned as the final solution. Besides, it is possible to implement local search methods here, but this aspect is not exploited in this current work, since we look firstly for low computational complexity.

2.6. Numerical Results

This subsection is divided in two parts. The first one deals with ACO typical performance and its input parameters optimization; the second part numerical simulations results for both power and rate allocation problems are presented and the NMSE is compared with the RA-PSO algorithm.

The simulations were carried out in the MatLab 7.0 platform and the scenario parameters are presented on Table 1. We assumed a rectangular cell with one base station in the center and users uniformly spread across all the cell extension. We considered that all mobile terminals experience slow fading channels, i.e. the following relation is always satisfied:

$$T_{\text{slot}} < (\Delta t)_c \tag{26}$$

where $T_{\text{slot}} = R_{\text{slot}}^{-1}$ is the time slot duration, R_{slot} is the transmitted power vector update rate, and $(\Delta t)_c$ is the coherence time of the channel[2]. This is part of the SNIR estimation process, which means that the channel is constant in each optimization window, assumed herein equal to $667\mu s$. Thus, the ACO algorithm must converge to the solution within each $667\mu s$ interval.

2.6.1. RA-ACO Input Parameters Optimization

Simulation experiments were carried out in order to determine the suitable values for the ACO input parameters for each problem, such as file size (Fs), pheromone evaporation coefficient (ξ), population (m) and the diversity parameter (q). The best parameters combination was chosen considering the solutions quality measured by the normalized mean squared error (NMSE), defined as:

$$\text{NMSE} = \mathbb{E}\left[\frac{||\mathbf{p} - \mathbf{p}^*||^2}{||\mathbf{p}^*||}\right] \tag{27}$$

where $\mathbf{p}$ is the solution found through the ACO algorithm, $\mathbf{p}^*$ the analytical (optimal) solution and $\mathbb{E}$ the mathematical expectation operator. In order to find the best parameters for both problems non-exhaustive tests were conducted.

A typical convergence behavior for the RA-ACO under equal-rate power control problem is shown in Fig. 2. At a first glance, power allocation for $U = 5$ users (lightly loading system) is performed by a) RA-ACO algorithm, b) RA-PSO algorithm from [1]. One can

[2] Corresponds to the time interval in which the channel characteristics do not suffer expressive variations.

Parameters	Adopted Values
	DS/CDMA Power-Rate Allocation System
Noise Power	$P_n = -63$ [dBm]
Chip rate	$r_c = 3.84 \times 10^6$
Min. Signal-noise ratio	$SNR_{\min} = 4$ dB
Max. power per user	$P_{\max} = 1$ [W]
Min. Power per user	$P_{\min} = 0$ [W]
Time slot duration	$T_{\text{slot}} = 666.7\mu s$ or $R_{\text{slot}} = 1500$ slots/s
# mobile terminals	$U \in \{5; 10; 20; 100; 250\}$ users
# base station	BS = 1
Cell geometry	rectangular, with $x_{\text{cell}} = y_{\text{cell}} = 5$ Km
Mobile terminals distrib.	$\sim \mathcal{U}[x_{\text{cell}}, y_{\text{cell}}]$
	Fading Channel Type
Path loss	$\propto d^{-2}$
Shadowing	uncorrelated log-normal, $\sigma^2 = 6$ dB
Fading	Rice: [0.6; 0.4]
Time selectivity	slow
	User Features and QoS
User Services	[voice; video; data]
User Rates	$r_{i,\min} = [\frac{r_c}{128}; \frac{r_c}{32}; \frac{r_c}{16}]$ [bps]
User BER	$BER = [5 \times 10^{-3}; 5 \times 10^{-5}; 5 \times 10^{-8}]$
	RA-ACO Algorithm
Problem Dimensionality	$U \in \{5; 10; 20; 100; 250\}$ users
File Size	$Fs \in [8, 25]$
Diversity Factor	$q \in [0, 1]$;
Pheromone Evaporation Rate	$\xi \in [0, 1]$;
Population Size	$m \in [7, 35]$;
Max. # iterations	$N = 1000$
	Monte-Carlo Simulation
Trials number	$\mathcal{T} = 1000$ realizations

Table 1. Multirate DS/CDMA system, channel and ACO input parameters

see the smooth-monotonic convergence of the RA-ACO algorithm toward the optimal power solution, in this case given by (16), in contrast to the non-monotonic oscillated convergence behavior presented by the RA-PSO algorithm. Besides, for $U = 5$ users power allocation problem, the ACO was able to achieve convergence after ≈ 250 iterations in contrast to the ≈ 450 iterations necessary for the RA-PSO convergence.

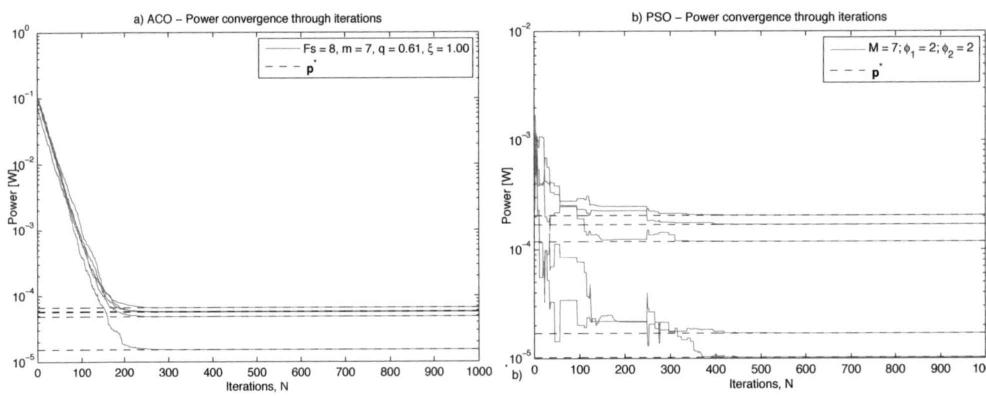

Figure 2. Power allocation for $U = 5$ users. Equally information rate among users is adopted. a) RA-ACO; b) RA-PSO algorithm from [1].

http://dx.doi.org/10.5772/53338

Power Allocation (PA) Problem. Under the first resource allocation problem posed by Eq. (9) or (17), Fig. 3 depicts the associated NMSE under different ACO input parameter values combination taking into account different loading system, i.e., $U = 5$, 10 and 20, respectively.

Note that the population size m and file size Fs parameters ($m, Fs \in \mathbb{N}$), both with entry values common for all the different $\{q, \xi\}$ input parameters configurations, where chosen based on the problem dimensionality. Numerical experiments have shown that different entries around the ones chosen do not affect substantially the NMSE results as the different entries for q and ξ parameters do. It is worth noting that the PA problem in (9) presents a non-convex characteristic; hence, the value entries for the population size m and file size Fs parameters assume relative high values regarding the dimensions of the problem, meaning that both parameters are of the order of problem dimension, $\{m, Fs\} \approx \mathcal{O}[U]$. It means that RA-ACO can solve the non-convex PA problem in DS/CDMA systems but with input parameter loads relatively high.

Herein, a parameter calibration strategy was adopted in order to find the best tradeoff for the $\{q; Fs\}$ set, given in Eq. (23). Since the parameters Fs and m are directly related to the computational complexity of the algorithm, finding a suitable parameter set with Fs entries as low as possible is of great interest.

On the other hand, the population size m parameter has a small or even no influence on the any other ACO input parameter (as q and Fs interfere each other). Although the m entries values directly increases the algorithm computational complexity. Therefore, the parameters m and Fs were fixed at low values and then the best q and ξ combination for it was sought. Hence, based on the NMSE *versus* convergence speed results obtained in Fig. 3, the optimized RA-ACO input q and ξ parameters for the power control problem in DS/CDMA networks under different level of interference could be found, as summarized in Table 2.

U (users)	5	10	20
q	0.61	0.40	0.40
ξ	1.00	0.82	0.75
m	7	15	35
Fs	8	4	25
Robustness, $\mathcal{R}$	100 %	100 %	30 %

Table 2. Optimized RA-ACO input parameters and respective robustness for the Problem of Eq. (17).

Also, the robustness achieved by the RA-ACO for the power allocation problem is added to the Table 2. Herein, the success of convergence is reached when the NMSE of the algorithm's solution goes less than 10^{-2}. Due to the non-convexity of the PA problem in (17), when the number of users grows from 10 to 20, the needed robustnes grows exponentially, thus, the algorithm's performance have a critical decay of 70%.

Weighted Throughput Maximization (WTM) Problem. For the weighted throughput maximization (WTM) problem posed in Eq. (21), Figure 4 shows different cost function evolutions when parameters q and ξ are combined under three distinct system loading, $U = 20$, 100 and 250 users. The average cost function evolution values where taken over $\mathcal{T} = 1000$ trials. Also, the correspondent sum rate difference ($\Delta \sum_{\rm rate}$) is zoomed in.

From Fig. 4-a it is clear that for $U = 20$ users, the $q = 0.10$ and $\xi = 1.00$ choice results in an average cost function value higher than the other ones. Besides, even in a

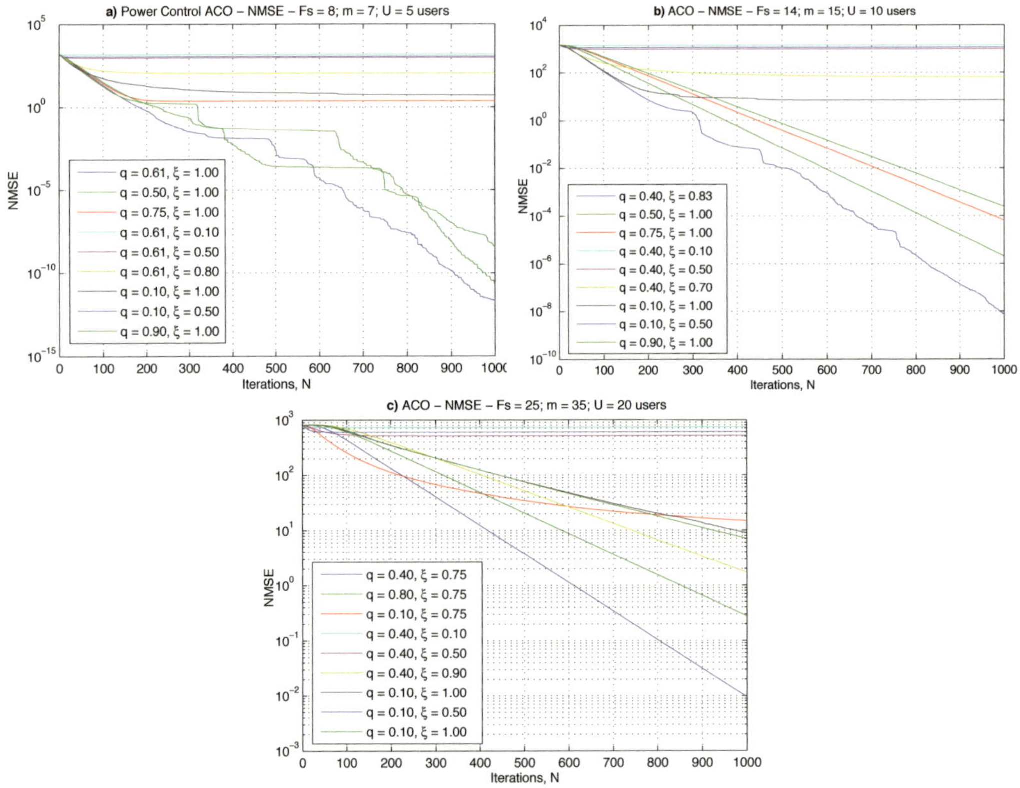

Figure 3. NMSE for different ACO input parameters. a) $U = 5$ users; b) $U = 10$ users and c) $U = 20$ users;

relative course optimization scenarios for q and ξ parameters it is clear the importance of deploying optimized RA-ACO input parameters; for instance, the best ACO input parameters configuration in Figure 4.a shows a difference of $\Delta \sum_{\text{rate}} = 76.8Kb/s$ on the total achievable system throughput regarding the second best parameters set choice, and a difference of $1.3Mb/s$ to the worst parameters set.

On Fig. 4.b, a lightly cost function value difference shows that the best parameter configuration for the system load of $U = 100$ users is $q = 0.20$ and $\xi = 1.00$. In terms of system throughput, the best parameter configuration shows a difference of $38.4Kb/s$ to the second one, and of $2Mb/s$ to the worst one.

Finally, the best input parameters configuration set for $U = 250$ users in Fig. 4.c is obtained as $q = 0.20$ and $\xi = 1.00$. Again, the associated total system throughput variations due to the different ACO input parameter configurations was not significant, ranging from $96Kb/s$ to $1.24Mb/s$. This confirms a certain robustness to the input $\{q; \xi\}$ deviation values, thanks to the convexity of the optimization problem formulated in (21). In summary, the RA-ACO algorithm was able to attain reasonable converge in less than $n = 150$ iterations for the WTM problem with dimension up to $U = 250$ users.

Note that in the WTM problem given the convex characteristic of the objective function, Eq. (21), the robustness of the RA-ACO approaches to $\mathcal{R} \approx 100\%$, and the value entries for the population size m and file size Fs parameters are impressively less than the number of dimensions of the problem, i.e., $\{m, Fs\} \ll U$. It means that RA-ACO can solve the WTM problem with soft parameter loads. The best input parameter configuration for the RA-ACO algorithm in order to solve the WTM problem is summarized in Table 3.

U (users)	20	100	250
q	0.10	0.20	0.20
ξ	1.00	1.00	1.00
m	3	7	7
Fs	5	5	5

Table 3. Optimized RA-ACO input parameters for the WTM Problem, Eq. (21).

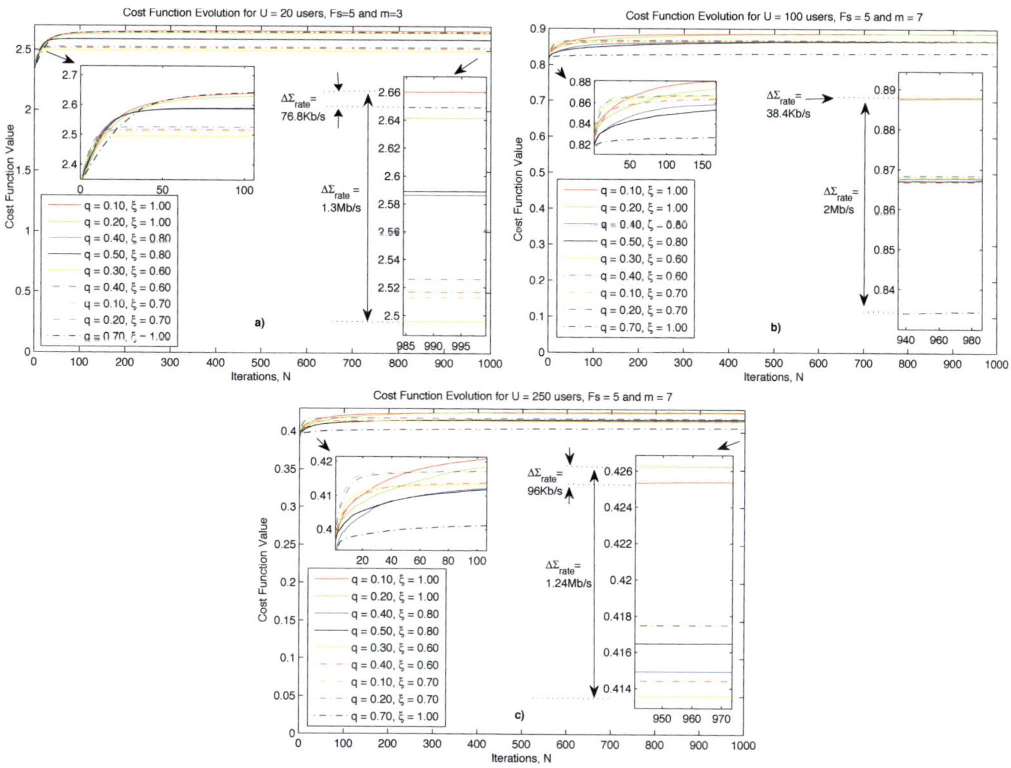

Figure 4. Cost function J evolution from Eq. (21) across $N = 1000$ iterations for different ACO input parameters values combination. The correspondent sum rate difference ($\Delta \sum_{rate}$) is zoomed in. a) $U = 20$; b) $U = 100$, and c) $U = 250$ users.

2.6.2. WTM RA-ACO Performance Results

Numerical results for the WTM problem with RA-ACO algorithm under optimized input parameters are shown in Figure 5. Here, its clear that ACO can evolve pretty fast to the three different system loads, finding a good solution in less than 100 iterations. Besides, one can note the great increase on the total system power from the 100 users case to the 250. It is due to the interference increase given the high number of users in the system. Nevertheless, a good system throughput result is found. For the 20 users results, a system total throughput of $200 Mb/s$ is found. This results in an remarkable average user rate of $10 Mb/s$.

On the $U = 100$ users results, $\approx 340 Mb/s$ of system throughput is reached, with a total power consumption of $\approx 55W$. Herein, the average user rate is $\approx 3,4 Mb/s$. This is due to the higher interference values given the medium system load.

Finally, for the $U = 250$ users results, a total system throughput of $\approx 400 Mb/s$ is reached, with a total power consumption of $\approx 125W$. Again, the average user rate decays regarding the low and medium system loads, reaching $\approx 1,6 Mb/s$.

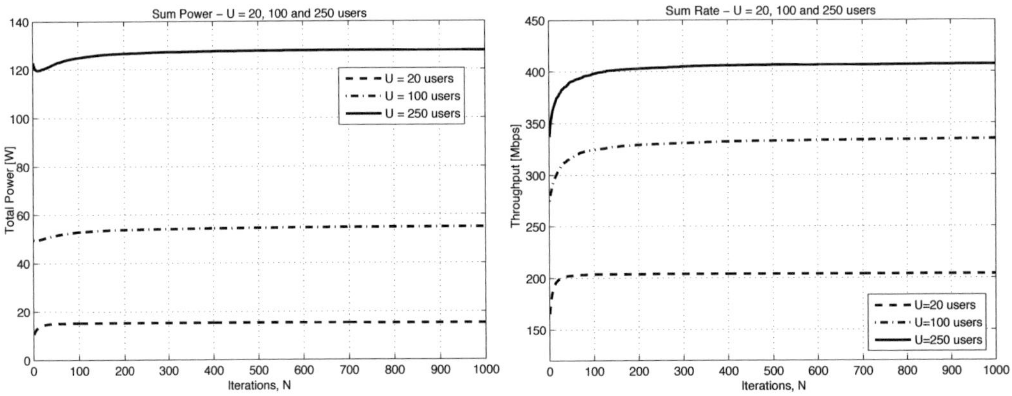

Figure 5. Sum Power and Sum Rate evolution for $U = 20, 100$ and 250 users under RA-ACO algorithm.

2.6.3. RA-ACO and RA-PSO Simulation Performance Results

The main objective in this analysis is put in perspective the both RA-heuristic algorithm performance regarding the non-convexity of the power allocation problem posed in (17). Simulations were conducted on different system loadings according to the best input RA-ACO parameters presented in Section 2.6.1 and those best parameters obtained in [1] for the resource (power) allocation using particle swarm optimization (RA-PSO) algorithm.

Figure 6, shows the NMSE evolution for the power control problem with $U = 5$, 10 and 20 users, respectively, for the algorithms RA-PSO [1] and RA-ACO. Clearly, the NMSE $\approx 10^{-12}$, 10^{-10} and 10^{-2} attainable by the RA-ACO is much more lower than that values reached by RA-PSO (NMSE $\approx 10^{-5}$, 10^{1} and 10^{1}) after $N = 1000$ iterations. This means the ACO could surpass the various convergence problems in solving the non-convex power control problem. The associated robustness shown that the RA-ACO achieves near to total convergence success, while the RA-PSO was not able to do. Fig. 6 also shows a table containing the percentage of algorithm success, i.e. the percentage of trials in which the algorithm ended with a NMSE less or equal to 10^{-2}, showing a clearly superiority of the RA-ACO scheme. Nonetheless, this robustness comes with a computational complexity increasing.

3. Anomaly Detection in Computer Networks

This section, presents the main concepts and related work in computer networks anomaly. It is presented anomalies type and their causes, the different techniques and methods applied into anomaly detection, as well as a compilation of recent proposals for detecting anomalies in networks using ACO and a case study.

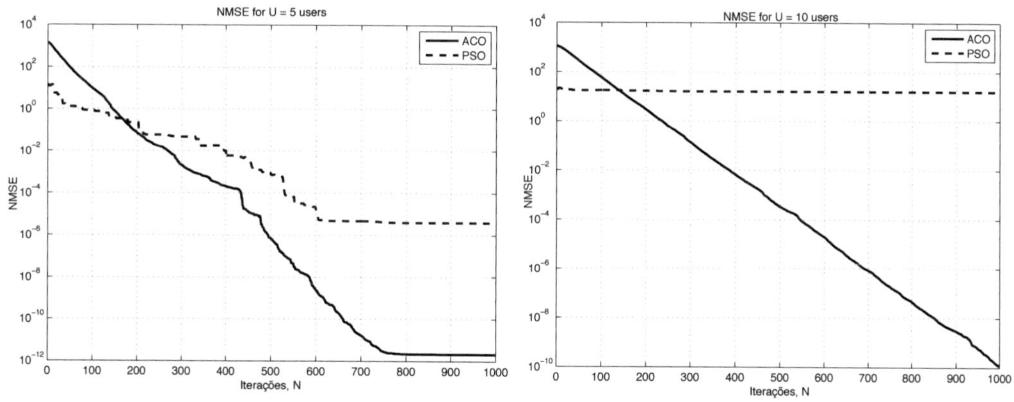

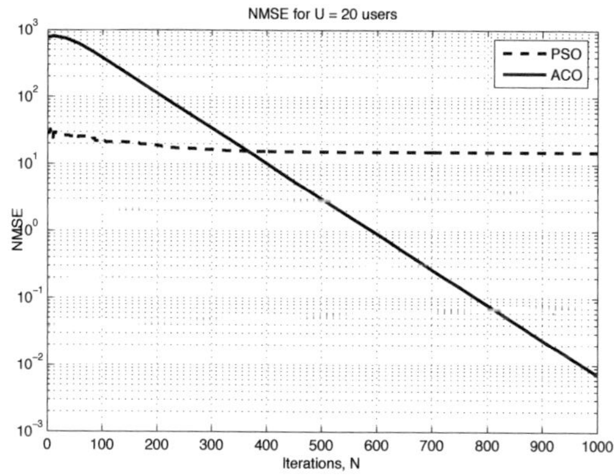

# Users U	Robustness, $\mathcal{R}$ RA-ACO	RA-PSO
5	100 %	40 %
10	100 %	0 %
20	30 %	0 %

Figure 6. NMSE attainable by RA-ACO and RA-PSO [1] algorithms, for $U = 5, 10$, and 20 users.

Management is an essential activity in enterprizes, service providers and other elements for which the networks have become a necessity because the continued growth of networks have introduced an extensive options of services. The main goal of management it is to ensure the fewest possible flaws and vulnerabilities to not affect the operation of networks. There are several factors that can lead to an anomaly such as configuration errors, improper use by users and programming errors and malicious attacks among many other causes [20].

A tool to help the network management task are the Anomaly Detection System (ADS), which consist of a technique or a set of techniques to detect anomalies and report to the network administrator, helping to decide which action perform in each situation. Anomaly detection is not an easy task and brings together a range of techniques in several areas. The quality and safety of the service provided to end users are ongoing concerns and receive special attention in network traffic.

3.1. Anomaly

Thottan et. al [21] present two anomalies categories. The first one is related to network failures and performance problems, where the role of a malicious agent does not exist. The flash crowd is a typical example of this category, where a server receives a huge amount of not malicious client requests in the same period of time, congesting the server, i.e., a webstore promotes a product at a lower price at a certain time. If the webstore does not prepare a infrastructure to support the client access, the server may interrupt the operations and do not operate all sales transactions, causing congestion on the server and possible financial loss to the webstore. The congestion itself is another anomaly type within the first category, due to the abrupt increase in traffic at some point in the network, causing delays in the delivery of packages until the saturation of the link, with packet discards. Congestion can also be generated by configuration errors, the server does not respond adequately to requests sent by clients by wrong settings.

In the second category, it is found anomalies that arise from problems related to security. Denial of Service (DoS) is a main example of a anomaly in this category. DoS occurs when a user is unable to obtain a particular service because of some malicious agent used methods of attack that occupied the machine resources such as CPU, RAM. Besides DoS attacks, also have Distributed Denial of Service (DDoS) where a master machine dominates other machine, called zombies, to perform a DoS [22]. The flash crowd is differentiate from DoS and DDoS because of the malicious agent. Worms, port scan and others usually are programmed to discover vulnerabilities in networks and perform attacks [23].

3.2. Anomaly Detection Techniques

The techniques implemented in ADS are present in diverse areas such as intrusion detection, fraud detection, medical anomaly detection, prevention of damage to industrial image processing, sensor networks, among others [24]. Presenting so many different application domains, many tools have been developed specifically for some activity and other solutions are more general. Chandola et. al [24] group the techniques into: based on classification, clustering, information theory, statistical and spectral theory. [24, 25] are surveys with a general content in anomaly detection field, but it is found surveys toward the area of computer network, [20, 26]. The nomenclature and some categories may differ, but the concept presented is consistent with each other.

Patcha et. al [20] divides the techniques of detecting anomalies in three categories: based on signature, based on the characterization of normal behavior and hybrid techniques. The signature-based techniques are based on the number of signatures of known or constructed attack patterns. The strength of this kind of detection is the low rate of false positives. The techniques based on characterization of normal traffic build the profile of network traffic, and any event that deviates from the normal behavior is considered an anomaly. The hybrid techniques are a junction of the two previous techniques [20].

Many authors consider the proposed work by Denning [27] as a watershed between the methods based on signature and the methods used to characterize the normal traffic behavior, and these methods consist of two phases: training phase and test phase. The training phase is generated from the network profile and the test phase is applied the profile obtained to evaluation.

3.2.1. Detection Based on Signature

Detection based on signature requires the construction of a database of events related to certain anomalies, thereby generating the signature. The signatures describes specific events that form a specific attack or anomaly; this way, when the tool monitors the traffic behavior,

http://dx.doi.org/10.5772/53338

the comparison with the signatures is performed, and if a match occurs as described the event in the signature, an alarm is generated [20].

Using signature, the method offers low rates of false positive, since the signature clearly describes what is required to be considered an anomaly, however, unknown attacks characteristics and not formulated signatures might pass unnoticed. Another negative point is the need to constantly update the signatures database [20].

3.2.2. Detection Based on the Characterization of Normal Behavior

Unlike the signature-based detection, the focus of this method is to detect anomalies based on the characterization of normal behavior. The first and fundamental step is to generate the normal behavior profile of traffic or adopt a model that more accurately describes the traffic. Consequently, any activity that deviates from the monitored normal profile built will be considered anomaly. The construction of the profile can be static or dynamic. It is static when the profile is built and replaced only when a new profile is constructed, and it is dynamic when the profile is updated according to the network behavior changes

A positive point is the possibility of detecting new anomalies, whereas these new anomalies describe a behavior different from normal. Another aspect is the difficulty created for the malicious agent devise an attack, because it ignores the profile and the possibility exists that it can not simulate an attack describing the profile and generates an alarm [20]. But there are disadvantages in profile construction as the required training period or the information amount on the basis of historical data. The difficulty in characterizing the traffic itself generates a high percentage rate of false positives, since the ADS can point to many natural variations of the network as an anomalous behavior.

There are several techniques, below are listed some relevant techniques that enrich the discussion with several different proposals:

- **Machine Learning:** The machine learning solutions have the ability to learn and improve the performance over time because the system changes the implementation strategy based on previous results. Bayesian networks, Markov chains, neural network are techniques applied to the generation of the normal profiles and detection of the anomalies. The main advantage of this approach is the ability to detect anomalies unknown and adapt to changes in the behavior of a monitored environment, however, this adjustment requires a large amount of data to generate a new profile [20].
- **Based on Specification:** These solutions are constructed by an expert, since the specifications of the normal behavior of the system are carried out manually. If the system is well represented, the false negative rate will be minimized by avoiding any behavior not predicted, but may increase if some behavior is overlooked or not well described. The most widely used technique for this task are finite state machines. A drawback of this approach point is the time and complexity to the development of the solutions [26].
- **Signal Processing:** The most commonly used techniques are the Fourier transforms, wavelet and algorithms such as ARIMA (Autoregressive Integrated Moving Average). It presents the advantage to adapt to the monitored environment and detecting unknown anomalies and low training period. The complexity is presented as a disadvantage of this approach [28].

- **Data Mining:** The Data Mining techniques usually deal with a huge amount of data, looking for patterns to form sets of normal data. Principal Component Analysis (PCA), clustering algorithms, Support Vector Machine (SVM) and others statistical tools are commonly employed in these solutions [20].

3.3. Recent Proposals Using ACO in Computer Networks Field

Since Dorigo et. al [29, 30] proposed the Ant System (AS) from the first time, several applications have emerged using AS itself or others algorithms arising from the ACO approach. One algorithm proposed to the networks routing problem is the AntNet, proposed by Di Caro et al. [31], a different approach to the adaptive learning of routing tables in communications networks. To the information to travel from point A to point B, it is necessary to determine the path that will be covered. The construction process itself and the path is known as routing, and it is one at the core of the To the network control system together with congestion control components, admission control, among others [31]. The AntNet is close to the real ants' behavior that inspired the ACO metaheuristic, because the routing problem can be characterized as a directed weighted graph, where the ants move on the graph, building the paths and loosing the pheromone trails.

Information Retrieval is another field where ACO found application, as proposed by [32, 33]. The problem in the information retrieval system consists in finding a set of documents including information expressed in a query specifying user needs. The process involves a matching mechanism between the query and the documents of the collection. In [32], Drias et. al designed two ACO algorithms, named AC-IR and ACS-IR. Each term of the document has an amount of pheromone that represents the importance of its previous contribution in constructing good solutions, the main difference between AC-IR and ACS-IR is mainly in the design of the probabilistic decision rule and the procedure of building solutions. In [33], the ACO algorithm is applied to retrieve relevant documents in the reduced lower-dimensionality document feature space, the probability function is built using the frequency of the terms and the total number of documents containing the term.

In [34], the autors make use of a Fuzzy Rule Based System (FRBS), Naive Bayes Classifier (NBC) and Support vector machine (SVM) to increase the interpretability and accuracy of intrusion detection model for better classification results. The FRBS system is a set of IF-THEN rules, whose antecedents and consequents are composed of fuzzy statements, related by the dual concepts of fuzzy implication and the compositional rule of inference. The NBC method based on the "Bayes rule" for conditional probability as this rule provides a framework for data assimilation. The SVM is a statistical tool for data classification which is one of the most robust and accurate methods among all well-known algorithms. Its basic idea is to map data into a high dimensional space and find a separating hyper plane with the maximal margin. Then, the authors proposed NBC with ACO, linking a Quality computation function, ranking the best rule between discovered ones, to the pheromone updating.

A commonly approach to network intrusion detection is to produce cluster using a swarm intelligence-based clustering. Therefore, in the traditional clustering algorithms it is used a simple distance-based metric and detection based on the centers of clusters, which generally degrade detection accuracy and efficiency because the centers might not be well calculated or the data do not associate to the closest center. Using ACO, it is possible to surround the local optimum and find the best or the most close to the best center. This technique is used in [35],

Feng et. al present a network intrusion detection, assuming two assumptions: the number of normal instances vastly outnumbers the number of intrusions and the intrusions themselves are qualitatively different from the normal instances. Then, three steps are followed: 1) Clustering, 2) Labeling cluster and 3) Detection.

In [36], it is found the use of data mining to intrusion detection. Abadeh et. al proposed an extract fuzzy classification rules for misuse intrusion detection in computer networks, named Evolutionary Fuzzy System with an Ant Colony Optimization (EFS-ACO). It consists of two stages, in the first stage, an iterative rule learning algorithm is applied to the training data to generate a primary set of fuzzy rules. The second stage of the algorithm employs an ant colony optimization procedure to enhance the quality of the primary fuzzy rule set from the previous stage.

3.4. Applying ACO for Anomaly Detection - A Study Case

This sections provides an application of ACO for anomaly detection. The proposed approach is under the categorie of the detection methods based on the characterization of normal behavior, and follow two steps: 1) Training Phase, 2)Detection Phase. The dataset used for evaluation of the method is the KDD'99 [37]. In the Training Phase, it is used the training dataset and it is generated the centroids for each class of attack, and in the Detection Phase, it is used the generated centroids in each conection to classify it and generate the final resuts.

3.4.1. KDD Cup 99 Data Set Description

For evaluation for anomaly detection methods it is commonly used the KDD'99 dataset, used for The Third International Knowledge Discovery and Data Mining Tools Competition, which was held in conjunction with KDD-99 The Fifth International Conference on Knowledge Discovery and Data Mining [37]. This dataset was built based on the data captured in DARPA'98 IDS evaluation program, used for The Second International Knowledge Discovery and Data Mining Tools Competition, which was held in conjunction with KDD-98 The Fourth International Conference on Knowledge Discovery and Data Mining [38].

KDD training dataset consists of approximately 490,000 single connection vectors each of which contains 41 features and it is labeled as: back, buffer overflow, ftp write, guess passwd, imap, ipsweep, land, loadmodule, multihop, neptune, nmap, perl, phf, pod, portsweep, rootkit, satan, smurf, spy, teardrop, warezclient, warezmaster, normal. Depending on the label, the connection fall in one of the following four attack categories:

1. **Denial of Service (DoS)**: is an attack in which the attacker makes some computing or memory resource too busy or too full to handle legitimate requests, or denies legitimate users access to a machine.
2. **User to Root (U2R)**: is a class of exploit in which the attacker starts out with access to a normal user account on the system and is able to exploit some vulnerability to gain root access to the system.
3. **Remote to Local (R2L)**: occurs when an attacker who has the ability to send packets to a machine over a network but who does not have an account on that machine exploits some vulnerability to gain local access as a user of that machine.

4. **Probing** : is an attempt to gather information about a network of computers for the apparent purpose of circumventing its security controls.

In Table 4 it is present the labels related to the attack categories. From all the attack categories, the study subject will be the DoS attacks. From all the 41 features, for this study case, it is adopt the source bytes and destiny bytes, because the main idea of the approach is to detect volume anomaly.

Attack Categorie	Labels	Samples
Denial of Service (DoS)	back, land, neptune, pod, smurf, teardrop	391458 (79.2391%)
User to Root (U2R)	buffer overflow,perl, loadmodule, rootkit	52 (0.0105%)
Remote to Local (R2L)	ftp write, guess passwd, imap, multihop, phf, spy, warezclient, warezmaster	1126 (0.2279%)
Probing	ipsweep, nmap, portsweep, satan	4107 (0.8313%)

Table 4. Labels related to the attack categories

3.4.2. The ACO Clustering

ACO is composed of a population of agents competing and globally asynchronous, cooperating to find an optimal solution. Although each agent has the ability to build a viable solution, as well as a real ant can somehow find a path between the colony and food source, the highest quality solutions are obtained through cooperation between individuals of the whole colony. Like other metaheuristics, ACO is compound of a set of strategies that guide the search for the solution. It makes use of choices based on the use of stochastic processes, verifying the information acquired from previous results to guide it through the search space [39].

Artificial ants travel the search space represented by a graph $G(V, E)$, where V is a finite set of all nodes and E is the set of edges. The ants are attracted to more favorable locations to optimize an objective function, in other words, those in which the concentration of pheromone deposited by ants that previously went through the same path is higher [40]. While real ants deposit pheromone on the place they visit, artificial ants change some numeric information stored locally which describe the state of problem. This information is acquired through the historical and current performance of the ant during construction of solutions [39].

The responsibility of hosting the information during the search of the solution lies with the trail pheromone, τ. In ACO, the tracks are channels of communication between agents and only they have access to the tracks, i.e., only ants have the propriety of reading and modifying the numeric information contained in the pheromone trails. Every new path selection produces a solution, and each ant modifies all local information in a given region of the graph. Normally, an evaporation mechanism modifies the pheromone trail information over time. This characteristic allows agents slowly forget their history, allowing their search for new directions without being constrained by past decisions, thereby, avoiding the problem of precipitated convergences and resulting in not so great solutions.

The technique of clustering is a data mining tool used to find and quantify similarities between points of determined group of data. This process seeks to minimize the variance between elements of a given group and maximize them in relation to other groups [41]. The similarity function adopted is the Euclidean distance described in Eq. (28).

$$d(x,y) = \sqrt{\sum_{i=1}^{m} |x_i - y_i|^2} = \|\mathbf{x}_i - \mathbf{y}_i\| \tag{28}$$

The equation that measures the similarity between the data is called the objective function. The purpose of the use clustering is to create a template from which to extract a pattern of information. Thus, when a distance of data is found in smaller quantities in relation to this standard, you can group them into clusters of different sets of greater representation. The most classical algorithm in the literature is the K-means (KM) algorithm. It is a partitional center-based clustering method and the popularity is due to simplicity of implementation and competence to handle large volumes of data.

The problem to find the K center locations can be defined as an optimization problem to minimize the sum of the Euclidean distances between data points and their closest centers, described in Eq. (29). The KM randomly select k points and make them the initial centres of k clusters, then assigns each data point to the cluster with centre closest to it. In the second step, the centres are recomputed, and the data points are redistributed according to the new centres. The algorithm stop when the number of iterations is achieved or there is no change in the membership of the clusters over successive iterations [42]. One issue founded in KM is the initialization due to partitioning strategy, when in local density data results in a strong association between data points and centers [43].

$$\mathrm{KM}(\mathbf{x},\mathbf{c}) = \sum_{i=1}^{n}\sum_{j=1}^{k} \|x_i - c_j\|^2 \tag{29}$$

where $\mathbf{x}$ is the data, $\mathbf{c}$ is the center. The parameter n is total number of elements in $\mathbf{x}$ and k is the number of center in $\mathbf{c}$.

The ACO described in this section aims to optimize the efficiency of clustering minimizing the objective function value described by Eq. (29) Thus, this ensures that each point i will be grouped to the best cluster j where $j = 1, ..., K$. In addition, it enables the construction of solutions that are not givens by local optimal, which is the existing problem in most clustering algorithms. The ACO algorithm proposed is described in Algorithm 2, in which the functions performed into the Algorithm 2 are briefly described in following.

a) *CalculateFitnessFunction()*: For each ant m is calculated the Fitness Function based on Eq. 29. As each ant represent a possible solution, each ant will be a possible center to clusterize the data $\mathbf{x}$, and the ant m describing the lowest value of $\mathrm{KM}(\mathbf{x},\mathbf{c}_m)$.

b) *SortAnts()*: This function sort and rank the ants according to the *CalculateFitnessFunction()*.

Algorithm 2 ACO Clustering

Objective function $f(\mathbf{x}), \mathbf{x} = (x_m, ..., x_d)^T$
Initialize the ants population $\mathbf{x}_m (m = 1, 2, ..., n)$
Set the parameters γ, β, ρ
WHILE (*The end conditions aren't met*)
  FOR $m = 1$ to M
    CalculateFitnessFunction();
  end FOR
  SortAnts();
  UpdatePheromone();
end WHILE

c) *UpdatePheromone()*: This function directs the algorithm at the search for new solutions using promising path that were previously found. The links between point-cluster that showed better results are intensified and expected to be used in the construction of increasingly better solutions. In contrast, point-cluster unsuccessful links are expected to be forgotten by the algorithm through the evaporation process of the pheromone. The pheromone updating can be described by:

$$\tau_{ij}(t+1) = (1-\rho)\tau_{ij}(t) \tag{30}$$

where ρ is a constant suitably defined, which describes the evaporation rate of the pheromone and has value $0 < \rho < 1$. The variable t identifies the interaction running.

After the ACOClustering generates the centers from the training dataset, it is applied to the dataset. Then, it is adopted a parameter ϵ which is a value describing a range accepted to cluster the data in that center of not. The figure 7 illustrated the idea.

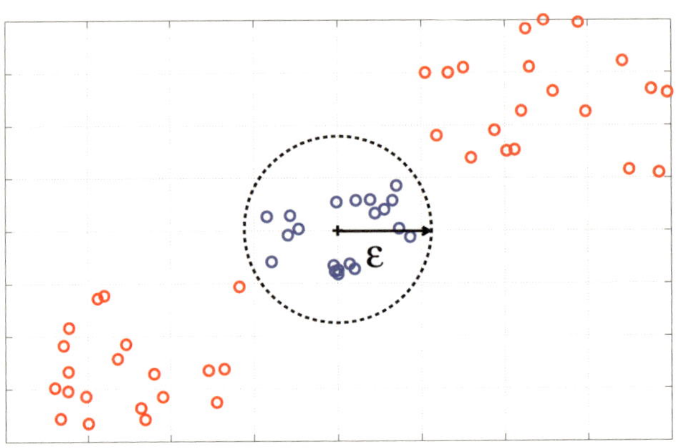

Figure 7. The area generated by the ϵ parameter.

3.4.3. Numerical Results

A paramount importance question when working with cluster is the optimal number of clusters to grouping the dataset in a good manner. We adopted the following clustering quality criteria: Dunn's index [44] and Davies-Bouldin index [45].

The Dunn's index is based on the calculation of the ratio between the minimal intracluster distance to maximal intercluster distance and the main idea is to identifying the cluster sets that are compact and well separated. The following Eq. (31) describes:

$$D = \frac{d_{\text{min}}}{d_{\text{max}}} \tag{31}$$

where, d_{min} is the smallest distance between two objects from different clusters, and d_{max} is the largest distance of two objects from the same cluster. D is limited to the interval $[0, \infty]$ and higher is the desirable value. In figure 8, it is presented the values for the tests for $K = [2, \ldots, 10]$.

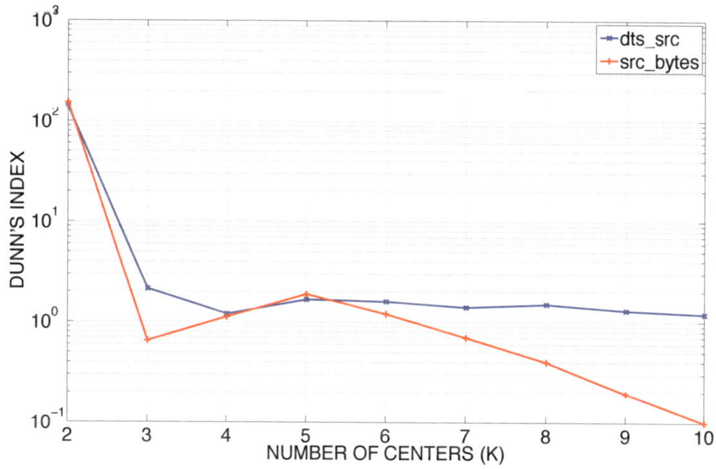

Figure 8. Dunn's index.

The Dunn's index aims to indentify clusters that are compact, well separeted and with a low variance between the members within the same cluster. The results indicates that $K = 2$ are the best number of centers to adopt, because it is the higher Dunn's index indicating a better clustering. But, before adopting $K = 2$, we tested another index, the Davies-Bouldin index [45]. It is a function of the ratio of the sum of within-cluster scatter to between-cluster separation and is describe by the Eq. (32):

$$DB = \frac{1}{n} \sum_{i=1, i \neq j}^{n} \max \left[\frac{\sigma_i + \sigma_j}{d(c_i, c_j)} \right] \tag{32}$$

where n is the number of clusters, σ_i is the average distance of all objects in cluster i to their cluster center c_i, σ_j is the average distance of all objects in cluster j to their cluster center c_j, and $d(c_i, c_j)$ is the distance of centers c_i and c_j. If DB result in low values, the clusters are compact and the centers are far from each other. In figure 9, it is presented the results for $K = [2, \ldots, 10]$.

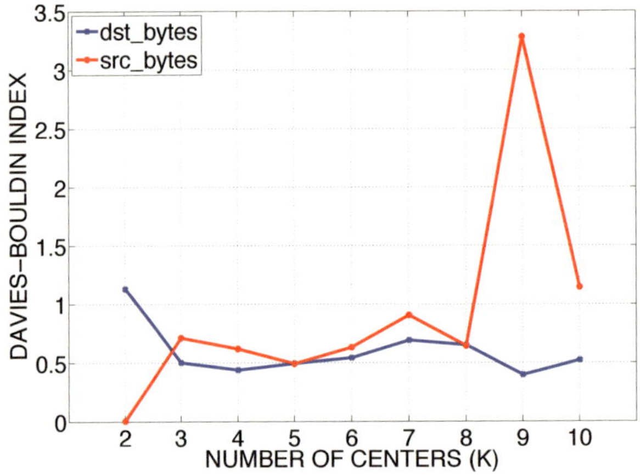

Figure 9. Davies-Bouldin index.

The ratio of the within cluster scatter to the between cluster separation will constrain the index to be symmetric and non-negative, thus it is expected a lower value for a fair clustering. From figure 9, the src_bytes (red line) present a value close to 0 at $K = 2$, and to the dst_bytes (blue line) for all the K tested values it had a regular behavior around 0.5. As $K = 2$ present a better result for Dunn's index and the best result for src_bytes, it is adopted for all the following tests.

Besides the number of centers, to measure the efficiency of the proposed case study, we adopted the following variables [46]:

- TRUE POSITIVE : If the instance is an anomaly and it is classified as an anomaly;
- FALSE NEGATIVE : If the instance is an anomaly and it is classified as normal;
- FALSE POSITIVE : If the instance is normal and it is classified as an anomaly;
- TRUE NEGATIVE : If the instance is an normal and it is classified as normal;

Hence, through the declaration of these variables the following equations can be calculated:

$$\text{False Alarm Rate (FAR)} = \frac{\text{FALSE NEGATIVE}}{\text{TOTAL OF NORMAL DATA}} \tag{33}$$

$$\text{Accuracy (ACC)} = \frac{\text{TRUE POSITIVE} + \text{TRUE NEGATIVE}}{\text{TOTAL NORMAL DATA + TOTAL ANOMALY DATA}} \tag{34}$$

$$\text{Precision (PRE)} = \frac{\text{TRUE POSITIVE}}{\text{TRUE POSITIVE} + \text{FALSE POSITIVE}} \tag{35}$$

http://dx.doi.org/10.5772/53338

Eq. (33) describes how much the method wrongly classified as anomalous from all the normal data, while Eq. (34) measures the closeness of the method measures in relation to the real values. The Eq. (35) describes the percentage of the corrected classified data among all the classified data.

It was decided to test four rules to capture the anomalies: 1)using only the src_bytes, 2)using only the dst_bytes, 3)using the src_bytes or dst_bytes and 4)using the src_bytes and dst_bytes. The figure 10 shows the results for the FAR.

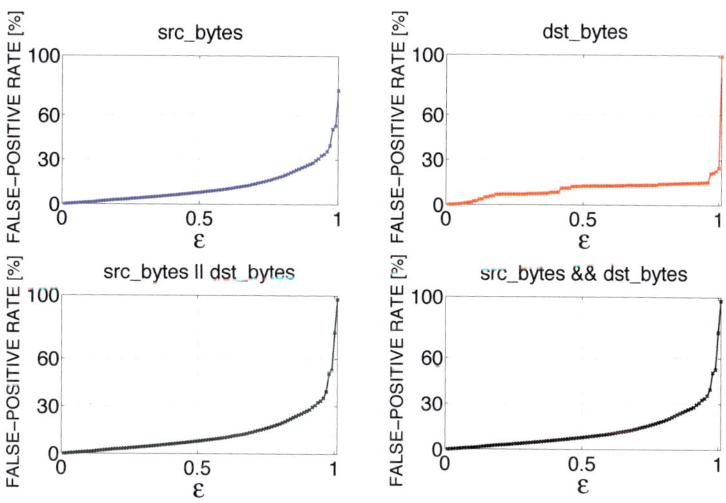

Figure 10. False alarm rate.

The figure 10 shows the method study achievies low rates for FAR, that means the method does not classify as anomalous the normal data. As we increase the value of ϵ, the FAR starts increasing, but only achieves high rates close to 1, therefore, the area created is large enough to capture anomalies data and wrongly classify.

In figure 11 shows the result for ACC. The method is not so close to the real value, expressed by the rate around 30%. This rate can be originated because the method is not classifying right the anomalies, in the other hand, it is classifying the normal data right.

To finally demonstrate that the method is not classifying the anomalies in a good manner, the figure 12 shows the precision results. It is observed higher rates when $\epsilon < 0.15$, meaning the method can classify anomaly from normal when the area adopted is small. This makes sense, because in computer networks the traffic behavior follows a regular action and the anomaly usually is a abrupt change in this behavior. When $\epsilon > 0.15$, the PRE rate decrease at a high pace, that means the method adopt more anomalies as normal data.

As for the four rules, we can conclude that when separate they show different results, the src_bytes describes better results. But when using then together, they express similar results as well, and the src_bytes results suppress the dst_bytes results.

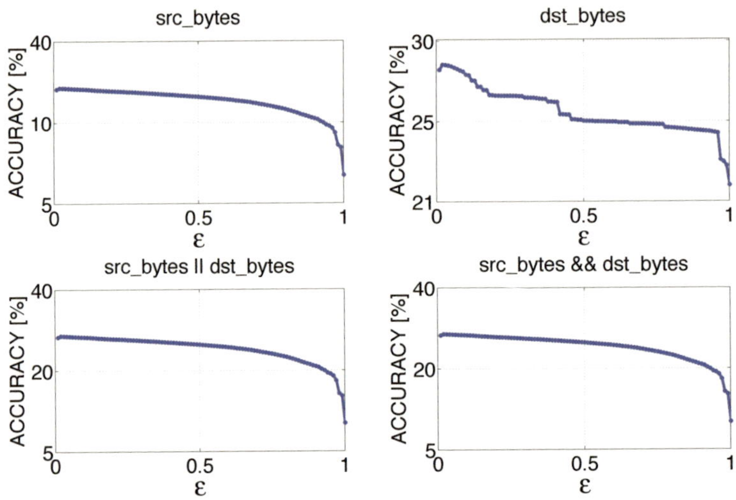

Figure 11. Accuracy rate.

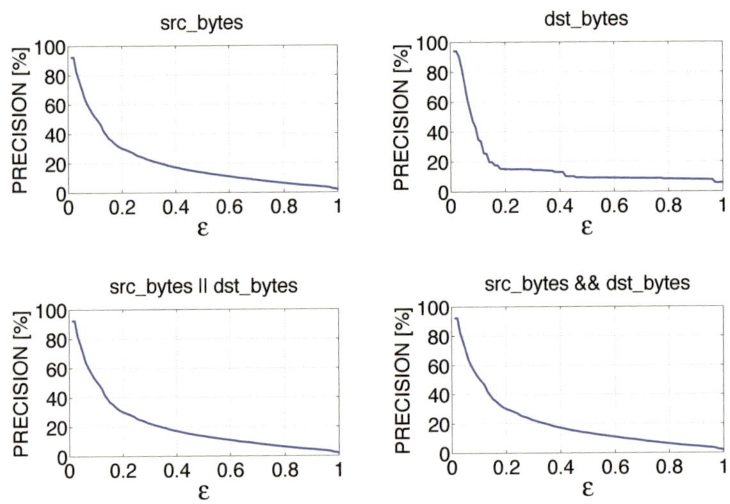

Figure 12. Precision rate.

4. Conclusion Remarks

4.1. ACO Resource Allocation in CDMA Networks

The ACO algorithm proved itself robust and efficient in solving both RA problems presented in this chapter. In fact, the ACO performance exceeded the PSO performance discussed in [1].

In terms of solution quality the ACO power control scheme was able to achieve much better solutions than the PSO approach. For the weighted throughput maximization problem, the numerical results show a fast convergence in the first hundred iterations and a solution

http://dx.doi.org/10.5772/53338

improvement after that. This fast convergence behavior is an important feature due to the system dynamics, i.e. the algorithm must update the transmission power every 666.7μs if we consider power control applications in 3G wireless cellular communication CDMA systems.

Future work includes analyzes over dynamic channels, i.e. the channel gain matrix is constant only over a single time slot.

4.2. ACO Anomaly Detection in Computer Networks

The method presented in the study case does not have excellent results, because for all the 41 features from the KDD dataset, it only works with 2: src_bytes and dst_bytes. From all the anomalies presented, we focus on the Denial of Service (DoS), and the method presented good False Alarm Rates (FAR), assuming the parameter ϵ normally is $0 < \epsilon < 2$, where the FAR is below 5%.

The Accuracy rate and Precision rate can be increased using the other features like flag, land. Flag is the status the connection and land assumes 1 if connection is from/to the same host/port; 0 otherwise. Adding more rules to the anomaly detection method, it is possible to increase the rates.

The ACOClustering have an important rule, helping to cluster the data, preventing to get stuck in local optimum centers. In the task of manage the computer network, it can handle a set of thousands of connections per second, thus it is possible to get stuck in some optimum local center. As each ant is a possible set of the centers, and the search is always guide by the best answer found so far.

Author details

Mateus de Paula Marques[1],
Mário H. A. C. Adaniya[1], Taufik Abrão[1],
Lucas Hiera Dias Sampaio[2] and Paul Jean E. Jeszensky[2]

1 State University of Londrina (UEL), Londrina, PR, Brazil
2 Polytechnic School of University of São Paulo (EPUSP), São Paulo, SP, Brazil

References

[1] T. Abrão, Sampaio L.D.H., M. L. Proença JR, B. A. Angélico, and P. J. E. Jeszensky. *Multiple Access Network Optimization Aspects via Swarm Search Algorithms*, volume 1, chapter 13, pages 261–298. InTech Open, 2011.

[2] Marco Dorigo and Gianni Caro. Ant colony optimization: A new meta-heuristic. In *Evolutionary Computation. CEC 99*. IEEE, 1999.

[3] Gerhard Fettweis and Ernesto Zimmermann. Ict energy consumption - trends and challenges. In *WPMC'08 – 11th International Symposium on Wireless Personal Multimedia Communications*, Sept. 2008.

[4] G. Foschini and Z. Miljanic. A simple distributed autonomous power control algorithm and is convergence. volume 42, pages 641–656. IEEE, November 1993.

[5] M. Moustafa, I. Habib, and M. Naghshineh. Genetic algorithm for mobiles equilibrium. MILCOM 200, October 2000.

[6] M.Moustafa, I. Habib, and M. Naghshineh. Wireless resource management using genethic algorithm for mobiles equilibrium. volume 37, pages 631–643, November 2011.

[7] H. Elkamchouchi, H. Elragal, and M. Makar. Power control in cdma system using particle swarm optimization. pages 1–8, March 2007.

[8] K. Zielinski, P. Weitkemper, R. Laur, and K. D. Kammeyer. Optimization of power allocation for interference cancellation with particle swarm optimization. volume 13, pages 128–150, February 2009.

[9] J.W. Lee, R.R. Mazumdar, and N. B. Shroff. Downlink power allocation for multi-class wireless systems. volume 13, pages 854–867. IEEE, August 2005.

[10] J. Dai, Z. Ye, and X. Xu. Mapel: Achieving global optimality for a non-convex wireless power control problem. volume 8, pages 1553–1563. IEEE, March 2009.

[11] T. J. Gross, T. Abrao, and P. J. E. Jeszensky. Algoritmo de controle de potência distribuido fundamentado no modelo populacional de verhulst. volume 20, pages 59–74. Revista da Sociedade Brasileira de Telecomunicacoes, 2010.

[12] J. H. Ping Qian and Ying Jun Zhang. Mapel: Achieving global optimality for a non-convex wireless power control problem. volume 8, pages 1553–1563, March 2009.

[13] Lucas Dias H. Sampaio, Moisés F. Lima, Bruno B. Zarpelão, Mario Lemes Proença Junior, and Taufik Abrão. Swarm power-rate optimization in multi-class services ds/cdma networks. 28th Brazilian Symposium on Computer Networks and Distributed Systems, May 2010.

[14] M. Elmusrati and H. Koivo. Multi-objective totally distributed power and rate control for wireless communications. volume 4, pages 2216–2220. VTC'03-Spring, Apr. 2003.

[15] C. E. Shannon. The mathematical theory of communication. *The Bell System Technical Journal*, 27((reprinted with corrections 1998)):379–423, 623–656, July, October 1948.

[16] M. Elmusrati, H. El-Sallabi, and H. Koivo. Aplications of multi-objective optimization techniques in radio resource scheduling of cellular communication systems. volume 7, pages 343–353. IEEE, Jan 2008.

[17] E. Seneta. *Non-Negative Matrices and Markov Chains*, volume 2. Springer-Verlag, 1981.

[18] N. T. H. Phuong and H. Tuy. A unified monotonic approach to generalized linear fractional programming. pages 229–259, 2003.

[19] Krzysztof Socha and Marco Dorigo. Ant colony optimization for continuous domains. In *European Jornal of Operational Research*, pages 1155–1173, Brussels, Belgium, 2008. Elsevier.

[20] Animesh Patcha and Jung-Min Park. An overview of anomaly detection techniques: Existing solutions and latest technological trends. *Computer Networks: The International Journal of Computer and Telecommunications Networking*, 51:3448–3470, August 2007.

[21] M Thottan and Chuanyi Ji. Anomaly detection in IP networks. *IEEE Transactions on Signal Processing*, 51(8):2191–2204, August 2003.

[22] Wentao Liu. Research on DoS attack and detection programming. In *Proceedings of the 3rd international conference on Intelligent information technology application*, volume 1 of *IITA'09*, pages 207–210, Piscataway, NJ, USA, November 2009. IEEE Press.

[23] J. Gadge and A.A. Patil. Port scan detection. In *16th IEEE International Conference on Networks*, ICON, pages 1–6, USA, December 2008. IEEE Press.

[24] Varun Chandola, Arindam Banerjee, and Vipin Kumar. Anomaly detection: A survey. *ACM Computing Surveys.*, 41(3), 2009.

[25] Victoria J. Hodge and Jim Austin. A survey of outlier detection methodologies. *Artif. Intell. Rev.*, 22(2):85–126, 2004.

[26] Juan M. Estévez-Tapiador, Pedro Garcia-Teodoro, and Jesús E. Díaz-Verdejo. Anomaly detection methods in wired networks: a survey and taxonomy. *Computer Communications*, 27(16):1569–1584, 2004.

[27] D.E. Denning. An intrusion-detection model. *Software Engineering, IEEE Transactions on*, SE-13(2):222–232, February 1987.

[28] Bruno Bogaz Zarpelão. *Detecção de Anomalias em Redes de Computadores*. PhD thesis, Universidade Estadual de Campinas (UNICAMP). Faculdade de Engenharia Eletrica e de Computação (FEEC)., 2010.

[29] Marco Dorigo, Vittorio Maniezzo, and Alberto Colorni. Positive feedback as a search strategy. Technical report, Technical Report No. 91-016, Politecnico di Milano, Italy, 1991.

[30] Marco Dorigo, Vittorio Maniezzo, and Alberto Colorni. Ant system: optimization by a colony of cooperating agents. *IEEE Transactions on Systems, Man, and Cybernetics, Part B*, 26(1):29–41, 1996.

[31] Gianni Di Caro and Marco Dorigo. Antnet: Distributed stigmergetic control for communications networks. *J. Artif. Intell. Res. (JAIR)*, 9:317–365, 1998.

[32] Habiba Drias, Moufida Rahmani, and Manel Khodja. Aco approaches for large scale information retrieval. In *World Congress on Nature and Biologically Inspired Computing (NaBIC)*, pages 713–718. IEEE, December 2009.

[33] Wang Ziqiang and Sun Xia. Web document retrieval using manifold learning and aco algorithm. In *Broadband Network Multimedia Technology, 2009. IC-BNMT '09. 2nd IEEE International Conference on*, pages 152–155, oct. 2009.

[34] Namita Shrivastava and Vineet Richariya. Ant colony optimization with classification algorithms used for intrusion detection. In *International Journal of Computational Engineering and Management, IJCEM*, volume 7, pages 54–63, January 2012.

[35] Yong Feng, Zhong-Fu Wu, Kai-Gui Wu, Zhong-Yang Xiong, and Ying Zhou. An unsupervised anomaly intrusion detection algorithm based on swarm intelligence. In *Machine Learning and Cybernetics, 2005. Proceedings of 2005 International Conference on*, volume 7, pages 3965–3969, aug. 2005.

[36] Mohammad Saniee Abadeh, Hamid Mohamadi, and Jafar Habibi. Design and analysis of genetic fuzzy systems for intrusion detection in computer networks. *Expert Syst. Appl.*, 38(6):7067–7075, 2011.

[37] The UCI KDD Archive. Kdd cup 1999 data, 1999.

[38] The UCI KDD Archive. Kdd cup 1998 data, 1998.

[39] Marco Dorigo and Thomas Stützle. *Ant colony optimization*. MIT Press, 2004.

[40] P.S Shelokar, V.K Jayaraman, and B.D Kulkarni. An ant colony approach for clustering. *Analytica Chimica Acta*, 509(2):187–195, 2004.

[41] Hui Fu. A novel clustering algorithm with ant colony optimization. In *Computational Intelligence and Industrial Application, 2008. PACIIA '08. Pacific-Asia Workshop on*, volume 2, pages 66–69, dec. 2008.

[42] D. T. Pham, S. Otri, A. A. Afify, M. Mahmuddin, and H. Al-Jabbouli. Data clustering using the bees algorithm. In *Proc. 40th CIRP Int. Manufacturing Systems Seminar*, Liverpool, 2007.

[43] Fengqin Yang, Tieli Sun, and Changhai Zhang. An efficient hybrid data clustering method based on k-harmonic means and particle swarm optimization. *Expert Syst. Appl.*, 36(6):9847–9852, 2009.

[44] J.C. Dunn. Well separated clusters and optimal fuzzy partitions. *Journal of Cybernetics*, 4:95–104, 1974.

[45] David L. Davies and Donald W. Bouldin. A Cluster Separation Measure. *Pattern Analysis and Machine Intelligence, IEEE Transactions on*, (2):224–227, 1979.

[46] Tom Fawcett. An introduction to ROC analysis. *Pattern Recognition Letters*, 27:861–874, 2005.

Chapter 6

Optical Network Optimization Based on Particle Swarm Intelligence

Fábio Renan Durand, Larissa Melo,
Lucas Ricken Garcia, Alysson José dos Santos and
Taufik Abrão

Additional information is available at the end of the chapter

http://dx.doi.org/10.5772/52225

1. Introduction

Modern optical communication networks are expected to meet a broad range of services with different and variable demands of bit rate, connection (session) duration, frequency of use, and set up time [1]. Thus, it is necessary to build flexible all-optical networks that allow dynamic resources sharing between different users and clients in an efficient way. The all-optical network is able to implement ultrahigh speed transmitting, routing and switching of data in the optical domain, presenting the transparency to data formats and protocols which increases network flexibility and functionality such that future network requirements can be met [2]. Optical code division multiplexing access (OCDMA) based technology has attracted a lot of interests due to its various advantages including asynchronous operation, high network flexibility, protocol transparency, simplified network control and potentially enhanced security [3]. Therefore, recent developments and researches on OCDMA have been experienced an expansion of interest, from short-range networks, such as access networks, to high-capacity medium/large networks.

The optical network presents two promising scenarios: the transport (backbone) networks with optical code division multiplexing/wavelength division multiplexing (OCDM/WDM) technology and the access network with OCDMA technology. In both, transport OCDM/WDM and access OCDMA networks, each different code defines a specific user or logic channel transmitted in a common channel. In a common channel, the interference that may arise between different user codes is known as multiple access interference (MAI), and it can limit the number of users utilizing the channel simultaneously [3]. In this work we have focus on hybrid OCDM/WDM systems. In this one, data signals in routing network configuration are carried on optical code path (OCP) from a source node to a destination node passing through nodes where the signals are optically routed and switched without regeneration in the electrical domain. Hence, in routing and channel (code/wavelength) assignment (RCA) problem, suitable paths and channels are carefully selected among the many possible choices for the required connections [2].

Establishing OCP with higher optical signal-to-noise plus interference ratio (SNIR) allows reducing the number of retransmissions by higher layers, thus increasing network throughput. Therefore, RCA techniques that consider physical layer impairments for the establishment of an OCP, namely Quality of Transmission-Aware (QoT-aware) RCA, could be much more practical [4-5]. For a dynamic traffic scenario the objective is to minimize the blocking probability of the connections by routing, assigning channels, and to maintain an acceptable level of optical power and adequate SNIR all over the network [6]. Furthermore, different channels can travel via different optical paths and also have different levels of quality of service (QoS) requirements. The QoS depends on SNIR, dispersion, and nonlinear effects [6]. Therefore, it is desirable to adjust network parameters in an optimal way, based on on-line decentralized iterative algorithms to accomplish such adjustment [7].

As a result, this dynamic optimization allows an increased network flexibility and capacity [6-7]. The SNIR optimization problem appears to be a huge challenge, since the MAI introduces the near-far problem [7]. Furthermore, if the distances between the nodes are quite different, like in real optical networks, the signal power received from various nodes will be significantly distinct. Thus, considering an optical node as the reference, the performance of closer nodes is many orders of magnitude better than that of far ones. Then, an efficient power control is needed to overcome this problem and enhance the performance and throughput of the network; this could be achieved through the SNIR optimization [6]. In this case, which is analogous to the CDMA cellular system, the power control (centralized or distributed) is one of the most important issues, because it has a significant impact on both network performance and capacity. It is the most effective way to avoid the near-far problem and to increase the SNIR [6-7].

The optical power control problem has been recently investigated in the context of access networks aiming at solving the near-far problem [7-8] and establishing the QoS at the physical layer [9-11]. In [7], the impact of power control on the random access protocol was investigated. In [8], the effect of near-far problem and a detailed review of the power control were presented including the use of distributed algorithms. On the other hand, in [9-12] the concept that users of various classes should transmit at different power levels was applied. Distinct power levels were obtained with power attenuators [10], adjustable encoders/decoders [11], and adjustable transmitters [12]. Furthermore, the optimal selection of the system's parameters such as the transmitted power and the information rate would improve their performances [9, 13-15]. In [13], optical power control and time hopping for multimedia applications using single wavelength was proposed. The approach accommodates various data rates using only one sequence by changing the time-hopping rate. However, in order to implement such system an optical selector device that consists of a number of optical hard-limiters is needed [13]. On the other hand, in [14] a multi rate and multi power level scheme using adaptive overlapping pulse-position modulator (OPPM) and optical power controller was proposed. The bit rate varies depending on the number of slots in the optical OPPM system and has the advantage that it is not required to change the code sequence depending on the required user's information rate. The power level can be achieved by accommodating users with the different transmitted power. The power controller requires only power attenuator, and the difference of the power does not cause the change of the bit rate. In [15] a hybrid power and rate control nonlinear programming algorithm for overlapped optical fast frequency hopping (OFFH) was proposed. The multi rate transmission is achieved by overlapping consecutive bits while coded using fiber Bragg grating (FBG). The intensity of the transmitted optical signal is directly adjusted from the laser source with respect to the transmission data rate. The proposed algorithm provides a joint transmission power and overlapping coefficient allocation strategy, which has been obtained via the solution of a constrained optimization problem, which maximizes the aggregate system throughput subject to a peak laser transmission power constraint. In [9], a control algorithm to solve the unfairness in the resource allocation strategy presented in [10] has been analyzed. Also, a unified framework for allocating and controlling the transmission rate and power in a way that it can be applied for any expression of the system capacity was implemented.

Besides, recently researches have showed the utilization of resource allocation and optimization algorithms such as Local Search, Simulated Annealing, GA, Particle Swarm optimization (PSO), Ant Colony optimization (ACO) and Game Theory to regulate the transmitted power, bit rate variation and the number of active users in order to maximize the aggregate throughput of the optical networks [16-17]. However, the complexity and unfairness in the strategies presented are aspects to be improved. On the other hand, resource allocation has not been largely investigated considering energy efficiency aspects. This issue has become paramount since energy consumption is dominated by the access segment due to the large amount of distributed network elements. The related works have showed the utilization of resource allocation and optimization algorithms to optimization of the access network; however, these issues have not been largely investigated considering routed OCDM/WDM networks [6]. In the case of the OCP networks optimization, it is necessary to consider the use of distributed iterative algorithms with high performance-complexity tradeoffs and the imperfections of physical layer, which constitute a new research area so far, which was investigated under an analytical perspective in [6].

It is worth noting the routed OCDM/WDM networks brings a new combination of challenges with the power control, like amplified spans, multiple links, accumulation, and self-generation of the optical spontaneous noise power (ASE) noise, as well as the MAI generated by the OCPs. On the other hand, the dispersive effects, such as chromatic or group velocity dispersion (GVD) and polarization mode dispersion (PMD), are signal degradation mechanisms that significantly affect the overall performance of optical communication systems [6, 18-21].

In this chapter, optimization procedures based on PSO are investigated in details, aiming to efficiently solve the optimal resource allocation for SNIR optimization of OCPs from OCDM/WDM networks under QoS restrictions and energy efficiency constraint problem, considering imperfections on physical constraints. Herein, the adopted SNIR model considers the MAI between the OCP based on 2-D codes (time/wavelength) [22, 23], ASE at cascaded amplified spans, and GVD and PMD dispersion effects.

The optimization method based on the heuristic PSO approach is attractive due to its performance-complexity tradeoff and fairness features regarding the optimization methods that deploy matrix inversion, purely numerical procedures and other heuristic approaches [9][17].

The chapter is organized in the following manner: in Section 2 the optical transport (OCDM/WDM) is described, while in Section 3 the SNIR optimization for the OCPs based on particle swarm intelligence is described in order to solve the resource allocation problem. In the network optimization context, figures of merit are presented and the PSO is developed in Section 4, with emphasis on its input parameters optimal choice and the network performance. Afterward, numerical results are discussed for realistic networks operation scenarios. Finally, the main conclusions are offered in Section 5.

2. Network architecture

2.1. OCDM/WDM transport network

The transport network considered in this work is formed by nodes that have optical core routers interconnected by OCDM/WDM links with optical code paths defined by patterns of short pulses in wavelengths, such as shown in Fig. 1. The links are composed by sequences of span and each span consists of optical fiber and optical amplifier. The transmitting and receiving nodes create virtual path based on the code and the total link length is given by $d_{ij} = \sum_i d_i^{tx} + \sum_j d_j^{rx}$, where d_i^{tx} is the span length from the transmitting node to the optical router and d_j^{rx} is the span between optical routers in the OCP route and the receiving node. The received power at the j-th node is given by $P_r = a_{star} p_i G_{amp} \exp(-\alpha_f d_{ij})$, where p_i is the transmitted power by the i-th transmitter node, α_f is the fiber attenuation (km^{-1}) and a_{star} is the star coupler attenuation (linear units), and G_{amp} is the total gain at the route. Considering decibel units, $a_{star} = 10\log(K) - [10\log_2(K)\log_{10}\delta]$, where, δ is the excess loss ratio [6]. A typical distance between optical amplifiers is about 60 km [20].

The optical core router consists of code converter routers in parallel forming a two-dimensional router node [23] and each group of code converters in parallel is pre-connected to a specific output performing routing by selecting a specific code from the incoming broadcasting traffic. This kind of router does not require light sources or optical-electrical-optical conversion and can be scaled by adding new modules [22]. This code is transmitted and its route in the network is determined by a particular code sequence. For viability characteristics, we consider network equipment, such as code-processing devices (encoders and decoders at the transmitter and receiver), star coupler, optical routers could be made using robust, lightweight, and low-cost technology platforms with commercial-off-the-shelf technologies [23-24]. For more details about transport networks the references [19],[25] should be consulted.

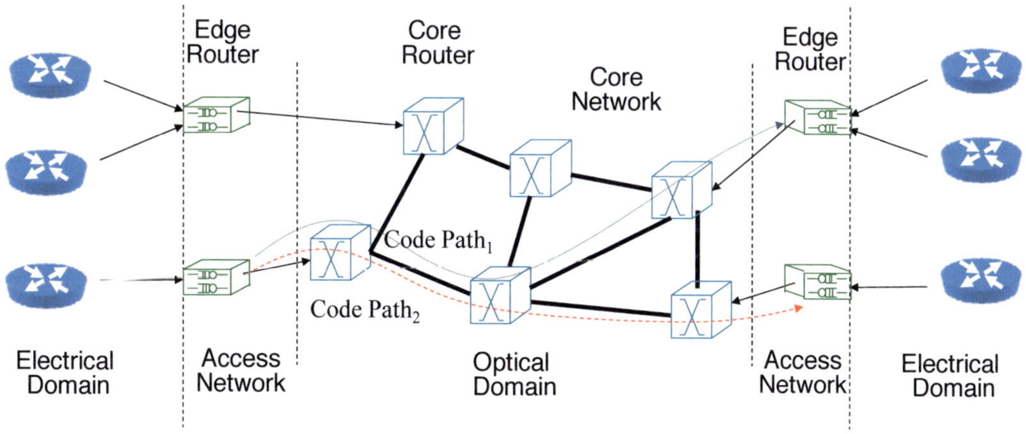

Figure 1. OCDM/WDM routed network architecture.

2.2. OCDMA codes

The OCDMA can be divided into a) non-coherent unipolar systems, based only on optical power intensity modulation [20], and b) coherent bipolar systems, based on amplitude and phase modulation [26]. As expected, the performance of coherent codes is higher than that of non-coherent ones when analyzing the SNIR [27]. This effect occurs, because the bipolar code is true-orthogonal, and the unipolar code is pseudo-orthogonal. However, the main drawback to the coherent OCDMA lies in the technical implementation difficulties, concomitant with the utilization of phase-shifted optical signals [20],[27]. In this work we adopt non-coherent codes because their technological maturity and implementation easiness when compared with coherent codes [28]. The non-coherent codes can be classified into one-dimensional (1-D) and two-dimensional (2-D) codes. In the 1-D codes, the bits are subdivided in time into many short chips with a designated chip pattern representing a user code. On the other hand, in the 2-D codes, the bits are subdivided into individual time chips, and each chip is assigned to an independent wavelength out of a discrete set of wavelengths. The 2-D codes have better performance than the 1-D codes, and they can significantly enhance the number of active users [29]. Besides, the 2-D codes have been applied only in access networks [2]; in this way, recently the utilization of the 2-D codes to obtain optical code path routed networks was proposed, which performance evaluated by simulation, considering coding, topology, load condition, and physical impairment [2][6][20][21][22].

The 2-D codes can be represented by $N_\lambda \times N_T$ matrices, where N_λ is the number of rows, that is equal to the number of available wavelengths, and N_T is the number of columns, that is equal to the code length. The code length is determined by the bit period T_B which is subdivided into small units namely chips, each of duration $T_c - T_B/ N_T$, as show Fig. 2(a). In each code, there are w short pulses of different wavelength, where w is called the weight of the code. An $(N_\lambda \times N_{T,}\ w,\ \lambda_a,\ \lambda_c)$ code is the collection of binary $N_\lambda \times N_T$ matrices each of code weight w; the parameters λ_a and λ_c are nonnegative integers and represent the constraints on the 2-D codes autocorrelation and cross-correlation, respectively [3]. The 2-D code design and selection is very important for good system performance and high network scalability with low bit error rate (BER). In [3] and [28] is presented an extensive list of code construction techniques, as well as their technological characteristics are discussed.

The OCDMA 2-D encoder creates a combination of two patterns: a wavelength-hopping pattern and a time-spreading pattern. The common technology applied for code encoders/ decoders fiber Bragg gratings (FBGs), as show Fig. 2(b). The losses associated with the encoders/ decoders are given by $C_{Bragg}(dB) = N_{\lambda} a_{Bragg} + a_{Circulator}$ [22], where a_{Bragg} is the FBG loss and $a_{Circulator}$ is the circulator loss. The usual value of losses for these equipments are a_{Bragg} = 0.5 dB and $a_{Circulator}$ = 3dB.

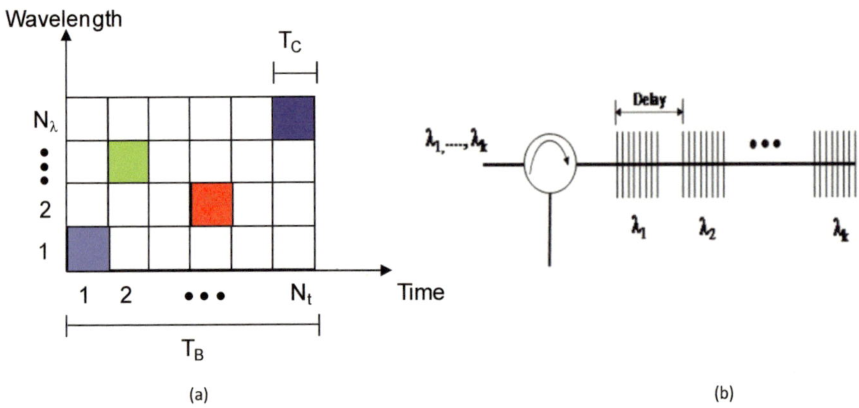

Figure 2. a) Representation of optical OCDMA codes. (b) Schematic of 2-D encoders/decoders based on fiber Bragg gratings (FBGs).

3. SNIR optimization procedures

In the present approach, the SNIR optimization is based on the definition of the minimum power constraint (also called sensitivity level) assuring that the optical signal can be detected by all optical devices. The maximum power constraint guarantees the minimization of nonlinear physical impairments, because it makes the aggregate power on a link to be limited to a maximum value. The power control in optical networks appears to be an optimization problem.

3.1. Problem description

Denoting Γ_i the carrier-to-interference ratio (CIR) at the required decoder input, in order to get a certain maximum bit error rate (BER) tolerated by the *i-th* optical node, and defining the *K*-dimensional column vector of the transmitted optical power p = $[p_1, p_2, \ldots, p_K]^T$, the optical power control problem consists in finding the optical power vector p that minimizes the cost function $J(\boldsymbol{p})$ can be formulated as [6],[8] :

$$\begin{aligned}
&\min_{\mathbf{p}\in\mathbb{R}_+^K} J(\mathbf{p}) = \min_{\mathbf{p}\in\mathbb{R}_+^K} \mathbf{1}^T\mathbf{p} = \min_{p_i\in\mathbb{R}_+} \sum_{i=1}^{K} p_i, \\
&\textit{subject to}: \Gamma_i = \frac{\left(G_{ii}p_iG_{amp}\right)/\sigma_D}{G_{amp}\sum_{j=1,j\neq i}^{K} G_{ij}p_j + 2N_{sp}^{eq}} \geq \Gamma^* \\
&P_{\min} \leq p_i \leq P_{\max}\ \forall i = 1,..,K, \\
&P_{\min} \geq 0, \quad P_{\max} > 0
\end{aligned} \tag{1}$$

where $1^T = [1,\ldots, 1]$ and $\Gamma^*$is the minimum CIR to achieve a desired QoS; G_{ii} is the attenuation of the OCP taking into account the power loss between the nodes, according to network topology, while G_{ij} corresponds to the attenuation factor for the interfering OCP signals at the same route, G_{amp} is the total gain at the OCP, $N_{sp}^{eq}$is the spontaneous noise power (ASE) for each polarization at cascaded amplified spans [29], p_i is the transmitted power for the i-OCP and p_j is the transmitted power for the interfering OCP; σ_D is the pulse spreading due to the combined effects of the GVD and the first-order PMD for Gaussian pulses [30]. Using matrix notations, (1) can be written as $[\boldsymbol{I}-\boldsymbol{\Gamma}^*\boldsymbol{H}]\boldsymbol{p}\geq\boldsymbol{u}$, where I is the identity matrix, H is the normalized interference matrix, which elements evaluated by $H_{ij}=G_{ij}/G_{ii}$for $i\neq j$ and zero for another case, thus $u_i=\Gamma^* N_{sp}^{eq}/G_{ii}$, where there is a scaled version of the noise power. Substituting inequality by equality, the optimized power vector solution through the matrix inversion $\boldsymbol{p}^*=[\boldsymbol{I}-\boldsymbol{\Gamma}^*\boldsymbol{H}]^{-1}\boldsymbol{u}$ could be obtained. The matrix inversion is equivalent to centralized power control, i.e. the existence of a central node in power control. The central node stores information about all physical network architecture, such as fiber length between nodes, amplifier position and regular update for the OCP establishment, and traffic dynamics. These observations justify the need for on-line SNIR optimization algorithms, which have provable convergence properties for general network configurations [6, 16, 29].

The SNIR and the carrier to interference ratio in eq. (1) are related to the factorN_T/σ, i. e., $\gamma_i\approx(N_T/\sigma)^2\Gamma_i$. The bit error probability (BER) is given by $P_b(i)=erfc\left(\sqrt{\gamma_i}/2\right)/2$, when the Gaussian approximation is adopted, and the signal-to-noise plus interference ratio (SNIR) at each OCP, considering the 2-D codes, is given by [6, 8],

$$\gamma_i = \frac{N_T^{\,2}\left(G_{ii}p_iG_{amp}\right)/\sigma_D}{\sigma^2 G_{amp}\sum_{j=1,j\neq i}^{K} G_{ij}p_j + 2N_{sp}^{eq}} \tag{2}$$

where the average variance of the Hamming aperiodic cross-correlation amplitude is represented by σ^2[3].

3.2. Physical restrictions

The physical impairments are signal degradation mechanisms that significantly affect the overall performance of optical communication systems [6]. For the data that are transmitted through a transparent optical network, degradation effects may accumulate over a large distance. The major linear physical impairments are group velocity dispersion (GVD), polarization mode dispersion (PMD), and amplifier spontaneous emission (ASE) noise [24]. On the other hand, the major nonlinear physical impairments are self phase modulation (SPM), cross-phase modulation (XPM), and four wave mixing (FWM), stimulated Brillouin scattering (SBS), and Raman scattering (SRS). The nonlinear physical impairments are excited with high power level [24]. However, the maximum power constraint guarantees the minimization of nonlinear physical impairments, because it makes the aggregate power on a link to be limited to a maximum value [6]. In the currently technology stage, besides GVD, the main linear impairment is the PMD, that must be considered in high capacity optical networks. Differently from GVD, PMD is usually difficult to accurately determine and compensate due to its dynamic nature and its fluctuations induced by external stress/strain applied to the fiber after installation [5] [21][22]. As a result, the signals quality in an OCDM/WDM network can be quickly evaluated by analyzing the GVD, PMD and MAI restrictions. PMD impairment establishes an upper bound on the length of the optical segment due to fiber dispersion which causes the temporal spreading of optical pulses. On the other hand, due to the advances in the fiber manufacturing process with a continuous reduction of the PMD parameter, the deleterious effect of PMD will not be an issue for 10 Gbps or lower bit rates, for future small and medium-sized networks [20] [21]. In this context, the dominant impairment in SNIR will be given by i) ASE noise accumulation in chains of optical amplifiers for future optical networks [29] and ii) ASE, GVD and PMD for currently stage of optical networks.

The dispersive effects, such as chromatic or group velocity dispersion (GVD) and polarization mode dispersion (PMD) constitute degradation mechanisms of the optical signal that significantly affect the overall performance of optical communication systems [21]. Currently, the PMD effect appears to be the only major physical impairment that must be considered in high capacity optical networks, which can hardly be controlled due to its dynamic and stochastic nature [5][21-22]. On the other hand, the GVD causes the temporal spreading of optical pulses that limits the product line rate and link length [6-30]. The pulse spreading effect due to the combined effects of the GVD and the first-order PMD for Gaussian pulses can be calculated as [30]:

$$\sigma_D = \left\{ \left(1 + \frac{C_p \beta_2 d_{ij}}{2\tau_0^2}\right) + \left(\frac{\beta_2 d_{ij}}{2\tau_0^2}\right) + x - \left(\frac{1}{2(1 + C_p^2)} \times \sqrt{1 + \frac{4}{3}(1 + C_p^2)x - 1}\right) \right\}^{1/2} \tag{3}$$

where C_p is the chirp parameter, $\tau_0 = \frac{T_C}{2\sqrt{2\ln 2}}$ is the RMS pulse width, T_c is the chip period at half maximum, $\beta_2 = -D\lambda_0^2 / 2\pi c$ is the GVD factor, D is the dispersion parameter, c is the speed of light in the vacuum, $x = \Delta\tau^2 / 4\tau_0^2$ and $\Delta\tau = D_{PMD}\sqrt{d_{ij}}$, D_{PMD} is the PMD parameter,

http://dx.doi.org/10.5772/52225

and d_{ij} is the link length. Although there is a difference in the GVD for each wavelength, resulting from time skewing between the wavelengths, the consideration of the same GVD value for the entire transmission window is reasonable for a small number of wavelengths, as for the present code [27], [28]. On the other hand, this approximation is utilized to obtain an analytical treatment of the GVD and the PMD, in the same and less complex formalism, rather than to apply a formalism based on numerical methods [6].

The ASE (N_{sp}^{eq}) at the cascaded amplified spans is given by the model presented in Fig. 3 [29].

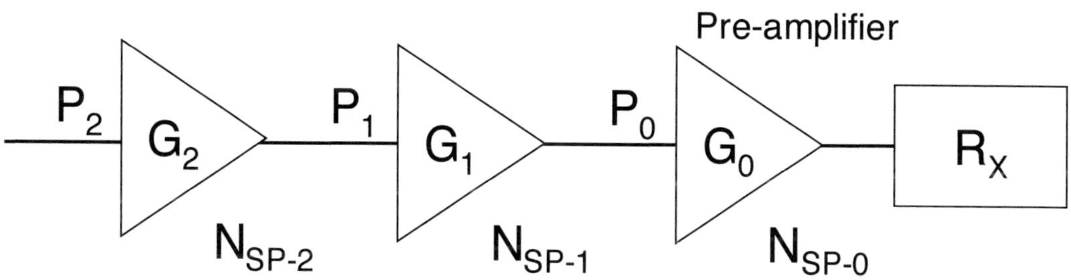

Figure 3. Cascading amplifiers.

This model considers that the receiver gets the signal from a link with cascading amplifiers, numbered as 1, 2,.., starting from the receiver. The pre-amplifier can be contemplated as the number 0 cascade amplifier. Let G_i be the amplifier gain, *i. e.* N_{sp}-i will be its spontaneous emission factor. The span between the i-*th* and the *(i – 1)-th* amplifier has the attenuation G_{ii}. Let P_{ti} be the mark power at the *i-th* amplifier input. The equivalent spontaneous emission factor is given by [24], [29]

$$N_{sp}^{eq} = \frac{N_{sp-1}\left(G_1 - 1\right)G_{ii}G_0 + N_{sp-0}\left(G_0 - 1\right)}{G_1 G_{ii} G_0 - 1} \tag{4}$$

Calculating recursively the N_{sp}^{eq} factor, one can find the noise at the cascading amplifiers. The noise for i-*th* amplifier is given by $N_{sp-i}=2n_{sp}hf\left(G_i - 1\right)B_0$, which take into account the two polarization mode presented in a single mode fiber [24]. Where n_{SP} is the spontaneous emission factor, typically around 2 – 5, *h* is Planck's constant, *f* is the carrier frequency, G_i is the amplifier gain and *Bo* is the optical bandwidth. Ideally, to reduce the ASE noise power, the optical bandwidth can be set to a minimum of *Bo* = 2*R*, where *R* is the bit rate. Without loss of generality, all employed optical amplifiers provide a uniform gain, setting the maximum obtainable Erbium-doped fiber amplifier (EDFA) to 20 dB across the transmission window. This is a reasonable assumption for the reduced number of wavelengths in the code transmission window (4 wavelengths), considering the optical amplifier gain profile, where the maximum difference of this gain is 0.4 dB for the wavelength, which is the most distant one from the central wavelength (1550 nm), with spectral spacing of 100 GHz [6][17].

3.3. Particle swarm optimization

3.3.1. PSO description

Particle Swarm Optimization (PSO) is a population-based stochastic optimization algorithm for global optimization that was presented first in 1995 [31]. It is based on the behavior of social groups like fish schools or bird flocks and it differs from other well-known Evolutionary Algorithms (EA). As in EA, a population of potential solutions is used to probe the search space, but no operators, inspired by evolution procedures, are applied on the population to generate new promising solutions [32]. The fact which is recursively exploited is that an improved performance can be gained by interactions between individuals, or more specifically by imitation of successful individuals. In a PSO system, particles fly around in multidimensional search space. During the flight, each particle adjusts its position according to its own experience, and the experience of neighboring particles, making use of the best position encountered by itself and its neighbors. The swarm direction of a particle is defined by the set of particles neighboring the particle and its history experience. Although PSO does not rely on the survival of the fittest principle, it is often classified as an evolutionary algorithm (EA) because the update equations, which control the movement of individuals, are similar to the evolutionary operators used in EAs.

In general, the PSO performance for resource allocation problem can guarantee fast convergence and fairness within fewer iterations regarding the genetic algorithm-based [16]. It is well known in the literature that the PSO performance for resource allocation problem is highly dependent on its control parameters and that recommended parameter settings from the literature often do not lead to reliable and fast convergence behavior for the considered optimization problem [33], [34], [35].

In the PSO process, each particle keeps track of its coordinates in the space of interest, which are associated with the best solution (fitness) it has achieved so far. Another best value tracked by the global version of the particle swarm optimizer is the overall best value, and its location, obtained so far by any particle in the population. At each time iteration step, the PSO concept consists of velocity changes of each particle toward local and global locations. Acceleration is weighted by a random term, with separate random numbers being generated for acceleration toward local and global locations. Let b_p and v_p denote a particle coordinates (position) and its corresponding flight speed (velocity) in a search space, respectively. In this strategy, each power-vector candidate $\mathbf{b}_p[t]$, with dimension K x 1, is used for the velocity-vector calculation of the next iteration [33]:

$$\mathbf{v}_p[t+1] = \omega[t] \bullet \mathbf{v}_p[t] + C_1 \bullet \mathbf{U}_{p1}[t]\left(\mathbf{b}_p^{best}[t] - \mathbf{b}_p[t]\right) + C_2 \bullet \mathbf{U}_{p2}[t]\left(\mathbf{b}_g^{best}[t] - \mathbf{b}_p[t]\right) \quad (5)$$

where $\omega[t]$ is the inertia weight of the previous velocity in the present speed calculation, the velocity-vector has K dimension $\mathbf{v}_p[t]=[\mathrm{v}_{p1}^t\,\mathrm{v}_{p2}^t\,\cdots\,\mathrm{v}_{pK}^t]^T$; the diagonal matrices $\mathbf{U}_{p1}[t]$ and $\mathbf{U}_{p2}[t]$ with dimension K have their elements as random variables with uniform distribution $\sim U\in[0,1]$, generated for the pth particle at iteration $t=1,2,\ldots,\mathrm{G}$; $\mathbf{b}_g^{best}[t]$ and $\mathbf{b}_p^{best}[t]$ are the best global vector-position and the best local vector-position found until the t_{th} iteration, respectively; C_1 and C_2 are acceleration coefficients regarding the best particles and the best global positions influences in the velocity updating, respectively. The p_{th} particle's position at the tth iteration is defined by the power candidate-vector $\mathbf{b}_p[t]=[\mathrm{b}_{p1}^t\,\mathrm{b}_{p2}^t\,\cdots\,\mathrm{b}_{pK}^t]^T$. The position of each particle is updated using the new velocity vector for that particle,

$$\mathbf{b}_p[t+1]=\mathbf{b}_p[t]+\mathbf{v}_p[t+1],\quad p=1,\;\ldots,\;\mathrm{P} \tag{6}$$

where P is the population size. In order to reduce the likelihood that the particle might leave the search universe, maximum velocity factor V_{max} factor is added to the PSO model, which will be responsible for limiting the velocity to the range $[\pm V_{max}]$. Hence, the adjustment of velocity allows the particle to move in a continuous but constrained subspace, been simply accomplished by:

$$\mathrm{v}_{pk}^t=\min\{V_{max};\max\{-V_{max};\,\mathrm{v}_{pk}^t\}\},\qquad k=1,\;\ldots,\;K;\;\;p=1,\;\ldots,\;\mathrm{P} \tag{7}$$

From (7) it's clear that if $|\mathrm{v}_{pk}^t|$ exceeds a positive constant value V_{max} specified by the user, the pth particle' velocity of kth user is assigned to be $\mathrm{sign}(\mathrm{v}_{pk}^t)V_{max}$, i.e. particles velocity on each of K-dimension is clamped to a maximum magnitude V_{max}. Besides, if the search space could be defined by the bounds $[P_{min};P_{max}]$, then the value of V_{max} typically is set so that $V_{max}=\tau(P_{max}-P_{min})$, where $0.1\leq\tau\leq1.0$; please refer to Chapter 1 within the definition of reference [35].

In order to elaborate further about the inertia weight it can be noted that a relatively larger value of ω is helpful for global optimum, and lesser influenced by the best global and local positions, while a relatively smaller value for ω is helpful for convergence, i.e., smaller inertial weight encourages the local exploration as the particles are more attracted towards $\mathbf{b}_p^{best}$ and $\mathbf{b}_g^{best}$ [31, 32]. Hence, in order to achieve a balance between global and local search abilities, a linear inertia weight decreasing with the algorithm convergence evolving was adopted, which has demonstrated good global search capability at beginning and good local search capability latter iterations:

$$\omega[t]=(\omega_{initial}-\omega_{final})\bullet\left(\frac{\mathrm{G}-t}{\mathrm{G}}\right)^m+\omega_{final} \tag{8}$$

where $\omega_{initial}$ and ω_{final} is the initial and final weight inertia, respectively, $\omega_{initial} > \omega_{final}$, G is the maximum number of iterations, and $m \in [0.6;\ 1.4]$ is the nonlinear index [36].

3.3.2. Optical code path resource allocation optimization

The following maximization cost function could be employed as an alternative to OCP resource allocation optimization [33]. This single-objective function was modified in order to incorporate the near-far effect [37], [38]

$$J_1(\mathbf{p}) = \max \begin{array}{l} \frac{1}{K}\sum_{k=1}^{K} \mathsf{F}_k^{th}\left(1 - \frac{p_k}{P_{max}}\right) + \frac{\rho}{\sigma_{rp}} \\ \gamma_k \geq \gamma_k^*,\ 0 < p_k^l \leq P_{\max},\quad R^l = R_{\min}^l \forall k \in K_l,\ and\ \forall l = 1,2,\ \ldots,\ L \end{array} \quad (9)$$

where L is the number of different group of information rates allowing in the system, and K_l is the number of user in the lth rate group with minimum rate given by $R_{\min}^l$. Important to say, the second term in eq. (9) gives credit to the solutions with small standard deviation of the normalized (by the inverse of rate factor, F^l) received power distribution:

$$\sigma_{rp}^2 = \mathrm{var}\left(F^1 p_1 G_{11},\ F^1 p_2 G_{22},\ \ldots,\ F^l p_k G_{kk},\ \ldots,\ F^L\ p_k G_{kk}\right) \quad (10)$$

i.e. the more close the normalized received power values are with other (small variance of normalized received power vector), the bigger contribution of the term $\frac{\rho}{\sigma_{rp}}$. For single-rate systems, $F^1 = \ldots = F^l = \ldots = F^L$. It is worth to note that since the variance of the normalized received power vector, σ_{rp}^2, normally assumes very small values, the coefficient ρ just also take very small values in order to the ratio $\frac{\rho}{\sigma_{rp}}$ achieves a similar order of magnitude of the first term in (9), been determined as a function of the number of users, K. Hence, the term $\frac{\rho}{\sigma_{rp}}$ has an effective influence in minimizing the near-far effect on OCDM/WDM systems, and at the same time it has a non-zero value for all swarm particles [33]. Finally, the threshold function in (9) is simply defined as:

$$\mathsf{F}_k^{th} = \begin{cases} 1, & \gamma_k \geq \gamma^* \\ 0, & otherwise \end{cases} \quad (11)$$

where the SNIR for the kth user, γ_k, is given by (2). The term $1 - \frac{p_k}{P_{max}}$ gives credit to those solutions with minimum power and punishes others using high power levels [33].

The PSO algorithm consists of repeated application of the updating velocity and position, eq. (5) and (6), respectively. The pseudo-code for the single-objective continuous PSO power allocation problem is presented in Algorithm 1.

http://dx.doi.org/10.5772/52225

Algorithm 1 Continuous PSO Algorithm for the Power Allocation Problem

Input: $\mathcal{P}, \mathcal{G}, \omega, C_1, C_2, V_{max}, R_{\min}$; Output: $\mathbf{p}^*$

begin

1. initialize the population at $t = 0$;
   $\mathbf{B}[0] \sim U[P_{min}; P_{max}]$
   $\mathbf{b}_p^{best}[0] = \mathbf{b}_p[0]\ and\ \mathbf{b}_g^{best}[0] = \mathbf{p}_{max}$;
   $\mathbf{v}_p[0] = 0$: null initial velocity;
2. while $t \leq \mathcal{G}$
   a. calculate $J(\mathbf{b}_p[t]), \forall \mathbf{b}_p[t] \in \mathbf{B}[t]$, using (9);
   b. update velocity $\mathbf{v}_p[t], p = 1, \ldots, \mathcal{P}$, through (5);
   c. update best positions:
      for $p = 1, \ldots, \mathcal{P}$
         If $J(\mathbf{b}_p[t]) < J(\mathbf{b}_p^{best}[t])$ & $R_p[t] \geq R_{\min}$,
            $\mathbf{b}_p^{best}[t+1] \leftarrow \mathbf{b}_p[t]$
         else $\mathbf{b}_p^{best}[t+1] \leftarrow \mathbf{b}_p^{best}[t]$
      end
      if $\exists\ \mathbf{b}_p[t] \mid [J(\mathbf{b}_p[t]) < J(\mathbf{b}_g^{best}[t])]$ & $R_p[t] \geq R_{\min}$ & $[J(\mathbf{b}_p[t]) \leq J(\mathbf{b}'_p[t]), \forall p' \neq p]$
            $\mathbf{b}_g^{best}[t+1] \leftarrow \mathbf{b}_p[t]$
      else $\mathbf{b}_g^{best}[t+1] \leftarrow \mathbf{b}_g^{best}[t]$
   d. Evolve to a new swarm population $\mathbf{B}[t+1]$, using (9);
   e. Set $t = t + 1$.
   end
3. $\mathbf{p}^* = \mathbf{b}_g^{best}[\mathcal{G}]$.

end

$\mathcal{P}$: population size.
$\mathbf{B} = [\boldsymbol{b}_1, \ldots, \boldsymbol{b}_p, \ldots, \boldsymbol{b}_{\mathcal{P}}]$ particle population matrix, dimension $K \times \mathcal{P}$.
$\mathcal{G}$: maximum number of swarm iterations.
$\mathbf{p}_{max}$: maximum power vector considering each mobile terminal rate class.
$R_{\min}$: minimum data rate common to all users

The quality of solution achieved by any iterative resource allocation procedure could be measured by how close to the optimum solution is the found solution, and can be quantified by the normalized mean squared error (NMSE) when equilibrium is reached. For power allocation problem, the NSE definition is given by,

$$NMSE[t] = \mathrm{E}\left[\frac{\|\mathbf{p}[t] - \mathbf{p}^*\|^2}{\|\mathbf{p}^*\|^2}\right] \tag{12}$$

where $\| \bullet \|^2$ denotes the squared Euclidean distance to the origin, and $\mathrm{E}[\bullet]$ the expectation operator.

3.3.3. Energy efficiency optimization in OCPs

Recent studies have showed the importance of the consideration of energy consumption in optical communications design [39], considering the transmission infrastructure (transmitters, receivers, fibers and amplifiers) [40] and network infrastructure (switchers and routers) [41] aspects. Researches in a global scale network have indicated that the energy consumption of the switching infrastructure is larger than the energy consumption of the transport infrastructure [39-41]. In this context, it is necessary to improving the energy efficiency of switching and optimizing the network design in order to reduce the quantity of switching and overheads. The energy necessary for 1 bit transmission on each OCP can be expressed as [40],

$$E_i = p_i T_{bit} \qquad [J/bit], i = 1,..,K \tag{13}$$

where T_{bit} =1/R is the time to transmit one bit over the network, with R is the bit rate. In our analysis, to determinate the energy is necessary define the individual OCPs transmitted power (p_i). The p_i is obtained by power control PSO algorithm given in Algorithm 1 and may be associated to a specific QoS, SNIR and maximum BER tolerated by the *i*-th optical node. In a power control situation, each optical node adjusts its transmitter power in an attempt to maximize the number of transmitted bits with minimum consumption of energy. This concept is formulated by the energy efficiency [42]:

$$\eta_i = \frac{R \cdot g(\gamma_i)}{p_i}, \quad i = 1,..,K \tag{14}$$

where $g(\gamma_i)=1-BER_i$ is the efficiency function, which represents the number of correct packets received for the for the *i*-th node, given a SNIR γ_i. In the same way this concept is used in a metric called *utility* that is the number of bits received per energy expended or the relation of the throughput and power dissipation [41].

For each *i*-th OCP, the maximum number of transmitted bits occurs at power level for which the partial derivative of energy efficiency function in (14) with respect to p_i is zero $\partial\eta_i / \partial p_i$=0. Considering a SNIR general formula for CDMA networks, given by [6]

$$\gamma_i = \frac{h_{ii} p_i}{I_i + N_i}, i = 1,..,K \tag{15}$$

where h_{ij} are the total loss in the path that connects *i*-th transmitter node to *j*-th receiver node, I_1 is the interference from the others transmitters nodes and N_i is the receiver noise. We can obtain the derivative of energy efficiency referring to efficiency function and (15),

http://dx.doi.org/10.5772/52225

$$\frac{\partial \eta_i}{\partial p_i} = \frac{R}{p_i^2}\left(\gamma_i \frac{\partial g(\gamma_i)}{\partial \gamma_i} - g(\gamma_i)\right), i = 1,..,K \tag{16}$$

From (16) we observe, for $p_i > 0$, the necessary condition to maximize the energy efficiency is

$$\gamma_i \frac{\partial g(\gamma_i)}{\partial \gamma_i} - g(\gamma_i) = 0, i = 1,..,K \tag{17}$$

To satisfy (17) it is necessary that the received node achieves the target SNIR, namelyγ_i^*. In this context, we propose the utilization of PSO power allocation algorithm in order to establish the lower energy per bit according to the OCDM/WDM network QoS requirements.

4. Numerical results

For all simulations, it is considered the transmission over a nonzero-dispersion shifted fiber (NFD)-ITU G.655 with fiber attenuation (α) of 0.2 dB/km, non-linear parameter (Γ) of 2 $(W.km)^{-1}$, zero-dispersion wavelength (λ_0) of 1550 nm, dispersion slope (S_0) of 0.07 ps/$(nm^2.km)$. The signal is placed at λ_0 and its peak power is P. Note that the nonlinear length [24] $L_{NL}= 1/(\Gamma P)$ is limited to 500 km, which is much longer than the considered fiber lengths; besides self-phase modulation (SPM) should not seriously affect the system performance. Furthermore, the threshold power for stimulated Brillouin scattering (SBS) is below a few mW; as a result, SBS should also not interfere in our results. Similarly, for these considerations, the physical impairments, such as stimulated Raman scattering (SRS) should not be relevant [24]. Typical parameter values for the noise power in all optical amplifiers were assumed [21]. So, it was adopted n_{SP}= 2, h=6.63 × 10–34 (J/Hz), f=193.1 (THz), G=20 (dB) and Bo=30 (GHz). Herein, it was considered an amplifier gain of 20 dB with a minimum spacing of 60 km, D_{PMD}=$0.1 ps/\sqrt{km}$, and D=15 ps/nm/km. Losses for encoder/decoder and router architecture of 5 dB and 20 dB, respectively, were included in the power losses model [22-24]. The parameters are code weight of 4 and code length of 101, thus the code is characterized by $(4 \times 101, 4, 1,0)$ and the target SNIR γ_i^*=20dB was adopted.

4.1. PSO parameters optimization for resource allocation problem

For power resource allocation problem, simulation experiments were carried out in order to determine the suitable values for the PSO input parameters, such as acceleration coefficients, C_1 and C_2, maximal velocity factor, V_{max}, weight inertia, ω, and population size, P, regarding the power optimization problem.

The continuous optimization for resource allocation problem was investigated in [33], [34], it indicates that after an enough number of iterations (G) for convergence, the maximization of cost function were obtained within low values for both acceleration coefficients. The V_{max} factor is then optimized. The diversity increases as the particle velocity crosses the limits established by $[\pm V_{max}]$. The range of V_{max} determines the maximum change one particle can take during iteration. With no influence of inertial weight (ω=1), it was obtained that the maximum allowed velocity V_{max} is best set around 10 to 20% of the dynamic range of each particle dimension [33]. The appropriate choose of V_{max} avoids particles flying out of meaningful solution space. Herein, for OCP power allocation problem, similar to the problem solved in [33], the better performance *versus* complexity trade-off was obtained setting the maximal velocity factor value as V_{max}=0.2 $(P_{max} - P_{min})$. For the inertial weight, ω, simulation results has confirmed that high values imply in fast convergence, but this means a lack of search diversity, and the algorithm can easily be trapped in some local optimum, whereas a small value for ω results in a slow convergence due to excessive changes around a very small search space. In this work, it was adopted a variable ω, as described in (8), but with m = 1, and initial and final weight inertia setting up to $\omega_{initial}$ = 1 and ω_{final} = 0.01. Hence, the initial and final maximal velocity excursion values were bounded through the initial and final linear inertia weight multiplied by V_{max}, adopted as a percentage of the maximal and minimal power difference values [33],

$$\omega_{initial} \bullet V_{max} = 0.2\,(P_{max} - P_{min}) \quad \omega_{final} \bullet V_{max} = 0.002\,(P_{max} - P_{min}) \tag{18}$$

Finally, stopping criterion can be the maximum number of iterations G (velocity changes allowed for each particle) combined with the minimum error threshold:

$$\left| \frac{J[t] - J[t-1]}{J[t]} \right| < \epsilon_{stop} \tag{19}$$

where typically $\epsilon_{stop} \in [0.001; 0.01]$. Alternately, the convergence test can be evaluated through the computation of the average percent of success, taken over T runs to achieve the global optimum, and considering a fixed number of iterations G. A convergence test is considered 100% successful if the following relation holds:

$$\left| J[\mathsf{G}] - J[\mathbf{p}^*] \right| < {}_1 J[\mathbf{p}^*] + {}_2 \tag{20}$$

where, $J[\mathbf{p}^*]$ is the global optimum of the objective function under consideration, $J[\mathsf{G}]$ is the optimum of the objective function obtained by the algorithm after G iterations, and ϵ_1, ϵ_2 are accuracy coefficients, usually in the range $[10^{-6}; 10^{-2}]$. In this study it was assumed that T = 100 trials and $\epsilon_1 = \epsilon_2 = 10^{-2}$.

The parameter ρ in cost function (9), was set as a function of the number of users OCPs (K), such that $\rho = K \times 10^{-19}$. This relation was adapted from [38] for the power-rate allocation

http://dx.doi.org/10.5772/52225

problem through non-exhaustive search [33]. The swarm population size was set by $P = K + 2$.

In power resource allocation problem for access network systems the parameters optimization, mainly acceleration coefficients C_1 and C_2, depend on the number of simultaneous transmitted users [16], [33]. In the case of realistic OCDM/WDM routed networks, the number of simultaneous transmitted OCPs is low, generally around or less than 10 [2], [5], [6], [25]. Fast convergence without losing certain exploration and exploitation capabilities could be obtained with optimization of acceleration parameters in relation to the standard values adopted in the literature [16]. Previous works have shown that the best convergence *versus* solution quality trade-off was achieved with $C_1 = 1$ and $C_2 = 2$ for number of codes less than 10 [16], [33]. On the other hand, the classical value adopted are $C_1 = C_2 = 2$ [32-35]. In this context, simulation experiments were carried out in order to determine the good choice for acceleration coefficients C_1 and C_2 regarding the power optimization problem. Fig. 4 illustrates different solution qualities in terms of the normalized mean squared error (NMSE), when different values for C_1 combining with $C_2 = 2$ in a system with number of OCPs equal to 7, considering 1 span. Previous simulations have shown the non poor convergence for different value of C_2 [33].The lengths of OCPs are uniformly distributed between 2 and 100 km. The NMSE values where taken as the average over T = 100 trials. Besides, the NMSE convergence values were taken after G = 800 iterations.

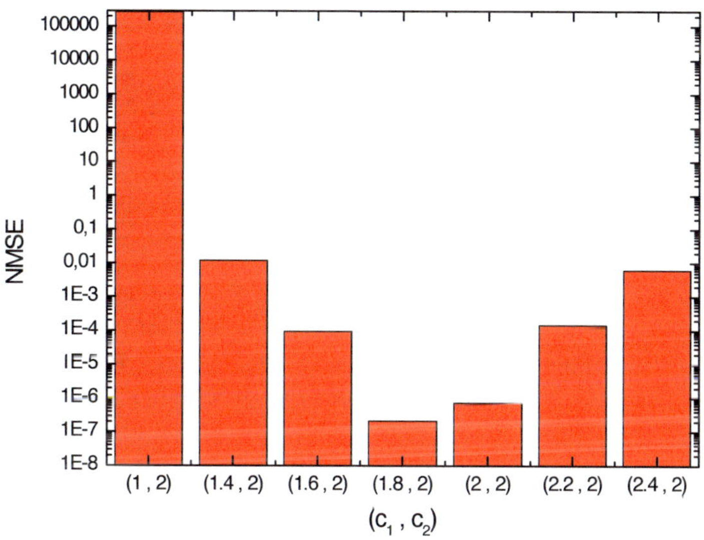

Figure 4. Normalized mean squared error (NMSE) for different values for C_1 combining with $C_2 = 2$ in a system with number of OCPs equal to 7, considering 1 span.

Numerical results have shown the solution quality for different values of acceleration coefficient C_1 and it was found $C_1 = 1.8$ presents the lower NMSE for the number of OCPs < 10. Hence, the best solution quality was achieved with $C_1 = 1.8$ and $C_2 = 2$. Fig. 5 shows the sum of power for the evolution through the t=1,…, 800 iterations for 7 OCPs under different acceleration value of C_1 and $C_2 = 2$, considering 1 span.

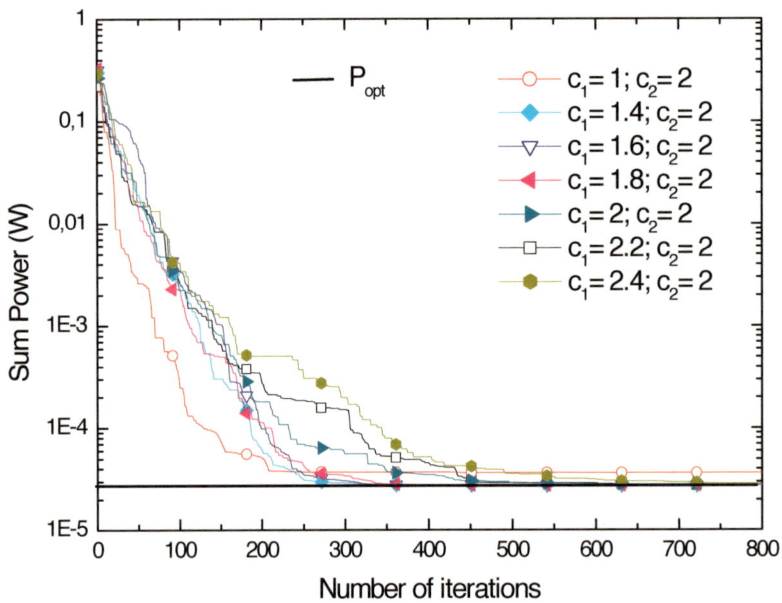

Figure 5. Sum of power for power vector evolution through the 800 iterations for 7 OCPs under different acceleration value of C_1 and $C_2 = 2$ for 1 spans.

The algorithm reaches convergence for $C_1 = 1.4$, 1.6, 1.8, 2, 2.2 and 2.4, however it doesn't reach acceptable convergence for $C_1 = 1$. Simulations revealed that increasing parameter C_1 results in a slower convergence with approximately 320, 343, 373, 639, 659 and 713, iterations, respectively. However, NMSE for faster convergence is higher than for the slower convergence parameters with minimum value for $C_1 = 1.8$, as indicated in Fig. 4. In this context, the best convergence *versus* solution quality trade-off was achieved with $C_1 = 1.8$ and $C_2 = 2$ for number of OCPs of 7.

It is worth to expand this analysis to other number of OCPs that are generally between 4 and 8 OCPs. For this purpose, Fig 6 shows the NMSE for the number of OCPs regarding two combination of acceleration coefficient: i) optimized herein ($C_1 = 1.8$ and $C_2 = 2.0$) and reported in the literature ($C_1 = 2.0$ and $C_2 = 2.0$).

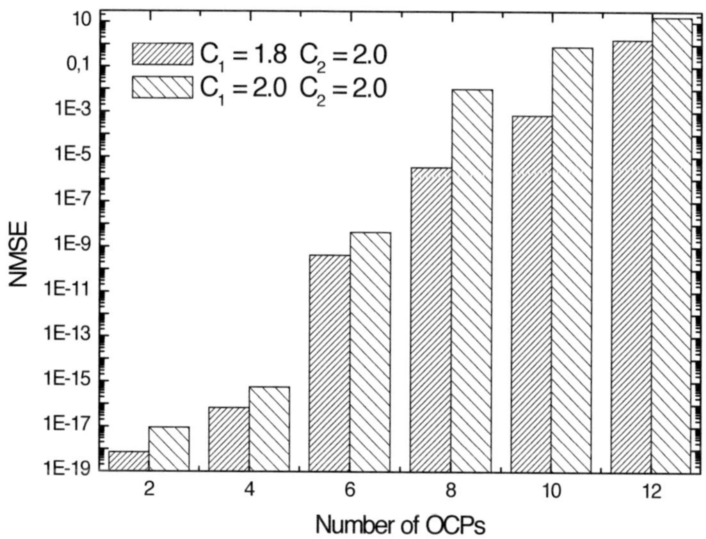

Figure 6. Normalized mean squared error (NMSE) for the number of OCPs for (C_1 = 1.8 and C_2 = 2.0) and (C_1 = 2.0 and C_2 = 2.0), considering 1 span.

The OCPs increasing affects the solution quality. This effect is directly related to the MAI rising which increases with the number of OCPs, i.e., the MAI effects are strongly influenced by the increase of the active OCPs; an error occurs when cross-correlational pulses from the ($K - 1$) interfering optical code paths built up to a level higher than the autocorrelation peak, changing a bit zero to a bit one.

In conclusion, our numerical results for the power minimization problem have revealed for low system loading that the best acceleration coefficient values lie on C_1 = 1.8 and C_2 = 2.0, in terms solution quality trade-off. This result was compared with C_1 = 2.0 and C_2 = 2.0 previously reported in the literature [32]-[35].

4.2. PSO optimization for OCPs

The solution quality *versus* convergence trade-off analysis presented in Figs. 4, 5 and 6 for the PSO's acceleration coefficients optimization in the case of OCDM/WDM networks with 1 span should be extended taking into account the use of more spans. The state-of-art for the number of spans without electronic regeneration is around 4 considering the ASE effect limiting applying fibers with low PMD effects for bit rate of 10 Gbps (lower than 40 Gbps) [29]. In Fig. 7 the analysis of subsection 4.1 is extended until 8 spans, showing the influence of the number of spans on the NMSE for 7 OCPs and the same parameters values optimized in that subsection.

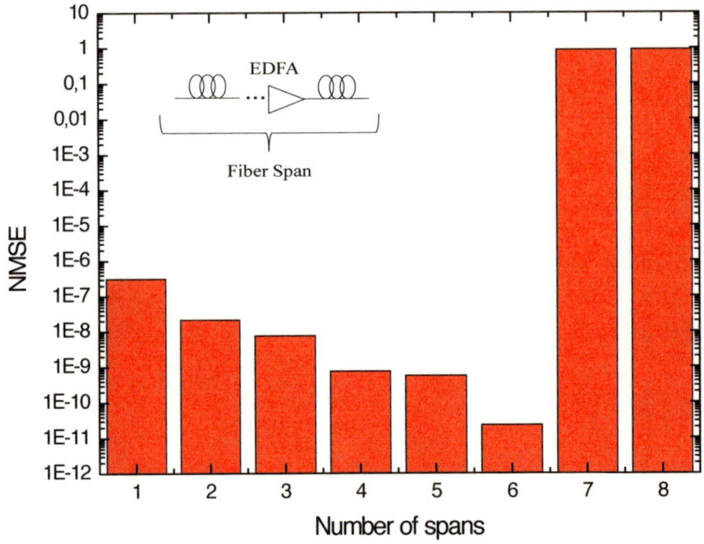

Figure 7. NMSE for the number of spans for 7 OCPs.

The results show that NMSE decreases when the number of spans increases until 6 spans, after this number of spans the NMSE alters the tendency and increases. This behavior shows the limitation of PSO convergence when the ASE increases. After 6 spans the PSO algorithm does not reach the total convergence. This fact occurs, directly by the limitations generate with the increase of the ASE. In other sense, the transmitted power needed to reach the target SNIR will overcome the maximum allowed transmitted power. The average number of spans increases slightly as K increases, as longer OCP routes become available. This increase is, however, not very significant, and on average, the path lengths are around four spans [22].

The convergence quality of the PSO algorithm presents variation with the increase in the number of spans. The figure of merit utilized as tool to this analysis is the rate of convergence (RC), which can be described as the ratio of PSO solution after the t-th iteration divided by the PSO solution after total convergence, which in this optimization context is given by the matrix inversion solution, as discussed in Section 3.1. Recalling eq. (19), the RC can be expressed in term of ϵ_{stop} as:

$$RC[t] = 1 - \left| \frac{J[t] - J[\mathbf{p}^*]}{J[\mathbf{p}^*]} \right| \qquad (21)$$

The reader interested in quality of solution metrics, a similar definition for RC and another figure of merit for the PSO, namely success cost, are presented in [35].

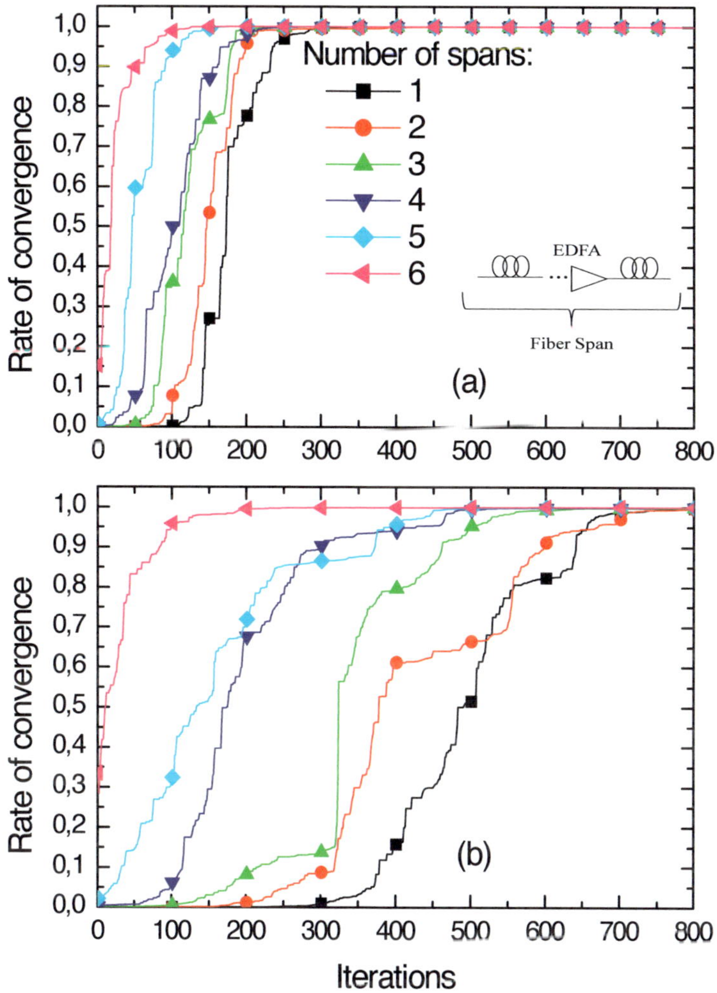

Figure 8. Rate of convergence versus the number of iterations for 1 until 6 spans for (a) 4 OCPs and (b) 8 OCPs.

Fig. 8 (a) and (b) shows the convergence rate of the sum of power for vector evolution through the 800 iterations for 4 a 8 OCPs, respectively, considering 1 until 6 spans. The results have shown that increasing span number results in a faster convergence. This fact occurs because until 6 spans the increase of the number of span increases the contribution of the amplifier with signal, besides for more than 6 spans the contribution of the amplifier is for the ASE noise. On the other hand, the increase in the number of OCPs results in a slow convergence that results from the MAI between the OCPs.

In summary, our numerical results for the power minimization problem considering different number of spans have revealed the viability of the PSO algorithm deployment to solve a power allocation in OCPs with until 6 spans in order to guarantee the solution quality in terms of NMSE. Furthermore, the numerical results revealed that increasing the number of spans results in a faster convergence. In this context, the PSO algorithm is quite suitable to solve a power allocation in OCPs that presents an average of 4 spans as reported in the literature [22].

In order to evaluate the impact of physical restrictions on the OCM/WDM network, further numerical results presented in Fig. 9 shows the sum power evolution of the PSO algorithm with respect to the number of iterations, considering a) 4 OCPs, and b) 8 OCPs. One span was considered as reference for bit rate of 10 Gbps taking into account two situations: i) with only ASE effects, ii) with ASE, GVD and PMD effects.

The target SNIR established for all the nodes is equal, and if the perfect power balancing with ideal physical layer (no physical impairments) is assumed, it could be demonstrated that the maximum SNIR and the transmitted power are defined by the number of OCPs in the same route. However, when the ASE, GVD and PMD effects are considered, there is a penalty. This penalty represents the received power reduction due to temporal spreading. Fig. 9 shows that when ASE, GVD and PMD effects are considered there is a power penalty of 3 decades compared with the situation where only ASE is considered (fibers with low PMD). Comparing Figs. 9(a) and 9(b), it could be noticed that the convergence velocity depends on the number of OCPs. The increase of OCPs from 4 to 8 affects the convergence velocity, from ≈200 to ≈500 iterations, respectively. This effect is directly related to the MAI increasing, which increases with the number of OCPs. The MAI effects are strongly influenced by the increase of the active number of OCPs; as explained before, an error occurs when cross-correlational pulses from the $(K-1)$ interfering optical code paths built up to a level higher than the autocorrelation peak, changing a bit zero to a bit one. The PMD effects degrade the performance when the link length and bit rate increase. This effect occurs because PMD impairment establishes an upper bound on the link length, which causes the temporal spreading of optical pulses. The upper bound for link distance depends on the chip-rate distance product $(d.R.N_T)$, where d is the link length, R is the bit rate, and N_T is the code length. The analysis of code parameters, MAI and PMD effects for 2D-based OCPs was previously reported in [20].

In OCDM/WDM networks, the OCPs with various classes of QoS are obtained with transmission of different power levels. Distinct power levels are obtained with adjustable transmitters and it does not cause the change of the bit rate. The intensity of the transmitted optical signal is directly adjusted from the laser source with respect to the target SNIR by PSO algorithm. Table 1 shows the optimization aspects of QoS regarding different levels of SNIR considering sum power and NMSE for 4 and 8 OCPs with 1 span.

http://dx.doi.org/10.5772/52225

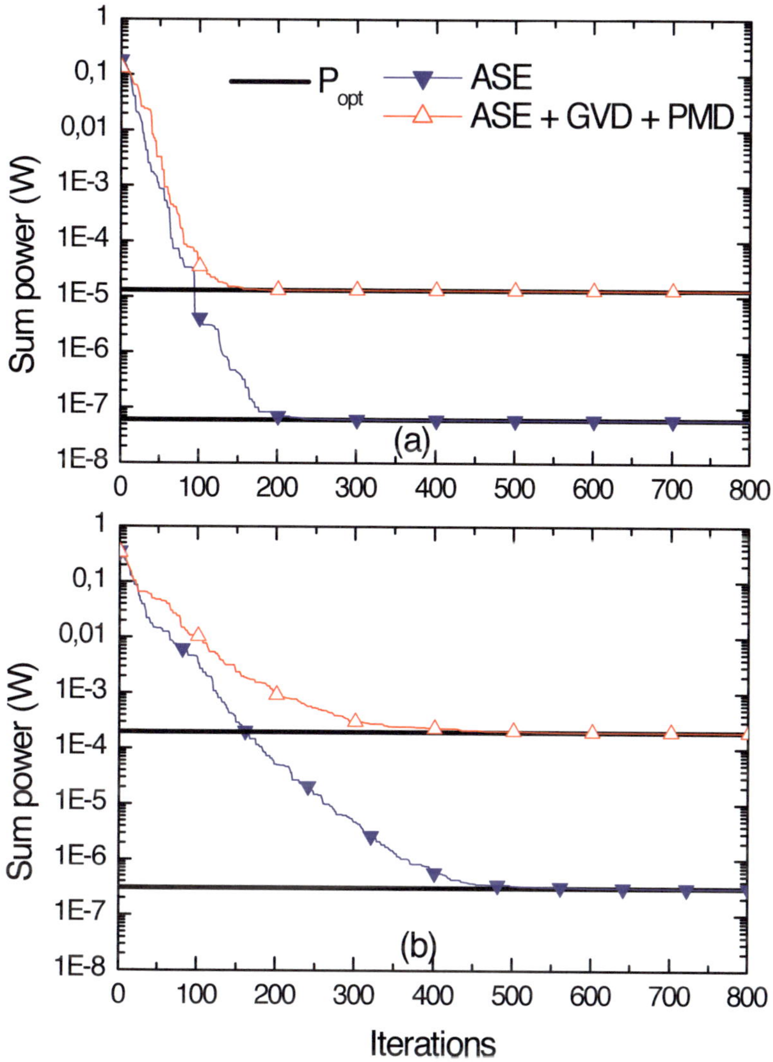

Figure 9. PSO sum power evolution for a) 4 OCPs; b) 8 OCPs. One span as reference, with R=10 Gbps. Two situations: i) ASE effects, ii) ASE, GVD and PMD effects.

The results in Table I show the necessary values for transmitted power, as well as the solution quality evaluation in terms of NMSE. The increase in the target SNIR results in the increase of the transmitted power, which is major for more OCPs. On the other hand, the solution quality (NMSE) decreases with the increase of SNIR target, since the number of the PSO iterations is fixed.

		4 OCPs		8 OCPs	
SNIR (dB)	BER	Sum power (W)	NMSE	Sum power (W)	NMSE
17	7.2×10^{-13}	3.3×10^{-8}	3.0×10^{-18}	1.2×10^{-7}	2.3×10^{-8}
20	7.6×10^{-24}	6.0×10^{-8}	6.2×10^{-16}	2.8×10^{-7}	1.2×10^{-3}
22	1.2×10^{-36}	9.5×10^{-8}	3.8×10^{-16}	4.3×10^{-7}	1.0×10^{-1}

Table 1. The optimization aspects of QoS.

4.3. PSO optimization for energy efficiency in OCPs

An efficient resource allocation algorithm is needed to overcome the problem of energy efficiency and to enhance the performance and QoS of the optical network. This could be achieved via signal-to-noise plus interference (SNIR) PSO optimization.

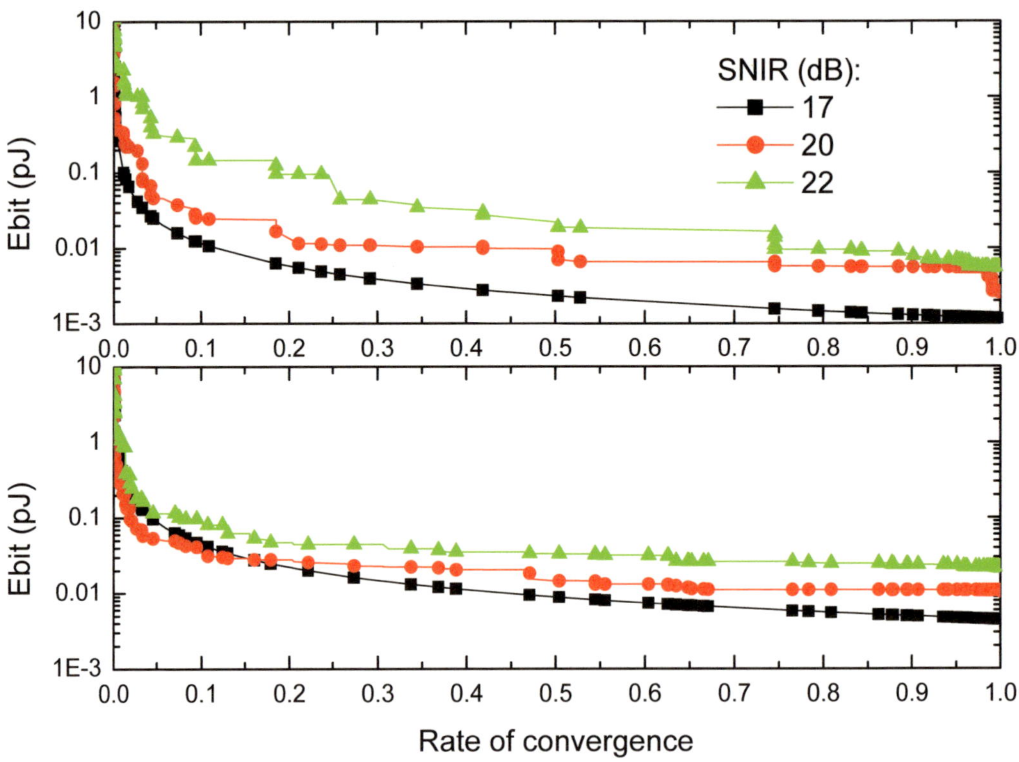

Figure 10. Energy per bit sum of the OCPs as a function of the rate of convergence using PSO algorithm. Three different SNIRs target of 17, 20 and 22 dB; a) 4 OCPs and b) 8 OCPs.

Fig. 10 shows the sum of energy per bit as a function of the rate of convergence of eq. (21) for the PSO optimization with different QoS requirements represented by SNIR target of 17, 20 and 22 dB, considering a) 4 OCPs and b) 8 OCPs, i.e., same scenario presented in the previous subsection. One can see when rate of convergence evolving, the energy per bit solution offered by the PSO algorithm convergences to the best lower values as predicted in (17).

It can be seen from Fig. 10 the impact of the PSO power allocation optimization procedure (in terms of transmitted energy per bit) on the energy efficiency improvement. The deployment of PSO with 100% of rate of convergence results in an enormous saving of energy. Indeed, with very low number of PSO iterations, rate of convergence is poor ($RC < 0.03$), the transmitted energy per bit is high because the MAI are strongly influenced by near-far effects. As expected, the increase of the active OCPs from 4 to 8, results in the increase of the transmitted energy per bit to reach the SNIR target. Furthermore, one can analyze the variation of saving energy for different levels of convergence rate; for instance, the variation of saving energy regarding the rate convergence in the range $RC \in [0.5; 1.0]$ remains in approximately from 40 to 60 % for different SNIR target and number of OCPs, as presented in Fig. 10. In this context, aiming to analyze the effect of the number of spans in the transmitted energy per bit for 4 and 8 OCPs, Table 2 presents the sum energy per bit considering SNIR target of 20 dB and rate of convergence of 0.5 and 1.0, for 2 and 4 spans.

	4 OCPs		8 OCPs	
Number of spans	Σ energy (pJ) RC=0.5	Σ energy (pJ) RC=1.0	Σ energy (pJ) RC=0.5	Σ energy (pJ) RC=1.0
2	0.0135	0.0120	0.3145	0.1100
4	1.0545	1.0108	2.2295	1.8199

Table 2. Sum energy per bit in [pJ] for SNIR target of 20 dB.

The results show the impact of the number of spans in the transmitted energy per bit for the variation of rate convergence of 0.5 and 1. As expected, the increase in the number of spans and the number of OCPs results in the increase of the transmitted energy per bit. Besides, the sum energy per bit variation, regarding the RC from 0.5 to 1.0, declines with the increase of the number of spans from 2 to 4. This results agree with the previous results illustrated in Fig. 8, meaning the increase of the number of spans accelerate the RC.

5. Conclusions

In this chapter, optimization procedures based on particle swarm intelligence are investigated in details, aiming to efficiently solve the optimal resource allocation for signal-to-noise plus interference ratio (SNIR) optimization of optical code paths (OCPs) from OCDM/WDM networks under quality of service (QoS) restrictions and energy efficiency constraint problem, considering imperfections on physical constraints. The SNIR model considers multiple access interference (MAI) between the OCP based on 2-D codes (time/wavelength), amplifier spontaneous emission (ASE) at cascaded amplified spans, and group velocity dispersion (GVD) and polarization mode dispersion (PMD) dispersion effects. The characteristic of the particle swarm optimization (PSO) is attractive due their performance-complexity tradeoff and fairness regarding the optimization methods that use numerical methods, matrix inversion and other heuristics. The resource allocation optimization based on PSO strategy allows the regulation of the transmitted power and the number of active OCPs in order to maximize the aggregate throughput of the OCDM/WDM networks considering QoS and energy efficiency constraint. For the network optimization context, system model was described, figures of merit were presented and a suitable model of PSO was developed, with emphasis in the optimization of input parameters and network performance. Afterward, extensive numerical results for the optimization problem are discussed taking into account realistic networks operation scenarios.

In order determine the suitable values for the PSO input parameters, such as acceleration coefficients, C_1 and C_2, maximal velocity factor, V_{max}, weight inertia, ω, and population size, P, simulation experiments were carried out in regarding the power optimization problem for OCDM/WDM networks. In these networks, the number of simultaneous transmitted OCPs is low, generally around or less than 10. For our specific problem, the optimized input parameters are different from the reported in the literature for similar problems. The numerical results considering the number of spans have revealed the viability of the PSO algorithm deployment in order to solve a power allocation in OCPs with until 6 spans to guarantee the solution quality and convergence. This result is adequate considering the average of 4 spans without electronic regeneration presented for this kind of network. Besides, the numerical results have shown a penalty when the ASE, GVD and PMD effects are considered. This penalty represents the received power reduction due to temporal spreading. Indeed, when ASE, GVD and PMD effects are considered there is a power penalty of 3 decades compared with the situation where only ASE is considered (fibers with low PMD). Finally, our numerical results reveal considerable variation of transmitted energy for different levels of convergence rate of PSO algorithm, in which the maximum energy efficiency is reached when the convergence of PSO algorithm is total. Interesting, even with only 10%-20% of the total PSO convergence, the network is able to operate within a remarkable energy efficiency gain region compared to network operation without power allocation policy.

Author details

Fábio Renan Durand[1], Larissa Melo[1], Lucas Ricken Garcia[1], Alysson José dos Santos[2] and Taufik Abrão[2]

1 Federal University of Technology, Parana, Campo Mourão, Brazil

2 State University of Londrina – Electrical Engineering Department, Brazil

References

[1] E. Wong, "Next-Generation Broadband Access Networks and Technologies," Journal of Lightwave Technology, vol. 30, no. 4, pp. 597 – 608, Feb., 2012

[2] H. Beyranvand and J. Salehi, "All-optical multiservice path switching in optical code switched GMPLS core network", Journal of Lightwave Technology, vol. 27, no. 17, pp. 2001 – 2012, Jun. 2009.

[3] H. Yin and D. J. Richardson, Optical code division multiple access communication networks: theory and applications. Berlin: Springer-Verlag and Tsinghua University Press, 2009.

[4] A. Rahbar, "Review of Dynamic Impairment-Aware Routing and Wavelength Assignment Techniques in All-Optical Wavelength-Routed Networks", IEEE Communications Surveys & Tutorials, ACCEPTED FOR PUBLICATION

[5] F. R. Durand, M. Lima and E. Moschim, "Impact of pmd on hybrid wdm/ocdm networks," IEEE Photonics Technology Letters, vol. 17, no. 12, pp. 2787–2789, December 2005.

[6] F. R. Durand and T. Abrão, "Distributed SNIR Optimization Based on the Verhulst Model in Optical Code Path Routed Networks With Physical Constraints", Journal of Optical Communications and Networking, vol. 3, no. 9, pp. 683–691, Sep. 2011. doi: 10.1364/JOCN.3.000683

[7] F. R. Durand, M. S. Filho and T. Abrão, "The effects of power control on the optical CDMA random access protocol", Optical Switching and Networking, (In press) doi: 10.1016/j.osn.2011.06.002

[8] N. Tarhuni, T. Korhonen, M. Elmusrati and E. Mutafungwa, "Power Control of Optical CDMA Star Networks", Optics Communications, vol. 259, pp. 655 – 664, Mar. 2006.

[9] E. Inaty, R. Raad, P. Fortier, and H. M. H. Shalaby, "A Fair QoS-Based Resource Allocation Scheme For a Time-Slotted Optical OV-CDMA Packet Networks: a Unified Approach," Journal of Lightwave Technology, vol. 26, no. 21, pp. 1-10, Jan. 2009.

[10] E. Inaty, H. Shalaby, P. Fortie, and L. Rusch, "Optical Fast Frequency Hopping CDMA System Using Power Control", Journal of Lightwave Tech., vol. 20, n. 2, pp. 166 – 177, March 2003.

[11] C. C. Yang, J. F. Huang, and T. C. Hsu, "Differentiated service provision in optical CDMA network using power control," IEEE Photon. Technol. Lett., vol. 20, no. 20, pp. 1664–1666, 2008.

[12] S. Khaleghi and Mohammad Reza Pakravan, Quality of Service Provisioning in Optical CDMA Packet Networks, Journal of Optical Communications and Networking, vol. 2, no. 5, pp. 283–292, Feb. 2010.

[13] H. Yashima, and T. Kobayashi "Optical CDMA with time hopping and power control for multirate networks," J. Lightwave Technol., vol. 21, pp. 695-702, March 2003.

[14] T. Miyazawa and I. Sasase "Multi-rate and multi-quality transmission scheme using adaptive overlapping pulse-position modulator and power controller in optical network," IEEE ICON, vol. 1, pp. 127-131, November 2004.

[15] R. Raad, E. Inaty, P. Fortier, and H. M. H. Shalaby, "Optimal resource allocation scheme in a multirate overlapped optical CDMA system," J. of Lightwave Technol., vol. 25, no. 8, pp. 2044 – 2053, August 2007.

[16] M. Tang, C. Long and X. Guan, "Nonconvex Optimization for Power Control in Wireless CDMA Networks," Wireless Personal Communications, vol. 58, n. 4, pp. 851-865, 2011.

[17] Q. Zhu. and L. Pavel, "Enabling Differentiated Services Using Generalized Power Control Model in Optical Networks", IEEE Transactions on Communications, vol. 57, no 9, p. 1 – 6, Sept. 2009.

[18] R. Ramaswami, K. Sivarajan and G. Sasaki, Optical Networks: A Practical Perspective, Morgan Kaufmann, Boston, MA, 2009.

[19] E. Mutafungwa, "Comparative analysis of the traffic performance of fiber-impairment limited WDM and hybrid OCDM/WDM networks", Photon Network Commun., vol. 13, pp.53–66, Jan. 2007.

[20] F R. Durand, L. Galdino, L. H. Bonani, F. R. Barbosa1, M. L. F. Abbade and Edson Moschim, "The Effects of Polarization Mode Dispersion on 2D Wavelength-Hopping Time Spreading Code Routed Networks", Photonics Network Communications, vol. 20, no. 1, pp. 27 – 32, Aug. 2010. DOI 10.1007/s11107-010-0242-6.

[21] F. R. Durand, M. L. F. Abbade, F. R. Barbosa, and E. Moschim, "Design of multi-rate optical code paths considering polarisation mode dispersion limitations," IET Communications, vol. 4, no. 2, pp. 234–239, Jan. 2010.

[22] Camille-Sophie Brès and Paul R. Prucnal, "Code-Empowered Lightwave Networks", J. Lightw. Technol. , vol. 25, n. 10, pp. 2911 – 2921, Oct. 2007.

[23] Yue-Kai Huang, Varghese Baby, Ivan Glesk, Camille-Sophie Bres, Christoph M. Greiner, Dmitri Iazikov, Thomas W. Mossberg, and Paul R. Prucnal, Fellow, "Novel Multicode-Processing Platform for Wavelength-Hopping Time-Spreading Optical CDMA: A Path to Device Miniaturization and Enhanced Network Functionality", IEEE Journal of Selected Topics in Quantum Electronics, vol. 13, no. 5, pp. 1471 – 1479, september/october 2007.

[24] G. P. Agrawal, Fiber-optic communication systems, John Wiley & Sons, 2002.

[25] K. Kitayama and M. Murata, "Versatile Optical Code-Based MPLS for Circuit, Burst and Packet Switching", J. Lightwave Technol, vol. 21, no. 11, pp. 2573 – 2764, Nov. 2003.

[26] S. Huang, K. Baba, M. Murata and K. Kitayama, "Variable-bandwidth optical paths: comparison between optical code-labeled path and OCDM path", J. Lightwave Technol., vol. 24, no. 10, pp. 3563 – 3573, Oct. 2006.

[27] Kerim Fouli e Martin Maier, "OCDMA and Optical Coding: Principles, Applications, and Challenges", IEEE Communications Magazine, vol. 45, no. 8, pp. 27 – 34, Aug. 2007.

[28] G.-C. Yang and W.C. Kwong, Prime codes with applications to CDMA optical and wireless networks, Artech House, Boston, MA, 2002.

[29] G. Pavani, L. Zuliani, H. Waldman and M. Magalhães, "Distributed approaches for impairment-aware routing and wavelength assignment algorithms in GMPLS networks", Computer Networks, vol. 52, no. 10, pp. 1905–1915, July 2008.

[30] A. L. Sanches, J. V. dos Reis Jr. and B.-H. V. Borges, "Analysis of High-Speed Optical Wavelength/Time CDMA Networks Using Pulse-Position Modulation and Forward Error Correction Techniques", J. Lightwave Technol., vol. 27, no. 22, pp. 5134 – 5144, Nov. 2009.

[31] J. Kennedy and R.C. Eberhart, "Particle swarm optimization", in Proceedings of IEEE International Conference on Neural Networks, Piscataway, USA, pp. 1942–1948, 1995.

[32] N. Nedjah and L. Mourelle, Swarm Intelligent Systems, Springer, Springer-Verlag Berlin Heidelberg, 2006.

[33] T. Abrão, L. D. Sampaio, M. Proença Jr., B. A. Angélico and Paul Jean E. Jeszensky, Multiple Access Network Optimization Aspects via Swarm Search Algorithms, In: Nashat Mansour. (Org.). Search Algorithms and Applications. 1 ed. Vienna, Austria: InTech, ISBN 978-953-307-156-5, 2011, v. 1, p. 261-298.

[34] K. Zielinski, P. Weitkemper, R. Laur, and K. Kammeyer, "Optimization of Power Allocation for Interference Cancellation With Particle Swarm Optimization", IEEE Transactions on Evolutionary Computation, vol. 13, no. 1, pp. 128 – 150, Feb. 2009.

[35] N. Nedjah and L. M. Mourelle. Swarm Intelligent Systems, Springer, Springer-Verlag Berlin Heidelberg, 2006.

[36] A. Chatterjee and P. Siarry, Nonlinear inertia weight variation for dynamic adaptation in particle swarm optimization, Computers & Operations Research, vol 33, no. 3, pp. 859–871.

[37] M. Moustafa, I. Habib, and M. Naghshineh, Genetic algorithm for mobiles equilibrium, MILCOM 2000. 21st Century Military Communications Conference Proceedings 2000.

[38] H. Elkamchouchi, H., Elragal and M. Makar, Power control in cdma system using particle swarm optimization, 24th National Radio Science Conference, pp. 1–8. 2007.

[39] S. Yoo, "Energy Efficiency in the Future Internet: the Role of Optical Packet Switching and Optical Label Switching", IEEE J Selected Topics in Quantum Electronics, vol. 17, no. 2, pp. 406 – 418, March-April 2011.

[40] Rodney S. Tucker, "Green Optical Communications - Part I: Energy Limitations in Transport", IEEE J Selected Topics in Quantum Electronics, vol. 17, no. 2, pp. 245 – 260, March-April 2011.

[41] Rodney S. Tucker, "Green Optical Communications - Part II: Energy Limitations in Networks", IEEE J Selected Topics in Quantum Electronics, vol. 17, no. 2, pp. 261 – 274, March-April 2011.

[42] D. Goodman and Narayan Mandayan "Power control for wireless data", IEEE Personal Communications, vol. 7, no. 2, pp. 48 – 54, April 2000.

Section 3

Power Systems and Industrial Processes Applications

Chapter 7

An Adaptive Neuro-Fuzzy Strategy for a Wireless Coded Power Control in Doubly-Fed Induction Aerogenerators

I. R. S. Casella, A. J. Sguarezi Filho, C. E. Capovilla,
J. L. Azcue and E. Ruppert

Additional information is available at the end of the chapter

http://dx.doi.org/10.5772/52090

1. Introduction

Renewable energy systems and specially wind energy have attracted governmental interests in opposition to energy sources that increase CO_2 emissions and cause enormous environmental impact. Recently, the concept of smart grid has been applied to power plants to enable and optimize the generation of energy by efficiently combining wind, solar and tidal. Moreover, the efforts for consolidation and implementation of this new concept through wind energy systems have attracted great interest from the technical community and has been the focus of several recent scientific works [1–3].

Smart grids are an evolution of conventional power grids to optimally manage the relationship between energy supply and demand in electrical systems to overcome the actual problem of contingency of energy of the modern world. They employ an interactive framework composed by integrated communication networks with real time monitoring, control and automatic intervention capability to use more efficiently the infrastructure of generation, transmission and distribution of energy.

Advances in wind power technology have greatly improved its system integration with smart grid, however, there are still some unsolved challenges in expanding its use. Due to the usual variations of the wind speed, its utilization entails undesirable fluctuations in the generated power that, if not compensated in real time, can lead to frequency imbalance and disturbance in the stability of the electrical system. Although smart grid can minimize this problem through a more precise demand response for load control and dispatch of other generation resources, it is still necessary to employ variable speed aerogenerators and a precise power control system to guarantee stability and maximum power generation.

Among the existing aerogenerators, Doubly-Fed Induction Generators (DFIG) are the most widely employed in wind power systems [4] due to their interesting main characteristics as, for instance, the ability to operate at variable speed and the capacity to control the active and reactive power into four quadrants by means of field orientation [5, 6]. The precise power control of the aerogenerators is essential to maximize the generated power. The conventional Proportional-Integral (PI) [7] and Proportional-Integral-Derivative (PID) [8, 9] have been widely used as the core of different power control systems due to their simplicity of implementation. However, the design of these fixed gain control systems is very cumbersome, since it depends on the exact mathematical model of the generator. Also, they are very sensitive to disturbances, parameter variations and system nonlinearities.

On the other hand, the design of intelligent control systems based on Computational Intelligence (CI) does not require the exact mathematical model of the generator. Among of the CI techniques, Fuzzy Logic (FL) and Artificial Neural Networks (ANN) appear as powerful options for identification and control of nonlinear dynamic systems as power control systems. However, each of these intelligent techniques has its own drawbacks which restrict its use. For instance, FL suffers from some limitations as the appropriate selection of Membership Functions (MF), the adequate selection of fuzzy rules and, furthermore, how to adjust both of them to achieve the best performance. ANN have also some limitations as their black-box nature, the selection of the best structure and size, and the considerable training time to solve a specific problem. In order to overcome these problems, it is possible to use a hybrid neuro-fuzzy system [10] that can combine the learning capability of ANN with the knowledge representation of FL based on rules. One of the most widely used neuro-fuzzy system is the Adaptive Neural Fuzzy Inference System (ANFIS), proposed by [11]. It can use a hybrid learning procedure to construct an input-output mapping based on both human-knowledge as fuzzy rules and approximate MF from the corresponding input-output data pairs as ANN learning.

Another very important issue in the deployment of smart grids is the application of a modern telecommunications system to guarantee an effective monitoring and control of the grid. Nevertheless, its development and operability require a fairly complex infrastructure and present several non-trivial questions due to the convergence of different areas of knowledge and design aspects. In this way, the wireless transmission appears as an interesting solution for presenting many benefits such as low cost of development, expansion facilities, possibility of using the technologies currently applied in mobile telephone systems, flexibility of use, and distributed management. However, the employment of wireless technologies for transmitting power control signals may cause apprehension due to the possibility of the occurrence of errors in the transmission process that can cause serious problems to the generators and, consequentially, to the energy system. Such behavior is different from what usually happens in telecommunications systems designed to voice and data transmissions, where small errors can be detected, initiate requests for retransmission (generating delays) or even, in some cases, be ignored without any significant impact to the network. It is worth noting that there are some works in the scientific literature referencing the application of wireless technology for monitoring wind energy systems [12, 13], but there has not been presented any deep research about the use of wireless technology for control applications in

http://dx.doi.org/10.5772/52090

these systems yet, making it difficult to estimate the real impact of its use or its advantages and difficulties.

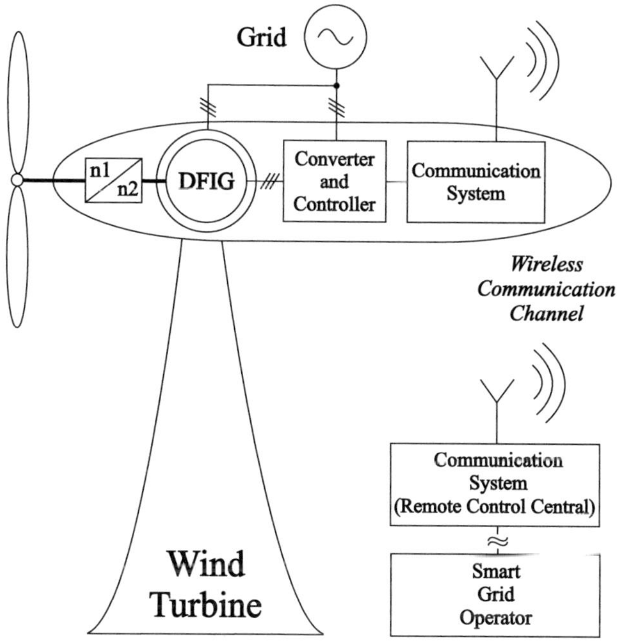

Figure 1. Wireless System Control Schematic.

Modern wireless digital communication techniques can be used to improve the robustness of the power control system and to minimize the mentioned problems through the application of Forward Error Correction (FEC) [14]. FEC is a coding technique used in all current digital wireless systems and it is essential to ensure the integrity of information, reducing significantly the Bit Error Rate (BER) and the latency of the information by adding controlled redundancy to the transmitted information [15]. In theory, the appropriate use of coding technology can offer the same reliability obtained by using fiber optic cables [16].

There are currently several different schemes of FEC that are used in commercial wireless communication systems [17], [18], [19–21]. Among them, LDPC is the one that presents the best performance and shows an excellent compromise between decoding complexity and performance [22, 23].The LDPC coding has recently been added to the IEEE 802.16e Standard, commonly known as Worldwide Interoperability for Microwave Access (WiMAX) for mobile applications [24].

In this context, this chapter will analyze the performance of a new wireless coded adaptive neuro-fuzzy power control system for variable speed wind DFIG, presented in Fig. 1. The system is based on a discrete dynamic mathematical model of the generator and uses the vector control technique to independently control the active and reactive power. The proposed adaptive neuro-fuzzy system is designed from an input and output data set collected from a DFIG with a deadbeat controller [25] operating at different conditions and considering the rotor current error as input. The wireless communication system, employed to send the power reference signals to the DFIG controller, uses LDPC coding to reduce the transmission errors and the overall latency of the system. The performance of the system is investigated in a frequency flat fading scenario, to evaluate the real impact of the wireless transmission in the wind energy control system. It is noteworthy that the errors generated in the wireless transmission cannot be easily removed without using advanced FEC coding techniques similar to those presented in this work.

2. Doubly-Fed Induction Generator model

The DFIG model in synchronous dq reference frame can be mathematically represented by [26]:

$$\vec{v}_{1dq} = R_1\vec{i}_{1dq} + \frac{d\vec{\lambda}_{1dq}}{dt} + j\omega_1\vec{\lambda}_{1dq} \tag{1}$$

$$\vec{v}_{2dq} = R_2\vec{i}_{2dq} + \frac{d\vec{\lambda}_{2dq}}{dt} + j\left(\omega_1 - N_P\omega_{mec}\right)\vec{\lambda}_{2dq} \tag{2}$$

and the relationships between fluxes and currents are:

$$\vec{\lambda}_{1dq} = L_1\vec{i}_{1dq} + L_M\vec{i}_{2dq} \tag{3}$$

$$\vec{\lambda}_{2dq} = L_M\vec{i}_{1dq} + L_2\vec{i}_{2dq} \tag{4}$$

so, the active and reactive power are given by:

$$P = \frac{3}{2}\left(v_{1d}i_{1d} + v_{1q}i_{1q}\right) \tag{5}$$

$$Q = \frac{3}{2}\left(v_{1q}i_{1d} - v_{1d}i_{1q}\right) \tag{6}$$

The subscripts 1 and 2 represent the stator and rotor parameters, respectively; ω_1 represents the synchronous speed, ω_{mec} represents machine speed, R represents winding per phase electrical resistance, L and L_M represent the proper and the mutual inductances of windings, $\vec{v}$ represents voltage vector, and N_P represents the machine number of pair of poles.

The DFIG power control aims independent stator active P and reactive Q power control by means of a rotor current regulation. For this purpose, P and Q are represented as functions of each individual rotor current. Using stator flux orientation, that decouples the dq axis, it has that $\lambda_{1d} = \lambda_1 = |\vec{\lambda}_{1dq}|$, thus, the equation (3) becomes:

$$i_{1d} = \frac{\lambda_1}{L_1} - \frac{L_M}{L_1} i_{2d} \tag{7}$$

$$i_{1q} = -\frac{L_M}{L_1} i_{2q} \tag{8}$$

Similarly, using stator flux oriented, the stator voltage becomes $v_{1d} = 0$ and $v_{1q} = v_1 = |\vec{v}_{1dq}|$. Hence, the active (5) and reactive (6) powers can be calculated by using equations (7) and (8):

$$P = -\frac{3}{2} v_1 \frac{L_M}{L_1} i_{2q} \tag{9}$$

$$Q = \frac{3}{2} v_1 \left(\frac{\lambda_1}{L_1} - \frac{L_M}{L_1} i_{2d} \right) \tag{10}$$

Thus, the rotor currents will reflect on the stator currents and on the active and reactive power of the stator, respectively. Consequently, this principle can be used to control the active and reactive power of the DFIG stator.

Thus, the rotor currents will reflect on the stator currents and the active and reactive power of the stator, respectively. Therefore, this principle can be used to control the active and reactive power of the DFIG.

3. Adaptive Neuro-Fuzzy Power Control

As mentioned before, an accurate power control system of the aerogenerators connected to the grid is essential to guarantee stability and maximum energy generation. In this way, ANFIS has been shown a powerful technique to control the nonlinear dynamic behavior of aerogenerators. Usually, it can use a hybrid learning procedure to construct an input-output mapping based on both human-knowledge as fuzzy rules and approximate MF from the stipulated input-output data pairs as ANN learning to precisely control the power generation.

3.1. ANFIS overview

An ANFIS is an adaptive network architecture whose overall input-output behavior is determined by the values of a collection of modifiable parameters [11, 27]. More specifically, the configuration of an adaptive network is composed by a set of nodes connected through directed links, where each node is a process unit that performs a static node function on its incoming signals to generate a single node output and each link specifies the direction of the signal flow from one node to another.

The parameters of an adaptive network are distributed into the network nodes, so each node has a local parameter set. The union of these local parameter sets forms the network overall parameter set. If a node parameter set is non-empty, then its node function depends on the parameter values, and it is used a square to represent this kind of adaptive node. On the other hand, if a node has an empty parameter set, then its function is fixed and it is used a circle to denote this type of fixed node [11].

This technique offers a method for learning information about a given data set during the fuzzy modeling procedure in order to adjust the MF parameters that best control the associated fuzzy inference system (FIS) to obtain the desired behavior.

3.1.1. ANFIS Architecture

In order to describe the ANFIS architecture in more detail, it is considered in the following, without loss of generality, a system with two inputs x_1 and x_2 and one output y, based on a first-order Takagi-Sugeno fuzzy model [28] with the fuzzy rules expressed as:

$$\text{Rule 1: If } x_1 \text{ is } A_1 \text{ and } x_2 \text{ is } B_1 \text{ then } f_1 = p_1x_1 + q_1x_2 + r_1 \quad (11)$$

$$\text{Rule 2: If } x_1 \text{ is } A_2 \text{ and } x_2 \text{ is } B_2 \text{ then } f_2 = p_2x_1 + q_2x_2 + r_2 \quad (12)$$

The ANFIS network structure is composed by a set of units and connections arranged into five layers [11, 27, 29], as shown in Fig. 2, where the output of the ith node of the lth layer is denoted as $O_{l,i}$.

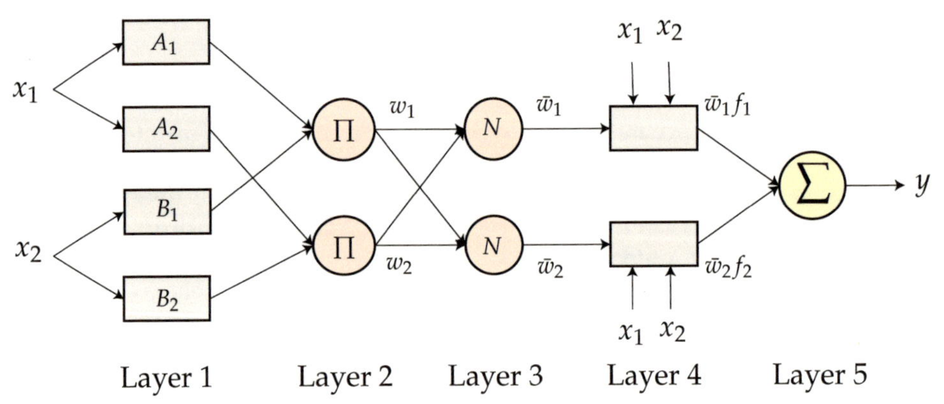

Figure 2. ANFIS Structure [11]

- Layer 1: Nodes in this layer are adaptive nodes and represent the membership grade of the inputs x_1 and x_2. Here, triangular or bell shaped MF [11] can be used. The parameters of this layer are called premise parameters and the nodes outputs are represented by:

$$O_{1,i} = \mu_{A_i}(x_1), \quad \text{for } i = 1,2 \text{ or}$$
$$O_{1,i} = \mu_{B_{i-2}}(x_2), \quad \text{for } i = 3,4 \tag{13}$$

where $\mu_{A_i}(x_1)$ and $\mu_{B_{i-2}}(x_2)$ are, respectively, the MF for the inputs x_1 and x_2.

- Layer 2: Each node in this layer, denoted as Π, is a non-adaptive node. This layer receives the input values x_1 and x_2 from the first layer and acts as a MF to represent the fuzzy sets of the respective input variables. Further, it computes the membership values that specify the degree to which the input values belongs to the fuzzy set. The nodes outputs are defined by the following product:

$$O_{2,i} = w_i = \mu_{A_i}(x_1) \cdot \mu_{B_i}(x_2), \quad \text{for } i = 1,2 \tag{14}$$

- Layer 3: Each node in this layer, denoted as N, is a non-adaptive node that normalizes the weight functions obtained from layer Π. This layer is usually called as the rule layer since it determines the activation level of each rule. The nodes outputs are given by:

$$O_{3,i} = \bar{w}_i = \frac{w_i}{w_1 + w_2}, \quad \text{for } i = 1,2 \tag{15}$$

- Layer 4: Every node i in this layer is an adaptive node. The parameters of this layer are called consequent parameters and the nodes outputs are expressed by:

$$O_{4,i} = \bar{w}_i f_i = \bar{w}_i(p_i x_1 + q_i x_2 + r_i), \quad \text{for } i = 1,2 \tag{16}$$

where $\bar{w}_i$ is the output of layer 3 and $\{p_i, q_i, r_i\}$ are the consequent parameters set.

- Layer 5: This layer, denoted by Σ, is non-adaptive and produces the output function by adding all inputs from the previous layers and transforming the fuzzy classification results into a crisp value, as expressed by:

$$O_{5,1} = \sum_i \bar{w}_i f_i = \frac{\sum_i w_i f_i}{\sum_i w_i} \tag{17}$$

3.1.2. Hybrid Learning Algorithm

The learning process can update the MF parameters by a hybrid learning procedure [29] composed by the Least Squares (LS) method, which is applied for tuning the linear output MF parameters, and the Backpropagation (BP) method, which is employed for tuning the nonlinear input MF parameters [30].

In the forward pass, LS estimates the consequent parameters, keeping the premise parameter fixed, and in the backward pass, the premise parameters are obtained by BP, keeping the consequent parameter fixed. When the values of the premise parameters are fixed, the overall output can be expressed as a linear combination of the consequent parameters.

Thus, considering the general LS approach, the output y of a given system to an input $\mathbf{x}$ can be represented by the following linearly parameterized equation:

$$y = \theta_1 f_1(\mathbf{x}) + \theta_2 f_2(\mathbf{x}) + \cdots + \theta_{N_i} f_{N_i}(\mathbf{x}) \tag{18}$$

where $\mathbf{x}$ is the input vector of the model with dimension N_x, f_i are known functions of $\mathbf{x}$, θ_i are the unknown parameters to be optimized and $i = 1, \cdots, N_i$.

Usually, to identify the unknown parameters, a training set of data pairs $(\mathbf{u}_m, v_m)$ is employed, where $\mathbf{u}$ is a vector with dimension N_x and $m = 1, \cdots, N_m$. Substituting each data pair into equation (18) yields the following set of N_m linear equations:

$$\begin{array}{ccccccc} f_1(\mathbf{u}_1)\theta_1 & + & f_2(\mathbf{u}_1)\theta_2 & + \cdots + & f_{N_i}(\mathbf{u}_1)\theta_{N_i} & = & v_1 \\ \vdots & & \vdots & & \vdots & & \vdots \\ f_1(\mathbf{u}_{N_m})\theta_1 & + & f_2(\mathbf{u}_{N_m})\theta_2 & + \cdots + & f_{N_i}(\mathbf{u}_{N_m})\theta_{N_i} & = & v_{N_m} \end{array} \tag{19}$$

These equations can also be expressed through the following vector representation:

$$\mathbf{A} \times \boldsymbol{\theta} = \mathbf{v} \tag{20}$$

where $\mathbf{A}$ is a (N_m) by (N_i) matrix:

$$\mathbf{A} = \begin{bmatrix} f_1(\mathbf{u}_1) & \cdots & f_{N_i}(\mathbf{u}_1) \\ \vdots & \ddots & \vdots \\ f_1(\mathbf{u}_{N_m}) & \cdots & f_{N_i}(\mathbf{u}_{N_m}) \end{bmatrix} \tag{21}$$

$\boldsymbol{\theta}$ is the (N_i) by (1) vector of unknown parameter:

$$\boldsymbol{\theta} = \begin{bmatrix} \theta_1 \\ \vdots \\ \theta_{N_i} \end{bmatrix} \tag{22}$$

and $\mathbf{v}$ is the (N_m) by (1) output vector:

$$\mathbf{v} = \begin{bmatrix} v_1 \\ \vdots \\ v_{N_m} \end{bmatrix} \tag{23}$$

Usually, $N_m > N_i$ and, instead of directly obtaining a solution for equation (20), an error vector $\mathbf{e}$ is introduced, resulting in the following equation:

$$\mathbf{A} \times \boldsymbol{\theta} + \mathbf{e} = \mathbf{v} \tag{24}$$

Whose solution $\boldsymbol{\theta} = \hat{\boldsymbol{\theta}}$ is obtained by minimizing the sum of the squared error:

$$\ell(\boldsymbol{\theta}) = \sum_{m=1}^{N_m} \left(v_m - \mathbf{a}_m^T \times \boldsymbol{\theta}\right)^2 = \mathbf{e}^T \times \mathbf{e} \tag{25}$$

where $\ell(\boldsymbol{\theta})$ is the objective function, $\mathbf{a}_m^T$ is the mth row vector of $\mathbf{A}$, $\mathbf{e} = \mathbf{v} - \mathbf{A} \times \boldsymbol{\theta}$ is the error vector produced by a specific choice of $\boldsymbol{\theta}$ and $()^T$ is the transpose operation.

The squared error in equation (25) is minimized when $\boldsymbol{\theta} = \hat{\boldsymbol{\theta}}$, where $\hat{\boldsymbol{\theta}}$ is obtained through a LS estimator that satisfies the following normal equation:

$$\left(\mathbf{A}^T \times \mathbf{A}\right) \times \boldsymbol{\theta} = \mathbf{A}^T \times \mathbf{v} \tag{26}$$

If $\mathbf{A}^T \times \mathbf{A}$ is non-singular, the LS solution $\hat{\boldsymbol{\theta}}$ can be obtained by:

$$\hat{\boldsymbol{\theta}} = \left(\mathbf{A}^T \times \mathbf{A}\right)^{-1} \times \mathbf{A}^T \times \mathbf{v} \tag{27}$$

where $\left(\mathbf{A}^T \times \mathbf{A}\right)^{-1} \times \mathbf{A}^T$ is the pseudo-inverse of $\mathbf{A}$ [30] and $()^{-1}$ is the inverse operation.

The LS solution is computationally very expensive, mainly for vectorial inputs, since it requires matrix inversion. Moreover, it can become ill-conditioned if $\mathbf{A}^T \times \mathbf{A}$ becomes singular. Thus, in practice, a recursive LS (RLS) algorithm can usually be employed [11, 29, 30].

In the backward pass, the premise parameters are estimated iteratively by a modified BP procedure that employs the Gradient Descent (GD) algorithm along with the chain rule to find the minimum of the error function in the weight space [29]. Considering a feed-forward adaptive network with N_L layers and N_i^l nodes in the lth layer, the output $O_{l,i}$ of the ith node of the lth layer can be represented by:

$$O_{l,i} = f_{l,i}\left(O_{l-1,1}, \cdots, O_{l-1,N_i^{l-1}}, \xi_{l,i}^1, \xi_{l,i}^2, \cdots, \xi_{l,i}^{N_\xi}\right) \tag{28}$$

where $f_{l,i}()$ is the node function of the ith node of the lth layer and $\xi_{l,i}^1, \xi_{l,i}^2, \cdots, \xi_{l,i}^{N_\xi}$ are the parameters associated with this specific node.

Assuming that the training data set has N_m terms, a measure of error at the output layer L corresponding to the mth term of the training set, where $1 \leq m \leq N_m$, can be defined as [11, 27, 29]:

$$\ell_m = \sum_{i=1}^{N_i^L} \left(v_m^i - O_{L,i}^m\right)^2 \tag{29}$$

where v_m^i is the mth desired output for the ith node of the output layer (L^{th} layer) and $O_{L,i}^m$ is the actual output of the ith node of the output layer produced by applying the mth training input vector to the network.

In this way, the main task of this step is to employ the GD algorithm to minimize the total measure of error defined as:

$$\ell_{total} = \sum_{m=1}^{N_m} \ell_m \tag{30}$$

In order to compute the gradient vector of the parameters, a form of the derivative information has to be passed backward, layer by layer, from the output layer to the input layer. This procedure is called BP because the gradient is obtained sequentially from the output layer to the input layer.

The corresponding error signal $e_{l,i}$ can be defined by the ordered derivative [31] of ℓ_m with respect to the output of the ith node of the lth layer:

$$e_{l,i}^m = \frac{\partial^+ \ell_m}{\partial O_{l,i}^m} \tag{31}$$

Particularly, for the output layer L, the error signal can be represented by the following simplified expression:

$$e_{L,i}^m = -2\left(v_m^i - O_{L,i}^m\right) \tag{32}$$

For the internal i^{th} node at the lth layer, the error signal can be derived iteratively by the following chain rule [11, 29]:

http://dx.doi.org/10.5772/52090

$$\begin{aligned} e_{l,i}^{m} &= \sum_{m=1}^{N_i^{l+1}} \frac{\partial^+ \ell_m}{\partial O_{l+1,i}^{m}} \frac{\partial f_{l+1,m}}{\partial O_{l,i}^{m}} \\ &= \sum_{m=1}^{N_i^{l+1}} e_{l+1,i}^{m} \frac{\partial f_{l+1,m}}{\partial O_{l,i}^{m}} \end{aligned} \tag{33}$$

where $1 \leq l \leq N_L - 1$ and $N_L = 5$ for the considered ANFIS model.

As the consequent parameters are fixed in the backward pass, the gradient vector can be defined as the derivative of the error measure with respect to the node parameters of the first layer $\varsigma_{1,i}$. The gradient vector elements corresponding to the ith node of 1st layer are given by:

$$\begin{aligned} \frac{\partial^+ \ell_m}{\partial \varsigma_{1,i}} &= \frac{\partial^+ \ell_m}{\partial v_{1,i}} \frac{\partial f_{1,i}}{\partial \varsigma_{1,i}} \\ &= e_{1,i} \frac{\partial f_{1,i}}{\partial \varsigma_{1,i}} \end{aligned} \tag{34}$$

The derivative of the total error measure ℓ_{total} with respect to $\varsigma_{1,i}$ is given by:

$$\frac{\partial^+ \ell_{total}}{\partial \varsigma_{1,i}} = \sum_{m=1}^{N_m} \frac{\partial^+ \ell_m}{\partial \varsigma_{1,i}} \tag{35}$$

Accordingly, the parameter $\varsigma_{1,i}$ can be updated using the following expression:

$$\Delta \varsigma_{1,i} = -\eta \frac{\partial^+ \ell_{total}}{\partial \varsigma_{1,i}} \tag{36}$$

where η is the learning rate [11, 29].

The parameter $\varsigma_{1,i}$ can be updated employing the following recursive expression:

$$\begin{aligned} \varsigma_{1,i}^{\text{new}} &= \varsigma_{1,i}^{\text{old}} + \Delta \varsigma_{1,i} \\ &= \varsigma_{1,i}^{\text{old}} - \eta \frac{\partial^+ \ell_{total}}{\partial \varsigma_{1,i}} \end{aligned} \tag{37}$$

In this type of learning procedure, the MF update occurs only after the whole set of training data pair is introduced. This process of introduction of the whole set of training data pairs is called epoch.

3.2. Proposed Adaptive Neuro-Fuzzy Power Control

The proposed adaptive neuro-fuzzy power control system, shown in Fig. 3, is composed by two ANFIS controllers based on a first-order Takagi-Sugeno fuzzy model [28] with one input and one output. The first controller has ε_{2q} as input and v_{2q} as output, and the second controller has ε_{2d} and v_{2d} as input and output, respectively, where $\varepsilon_{2q} = i_{2q_{ref}} - i_{2q}$ and $\varepsilon_{2d} = i_{2d_{ref}} - i_{2d}$.

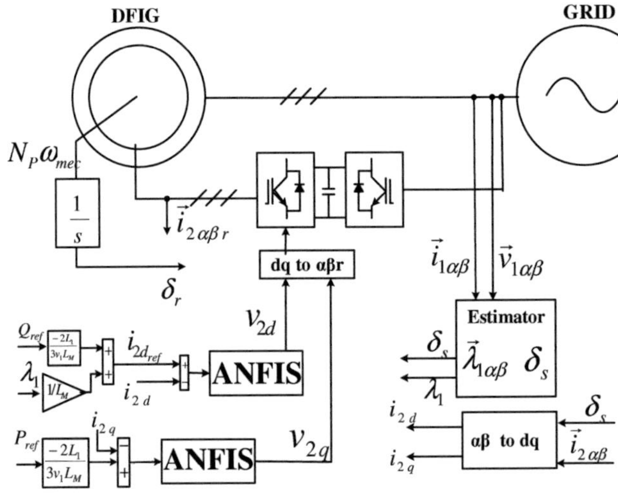

Figure 3. Neuro-Fuzzy DFIG Power Control.

Differently from what has been presented in the literature [10], the ANFIS controller employs just the rotor current error as input and it is designed from an input and output data set collected from a DFIG with a deadbeat controller [25] operating at different conditions.

In Fig. 4, it is shown a simplified representation of the substitution that was made in this work with the purpose of illustrating that the designed ANFIS presents the same behavior of a deadbeat controller.

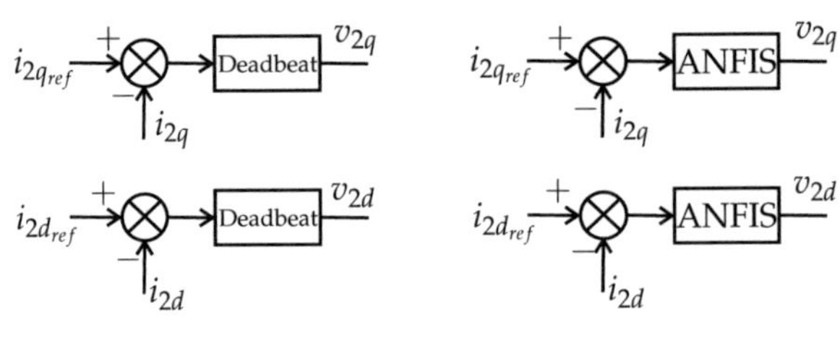

Figure 4. ANFIS Controller.

The tests used in this work for collect data set training are similar to the ones used in [10]:

http://dx.doi.org/10.5772/52090

- Operation at different wind speed profiles;
- Operation at different ramp increments in wind speed;
- Operation at different step increments in wind speed;
- Operation during voltage sag conditions.

The grid partition method is chosen to design the fuzzy controller structure, which usually involves just some few state variables as inputs to the controller. A generalized bell MF with parameters a, b, c, besides the input u, has been considered:

$$bell(u,a,b,c) = \frac{1}{1+|\frac{u-c}{a}|^{2b}} \quad (38)$$

where, the parameter b is usually positive.

The desired generalized bell MF can be obtained by a proper selection of the parameter set $\{a, b, c\}$. Specifically, we can adjust c and a to vary, respectively, the center and width of the MF and then use b to control the slopes at the crossover points [27].

To calculate the numerical value resulting from the activated rules, it was used the weighted average method for defuzzification. Thus, if the *direct* and *quadrature* voltage components are calculated according to the ANFIS controller and are applied to the generator, then the active and reactive power convergence to their respective commanded values will occur in a few sampling intervals. The desired rotor voltage in the rotor reference frame $(\delta_s - \delta_r)$ generates switching signals for the rotor side using either space vector modulation.

4. Flux estimation

For work properly, the ANFIS controller requires the estimation of stator flux, stator position and angle between stator and rotor flux.

The stator flux $\vec{\lambda}_{1\alpha\beta}$ in stationary reference frame can be estimated by:

$$\vec{\lambda}_{1\alpha\beta} = \int \left(\vec{v}_{1\alpha\beta} - R_1\vec{i}_{1\alpha\beta}\right) dt \quad (39)$$

The stator flux position can be obtained by using equation (39) as:

$$\delta_s = \arctan\left(\frac{\lambda_{1\beta}}{\lambda_{1\alpha}}\right) \quad (40)$$

And the angle between stator and rotor flux is given by:

$$\delta_s - \delta_r = \int \omega_{sl} dt \quad (41)$$

5. Wireless Coded Communication System

The wireless communication system, shown in Fig. 5, is responsible for transmitting the power references from the remote control unit to the ANFIS controller and it uses Quaternary Phase Shift Keying (QPSK) modulation and LDPC coding [18, 32, 33] to improve system performance and reliability.

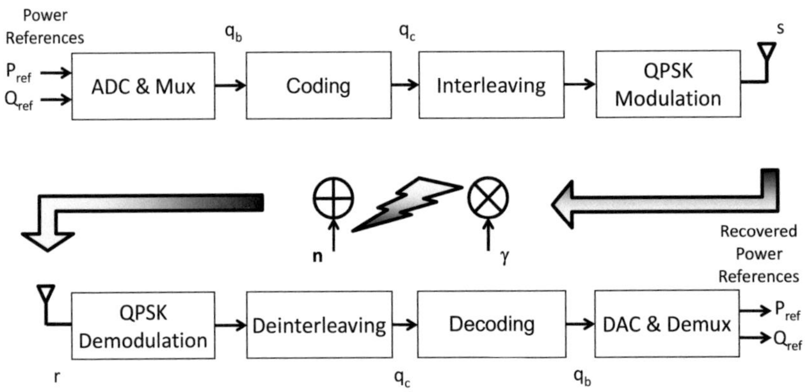

Figure 5. Wireless Coding Communication Diagram.

LDPC are (N_c, N_b) binary linear block codes that have a sparse parity-check matrix $\mathbf{H}$ that can be described in terms of a Tanner graph [34], where each bit in the codeword corresponds to a variable node and each parity-check equation corresponds to a check node. A check node j is connected to a variable node k whenever the element $h_{j,k}$ in $\mathbf{H}$ is equal to 1 [18, 34].

Extended Irregular Repeat Accumulate (eIRA) codes [35–39] are a special subclass of LDPC codes that improve the systematic encoding process and generate good irregular LDPC codes for high code rate applications. The eIRA parity-check matrix can be represented by $\mathbf{H} = [\mathbf{H}_1 \ \ \mathbf{H}_2]$, where $\mathbf{H}_1$ is a sparse (N_z) by (N_c) matrix, that can be constructed irregularly by density evolution according to optimal weight distribution [38], and $\mathbf{H}_2$ is the (N_z) by (N_z) dual-diagonal square matrix given by:

$$\mathbf{H}_2 = \begin{bmatrix} 1 & & & & \\ 1 & 1 & & & \\ & 1 & \ddots & & \\ & & \ddots & 1 & \\ & & & 1 & 1 \end{bmatrix} \tag{42}$$

where N_b is the number of control bits, N_c is the number of coded bits and N_z is the number of parity bits.

Given the constraint imposed on the $\mathbf{H}$ matrix, the generator matrix can be represented in the systematic form by the (N_b) by (N_c) matrix:

$$\mathbf{G} = [\mathbf{I} \ \ \mathbf{\Psi}] \tag{43}$$

where $\mathbf{I}$ is the identity matrix, $\mathbf{\Psi} = \mathbf{H}_1^T \times \mathbf{H}_2^{-T}$, $()^{-T}$ is the inverse transpose operation, and $\mathbf{H}_2^{-T}$ is the upper triangular matrix given by:

$$\mathbf{H}_2^{-T} = \begin{bmatrix} 1 & 1 & 1 & \cdots & 1 & 1 \\ & 1 & 1 & \cdots & 1 & 1 \\ & & 1 & & 1 & 1 \\ & & & \ddots & \vdots & \vdots \\ & & & & 1 & 1 \\ & & & & & 1 \end{bmatrix} \tag{44}$$

The encoding process can be accomplished by first multiplying the control information vector $\mathbf{q}_b = \left[q_{b,1} \cdots q_{b,N_b} \right]^T$ by the sparse matrix $\mathbf{H}_1^T$ and then differentially encoding this partial result to obtain the parity bits. The systematic codeword vector $\mathbf{q}_c = \left[q_{c,1} \cdots q_{c,N_c} \right]^T$ can be simply obtained by combining the control information and the parity bits.

In the transmission process, for each transmitted frame, the codeword vector is interleaved and mapped to QPSK symbols by using Gray coding [15], resulting in the symbol vector $\mathbf{s} = \left[s_1 \cdots s_{N_s} \right]^T$, where N_s is the number of transmitted coded control symbols. Afterward, the coded symbols are filtered, upconverted and transmitted by the wireless fading channel.

Assuming that the channel variations are slow enough that intersymbol interferences (ISI) can be neglect, the fading channel can be modeled as a sequence of zero-mean complex Gaussian random variables with autocorrelation function [15, 40]:

$$R_h(\tau) = J_0(2\pi f_D T_s) \tag{45}$$

where $J_0()$ is the $zero^{th}$ order Bessel function, T_s is the signaling time and f_D is the Doppler spread.

Thus, in the receive process, the complex low-pass equivalent discrete-time received signal can be represented by [15]:

$$\mathbf{r} = \boldsymbol{\gamma} \cdot \mathbf{s} + \mathbf{n} \tag{46}$$

where $\mathbf{r} = \left[r_1 \cdots r_{N_s} \right]^T$ is the received signal vector, $\boldsymbol{\gamma} = \left[\gamma_1 \cdots \gamma_{N_s} \right]^T$ is the vector of complex coefficients of the channel and $\mathbf{n} = \left[n_1 \cdots n_{N_s} \right]^T$ is the Additive White Gaussian

Noise (AWGN) vector. Note that the above vector multiplication is performed element by element.

Once the transmitted vector **s** is estimated, considering perfect channel estimation, the transmitted control bits can be recovered by performing symbol demapping, code deinterleaving and bit decoding. Bit decoding can be accomplished by a message passing algorithm [23, 41–43] based on the Maximum A Posteriori (MAP) criterion [18], that exchanges soft-information iteratively between the variable and check nodes.

The exchanged messages can be represented by the following Log-Likelihood Ratio (LLR):

$$L_{C_k} = \log\left[\frac{p(\mathbf{q}_{c,k} = 0|\mathbf{d})}{p(\mathbf{q}_{c,k} = 1|\mathbf{d})}\right] \tag{47}$$

where, **d** is the vector of coded bits obtained by the processes of demodulation and deinterleaving.

The LLR message from the jth check node to the kth variable node is given by [37]:

$$L_{r_{j,k}} = 2 \text{ atanh}\left[\prod_{k' \in V_{j\backslash k}} \tanh\left(\frac{L_{q_{k',j}}}{2}\right)\right] \tag{48}$$

On the other hand, the LLR message from the kth variable node to the jth check node is obtained by:

$$L_{q_{k,j}} = L_{C_k} + \sum_{j' \in C_{k\backslash j}} L_{r_{j',k}} \tag{49}$$

And the LLR for the kth code bit can be represented by:

$$L_{Q_k} = L_{C_k} + \sum_{j \in C_k} L_{r_{j,k}} \tag{50}$$

where the set V_j contains the variable nodes connected to the jth check node and the set C_k contains the check nodes connected to the kth variable node. $V_{j\backslash k}$ is the set V_j without the kth element, and $C_{k\backslash j}$ is the set C_k without the jth element.

At the end of each iteration, L_{Q_k} provides an updated estimate of the a posteriori LLR of the transmitted coded bit $q_{c,k}$. If $L_{Q_k} > 0$, then $\hat{q}_{c,k} = 1$, else $\hat{q}_{c,k} = 0$.

6. Performance of the Wireless Power Control System

In the presented simulations, the sampling time is of 5×10^{-5}s and the active and reactive power references are step changed, respectively, from -100 to -120 kW and from 60 to 0 kvar at 1.25s. At 1.5s, the references also are step changed from -120 to -60 kW and from 0 to -40 kvar. Again, at 1.75s, the references are step changed from -60 to -100 kW and from -40 to -60 kvar. These references are the inputs of the wireless coded power control shown in Fig. 5. The nominal power at each instant of time is calculated by means of the turbine parameters, presented in the Appendix, and the speed of the wind, monitored by the remote control unit (that can also be obtained directly from the turbine and sent to the remote control unit).

The wireless control system is evaluated in a frequency flat fading Rayleigh channel with a Doppler spread of 180 Hz. The employed LDPC coding scheme is the (64800, 32400) eIRA code specified in [44] and an ordinary Convolution Coding (CC) scheme with a (171, 133) generator polynomial with constrain length of 7 is used as a reference of performance [18]. Both schemes have code rate of 1/2 and employ a random interleaving of length 64,800. For simplicity, the number of iterations in the LDPC decoding is limited to 25. The bit duration is 16×10^{-5} s and each transmitted frame is composed by 32,400 QPSK coded symbols.

The training process of the adaptive neuro-fuzzy power control strategy employed in this work considered a maximum of 5,000 epochs and an error tolerance of 0.0001.

In Fig. 6 and 7, the step response of the active and reactive power and of the rotor currents of the wireless controller using CC scheme are presented, respectively, considering an E_b/N_0 of 10 dB.

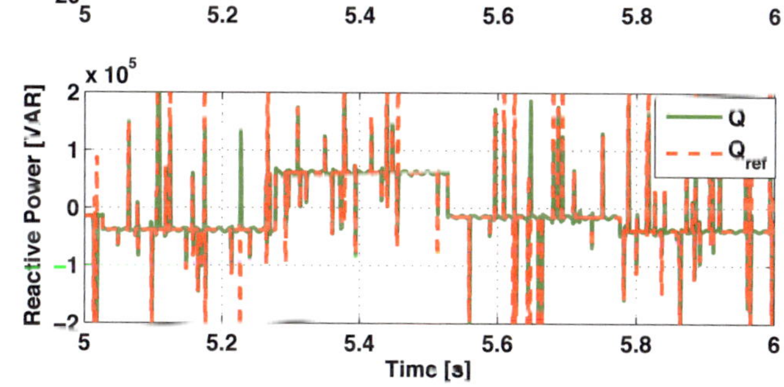

Figure 6. Step Response of Active and Reactive Power using CC in a Flat Fading Channel.

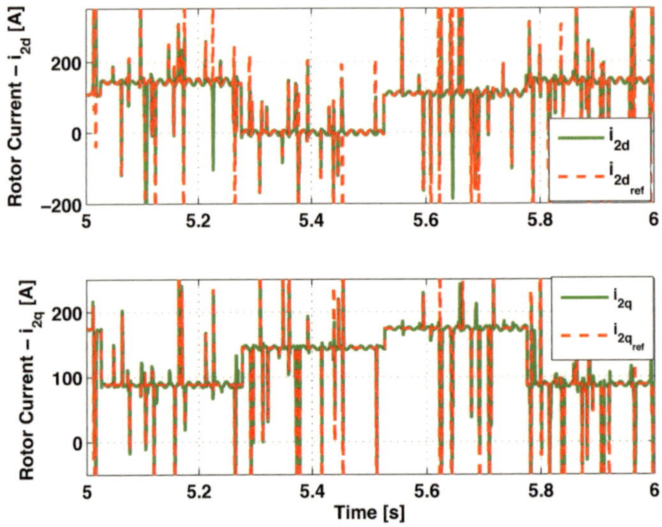

Figure 7. Step Response of Rotor Current $\vec{i}_{2_{dq}}$ using CC in a Flat Fading Channel.

The spikes presented in the responses of the system occur due to the errors in the wireless communication caused by the fading channel, even with the use of a very efficient error correction CC scheme. It can be observed that several of these spikes, presented in the recovered reference signals, are followed by the controller.

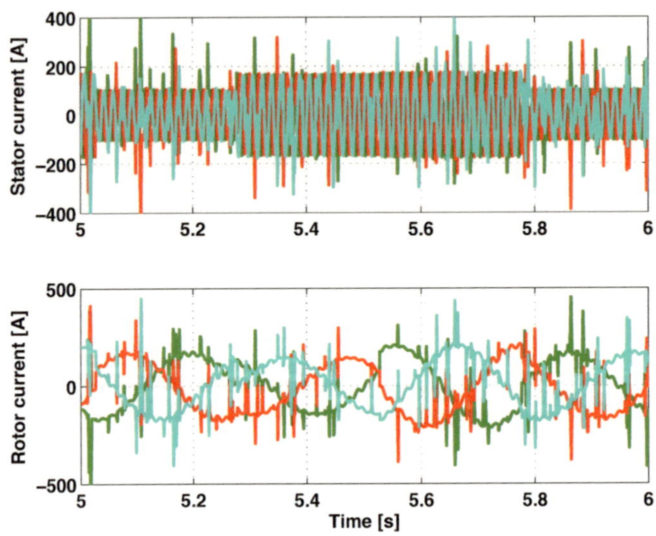

Figure 8. Stator and Rotor Currents using CC in a Flat Fading Channel.

These errors in the control system can permanently damage the aerogenerator or even cause a loss of system efficiency, since the machine will not generate its maximum power track at that moment. Additionally, they generate undesirable harmonic components to the power grid. As shown in Fig. 8, the damages can occur due to the high values of $\frac{di}{dt}$ that can completely deteriorate the Insulated Gate Bipolar Transistors (IGBTs) and, consequently, cause short circuits in rotor and/or stator of the generator through the power converter.

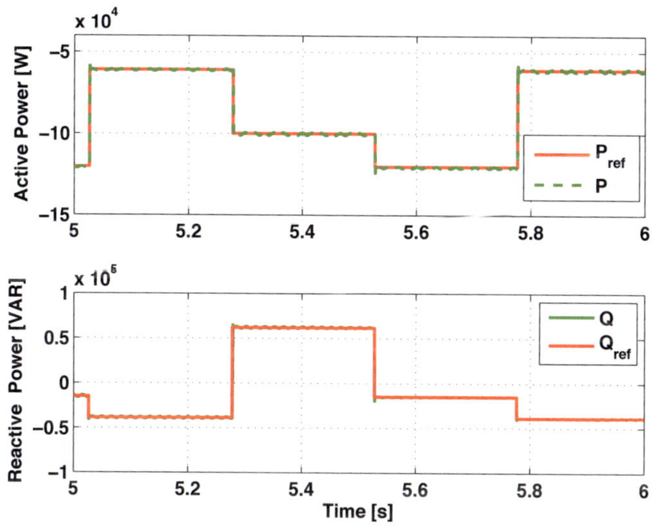

Figure 9. Step Response of Active and Reactive Power using LDPC in a Flat Fading Channel.

In this way, it turns out the necessity of employing a wireless coded control system capable of minimizing the occurrence of these spikes arising from errors caused by the channel distortions. With this finality, it is highlighted the proposal of using a more robust wireless neuro-fuzzy power control system based on LDPC coding.

In Fig. 9 and 10, the step response of the active and reactive power and of the rotor currents of the wireless controller using the LDPC scheme are presented, respectively, considering an E_b/N_0 of 10 dB. The satisfactory performance of the wireless control system can be seen due to the fact the references were perfectly followed by the controller and the inexistence of destructive spikes caused by errors in the wireless transmission system. Additionally, these good functionalities are shown in Fig. 11, where the stator currents present expected waveforms.

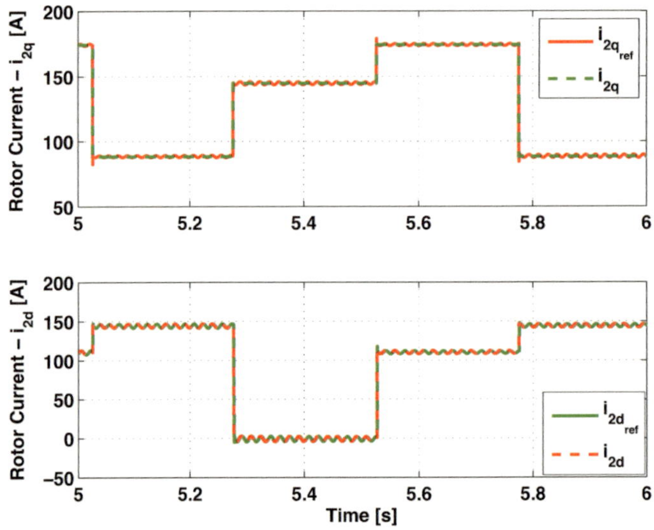

Figure 10. Step Response of Rotor Current $\vec{i}_{2_{dq}}$ using LDPC in a Flat Fading Channel.

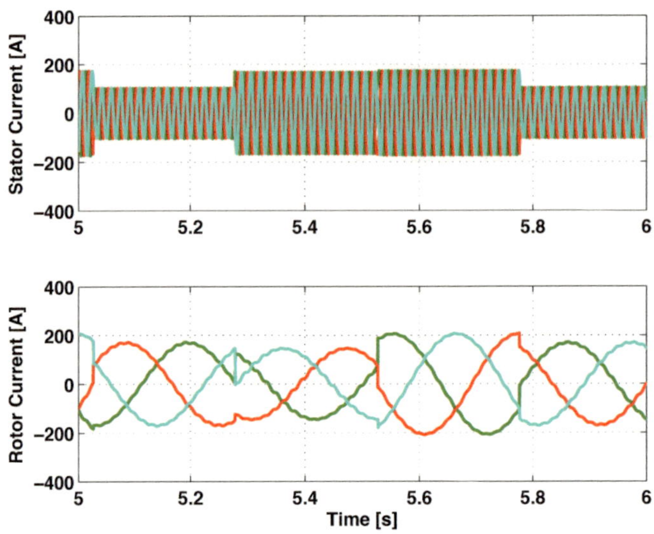

Figure 11. Stator and Rotor Currents using LDPC in a Flat Fading Channel.

To complete the analysis, it is presented in Fig. 12, a comparison of the BER performance for different values of E_b/N_0 for the proposed wireless coded neuro-fuzzy power control system using three different schemes: No Coding, CC, and LDPC. As expected, the performance of

http://dx.doi.org/10.5772/52090

LDPC is significantly superior than CC for flat fading channels. As pointed out in Fig. 12, for a BER of 10^{-5}, the performance improvement of LDPC over CC is approximately 26.8 dB and more than 30 dB over no coding. Table 1 shows a resume of the results presented in Fig. 12.

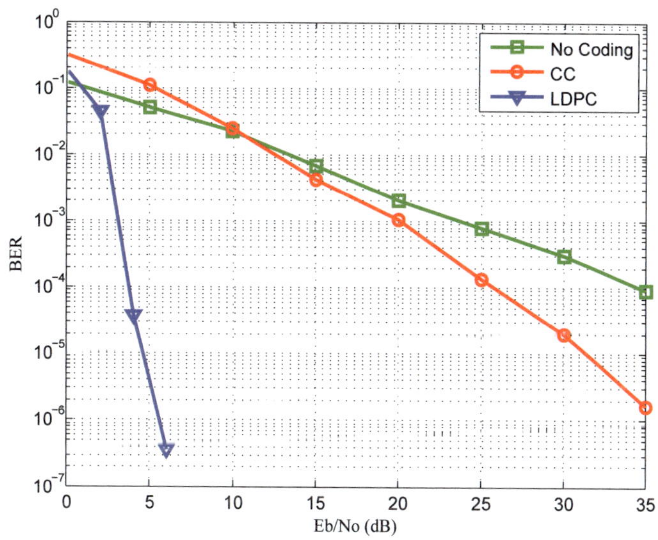

Figure 12. Performance Comparison for Different Coding Schemes in a Flat Fading Channel.

It can be seen that LDPC requires a significant lower E_b/N_0 to present the same order of performance of CC. It can be noted that, even for a low BER as 10^{-5}, it can occur some changes in the active and reactive power references that can cause serious problems in the generator, and consequentially, in the energy plant. However, the use of LDPC can reduce notably this number for a typical E_b/N_0 value and can improve considerably the system robustness to the channel impairments. For instance, a system operating with a typical E_b/N_0 of 10 dB employing CC will fail dramatically, while a system using LDPC coding will be free of errors.

Coding Scheme	Bit Error Rate	Eb/No (dB)
CC	10^{-3}	20.0
CC	10^{-4}	25.5
CC	10^{-5}	31.5
LDPC	10^{-3}	3.20
LDPC	10^{-4}	3.80
LDPC	10^{-5}	4.70

Table 1. Performance Comparison for Both Coding Schemes.

7. Conclusion

In this chapter, it was introduced a new wireless coded control system based on an adaptive neuro-fuzzy inference system applied to a doubly-fed induction aerogenerator. In order to improve the performance of the controller to the nonlinear characteristics of the wind system, the proposed adaptive neuro-fuzzy system uses a different design strategy based on the input and output data set collected from a DFIG with a deadbeat controller operating at different conditions and considering as input the rotor current error. The wireless communication system, employed to send the power reference signals to the DFIG controller, uses LDPC coding to reduce the transmission errors and the overall latency of the system by mitigating the necessity of retransmissions.

The presented analysis showed that the adaptive neuro-fuzzy strategy is a powerful technique to precisely control the dynamic nonlinear behavior of aerogenerators and to guarantee stability and maximum energy generation in smart grids applications. Furthermore, it was concluded that the use of LDPC coding can significantly improve the robustness of the wireless power control system in severe noise and fading channel conditions, being essential to maintain the integrity of the aerogenerators and to reduce the overall system response.

Appendix

Doubly-fed induction generator parameters [45]:

$R_1 = 24.75\ m\Omega$; $R_2 = 13.3\ m\Omega$; $L_M = 14.25\ mH$; $L_{l1} = 284\ \mu H$; $L_{l2} = 284\ \mu H$; $J = 2.6\ Kg \cdot m^2$; $N_P = 2$; $P_N = 149.2\ kVA$ and $V_N = 575\ V$.

Acknowledgments

This work was partially supported by Fundação de Amparo à Pesquisa do Estado de São Paulo (FAPESP), Conselho Nacional de Desenvolvimento Científico e Tecnológico (CNPq) and Coordenação de Aperfeiçoamento de Pessoal de Nível Superior (CAPES).

Author details

I. R. S. Casella[1,*], A. J. Sguarezi Filho [1],
C. E. Capovilla [1], J. L. Azcue [2] and E. Ruppert [2]

* Address all correspondence to: ivan.casella@ufabc.edu.br; alfeu.sguarezi@ufabc.edu.br; carlos.capovilla@ufabc.edu.br; jl.azcue@gmail.com; ruppert@fee.unicamp.br

1 Centro de Engenharia, Modelagem e Ciências Sociais Aplicadas - CECS, Universidade Federal do ABC - UFABC, Brazil

2 Faculdade de Engenharia Elétrica e de Computação - FEEC, Universidade de Campinas - UNICAMP, Brazil

References

[1] M. Glinkowski, J. Hou, and G. Rackliffe. Advances in wind energy technologies in the context of smart grid. *Proceedings of the IEEE*, 99(6):1083–1097, June 2011.

[2] J. Wang, X. Du, and X. Zhang. Comparison of wind power generation interconnection technology standards. *Asia-Pacific Power and Energy Engineering Conference*, March 2011.

[3] W. Xiwen, Q. Xiaoyan, X. Jian, and L. Xingyuan. Reactive power optimization in smart grid with wind power generator. *Asia-Pacific Power and Energy Engineering Conference*, March 2010.

[4] J. F. Manwell, J. G. McGowan, and A. L. Rogers. *Wind Energy Explained: Theory, Design and Application*. Wiley, 2 edition, 2010.

[5] R. G. de Almeida and J. A. Pecas Lopes. Participation of doubly fed induction wind generators in system frequency regulation. *IEEE Transactions on Power Systems*, 22(3):944–950, August 2007.

[6] R. Datta and V. T. Rangathan. Variable-speed wind power generation using doubly fed wound rotor induction machine - a comparison with alternative schemes. *IEEE Trans. on Energy Conversion*, 17(3):414–421, September 2002.

[7] A. Tapia, G. Tapia, J. X. Ostolaza, and J. R. Saenz. Modeling and control of a wind turbine driven doubly fed induction generator. *IEEE Transactions on Energy Conversion*, 18(194-204), June 2003.

[8] S. A. Shaheen, H. M. Hasanien, and M. A. Badr. Study on doubly fed induction generator control. *International Middle East Power Systems Conference*, pages 627–633, 2010.

[9] Jin-Sung Kim, Jonghyun Jeon, and Hoon Heo. Design of adaptive pid for pitch control of large wind turbine generator. *International Conference on Environment and Electrical Engineering*, pages 1–4, 2011.

[10] B. Singh, E. Kyriakides, and S. N. Singh. Intelligent control of grid connected unified doubly-fed induction generator. In *IEEE Power and Energy Society General Meeting*, pages 1–7, July 2010.

[11] J.-S.R. Jang. Anfis: adaptive-network based fuzzy inference system. *IEEE Transactions on Systems, Man, and Cybernetics*, 23(3):665–685, May-June 1993.

[12] M. Adamowicz, R. Strzelecki, Z. Krzeminski, J. Szewczyk, and L. Lademan. Application of wireless communication to small wecs with induction generator. *15th IEEE Mediterranean Electrotechnical Conference*, pages 944–948, June 2010.

[13] M. Adamowicz, R. Strzelecki, J. Szewczyk, and L. Lademan. Wireless short-range device for wind generators. *12th Biennial Baltic Electronics Conference*, pages 1736–3705, November 2010.

[14] T. J. Li. *Low complexity capacity approaching schemes: Design, analysis and applications*. Ph.D. dissertation, Texas AM Univ., 2002.

[15] J. G. Proakis. *Digital Communications*. MCGraw-Hill, 2008.

[16] H. Li, L. Lai, and W. Zhang. Communication requirement for reliable and secure state estimation and control in smart grid. *IEEE Transaction on Smart Grid*, 2(3):477–486, 2011.

[17] J. Jiang and K. R. Narayanan. Iterative soft decision decoding of reed solomon. *IEEE Communications Letters*, 8:244–246, 2004.

[18] S. Lin and D. J. Costello. *Error control coding*. Prentice Hall, 2004.

[19] C. Berrou, A. Glavieux, and P. Thitimajshima. Near shannon limit error-correcting coding and decoding: Turbo-codes. *IEEE International Communications Conference*, pages 1064–1070, 1993.

[20] J. Chen and A. Abedi. Distributed turbo coding and decoding for wireless sensor networks. *IEEE Communications Letters*, 15:166–168, 2011.

[21] I. R. S. Casella. Analysis of turbo coded ofdm systems employing space-frequency block code in double selective fading channels. *IEEE International Microwave and Optoelectronics Conference*, pages 516–520, November 2007.

[22] D. J. C. MacKay and R. M. Neal. Near shannon limit performance of low-density parity-check codes. *IET Electronics Letters*, 32:1645–1646, 1996.

[23] T. Richardson, A. Shokrollahi, and R. Urbanke. Design of capacity-approaching low-density parity-check codes. *IEEE Transactions on Information Theory*, 47:619–637, February 2001.

[24] IEEE. Standard for local and metropolitan area networks, part 16: Air interface for fixed and mobile wireless access systems. *IEEE Std. 802.16-2004*, 2004.

[25] A. J. Sguarezi Filho and E. Ruppert. A deadbeat active and reactive power control for doubly-fed induction generators. *Electric Power Components and Systems*, 38(5):592–602, 2010.

[26] W. Leonhard. *Control of Electrical Drives*. Springer-Verlag Berlin Heidelberg New York Tokyo, 1985.

[27] J.-S.R. Jang and C.-T. Sun. Neuro-fuzzy modeling and control. *Proceedings of the IEEE*, 83(3):378–406, March 1995.

[28] M. Sugeno and G. T. Kang. Structure identification of fuzzy model. *Fuzzy Sets and Systems*, 28, 1988.

[29] J. S. R. Jand, C.T. Sun, and E. Mizutani. *Neuro-Fuzzy and Soft Computing: A Computational Approach to Learning and Machine Intelligence*. Prentice Hall, 1997.

[30] S. Haykin. *Neural Networks: A Comprehensive Foundation*. Prentice Hall, 2 edition, 1998.

http://dx.doi.org/10.5772/52090

[31] P. Werbos. *Beyond regression: New tools for prediction and analysis in the behavioral sciences.* PhD thesis-Harvard University, 1974.

[32] R. G. Gallager. *Low-Density Parity-Check Codes*. Cambridge, 1963.

[33] Y. Zhang and W. E. Ryan. Toward low ldpc-code floors: a case stud. *IEEE Transactions on Communications*, 57(6):1566–1573, June 2009.

[34] R. M. Tanner. A recursive approach to low complexity codes. *IEEE Transactions on Information Theory*, 27(5):533–547, September 1981.

[35] H. Jin, A. Khandekar, and R. J. McEliece. Irregular repeat-accumulate codes. *Proc. Int. Symp. Turbo Codes and Related Topics*, pages 1–5, September 2000.

[36] M. Yang and W. E. Ryan. Lowering the error-rate floors of moderate length high-rate irregular ldpc codes. *Int. Symp. Information Theory*, 2:237, July 2003.

[37] M. Yang, W. E. Ryan, and Y. Li. Design of efficiently encodable moderate-length high-rate irregular ldpc codes. *IEEE Transactions on Communications*, 52(4):564–571, April 2004.

[38] Y. Zhang, W. E. Ryan, and Y. Li. Structured eira codes with low floors. *Proceedings of the International Symposium on Information Theory*, pages 174–178, September 2005.

[39] J. Kim, A. Ramamoorthy, and S. Mclaughlin. The design of efficiently-encodable rate-compatible ldpc codes. *IEEE Transactions on Communications*, 57:365–375, 2009.

[40] A. Barbieri, A. Piemontese, and G. Colavolpe. On the arma approximation for frequency-flat rayleigh fading channels. *IEEE International Symposium on Information Theory*, pages 1211–1215, June 2007.

[41] T. Richardson and R. Urbanke. The capacity of low-density parity check codes under message-passing decoding. *IEEE Transactions on Information Theory*, 47:599–618, February 2001.

[42] L. Dinoi, F. Sottile, and S. Benedetto. Design of versatile eira codes for parallel decoders. *IEEE Transactions on Communications*, 56(12):2060–2070, 2008.

[43] B. Shuval and I. Sason. On the universality of ldpc code ensembles under belief propagation and ml decoding. *IEEE 26th Convention of Electrical and Electronics Engineers*, pages 355–359, 2010.

[44] ETSI. Dvb-s.2. *Standard Specification*, pages 302–307, March 2005.

[45] A. J. Sguarezi Filho, M. E. de Oliveira Filho, and E. Ruppert. A predictive power control for wind energy. *IEEE Transactions on Sustainable Energy*, 2(1):97–105, January 2011.

Chapter 8

Application of Harmony Search Algorithm in Power Engineering

H. R. Baghaee, M. Mirsalim and G. B. Gharehpetian

Additional information is available at the end of the chapter

http://dx.doi.org/10.5772/55509

1. Introduction

1.1. On the use of harmony search algorithm in optimal placement of FACTS devices to improve power systems

With the increasing electric power demand, power systems can face to stressed conditions, the operation of power system becomes more complex, and power system will become less secure. Moreover, because of restructuring, the problem of power system security has become a matter of concern in deregulated power industry. Better utilization of available power system capacities by Flexible AC Transmission Systems (FACTS) devices has become a major concern in power systems too.

FACTS devices can control power transmission parameters such as series impedance, voltage, and phase angle by their fast control characteristics and continuous compensating capability. They can reduce flow of heavily loaded lines, resulting in low system losses, improved both transient and small signal stability of network, reduced cost of production, and fulfillment of contractual requirement by controlling the power flow in the network. They can enable lines to flow the power near its nominal rating and maintain its voltage at desired level and thus, enhance power system security in contingencies [1-6]. For a meshed network, an optimal allocation of FACTS devices allows to control its power flows and thus, to improve the system loadability and security [1].

The effect of FACTS devices on power system security, reliability and loadability has been studied according to proper control objectives [1-14]. Researchers have tried to find suitable location for FACTS devices to improve power system security and loadability [13-16]. The optimal allocation of these devices in deregulated power systems has been presented in

[17-18]. Heuristic approaches and intelligent algorithms to find suitable location of FACTS devices and some other applications have been used in [15-21].

In this chapter, a novel heuristic method is presented based on Harmony Search Algorithm (HSA) to find optimal location of multi-type FACTS devices to enhance power system security and reduce power system losses considering investment cost of these devices. The proposed method is tested on IEEE 30-bus system and then, the results are presented.

2. Model of FACTS devices

2.1. FACTS devices

In this chapter, we select three different FACTS devices to place in the suitable locations to improve security margins of power systems. They are TCSC (Thyristor Controlled Series Capacitor), SVC (Static VAR Compensator), and UPFC (Unified Power Flow Controller) that are shown in Fig. 1.

Power flow through the transmission line i-j namely P_{ij}, depends on the line reactance X_{ij}, the bus voltage magnitudes V_i, and V_j, and phase angle between sending and receiving buses δ_i and δ_j, expressed by Eq. 1.

$$P_{ij} = \frac{V_i V_j}{X_{ij}} \sin(\delta_i - \delta_j) \tag{1}$$

TCSC can change line reactance, and SVC can control the bus voltage. UPFC is the most versatile member of FACTS devices family and controls all power transmission parameters (i.e., line impedance, bus voltage, and phase angles). FACTS devices can control and optimize power flow by changing power system parameters. Therefore, optimal device and allocation of FACTS devices can result in suitable utilization of power systems.

2.2. Mathematical model of FACTS devices

In this chapter, steady-state model of FACTS devices are developed for power flow studies. TCSC is simply modeled to modify just the reactance of transmission lines. SVC and UPFC are modeled using the power injection models. Therefore, SVC is modeled as shunt element of transmission line, and UPFC as decoupled model. A power flow program has been developed in MATLAB by incorporating the mathematical models of FACTS devices.

2.2.1. TCSC

TCSC compensates the reactance of the transmission line. This changes the line flow due to change in series reactance. In this chapter, TCSC is modeled by changing transmission line reactance as follows:

http://dx.doi.org/10.5772/55509

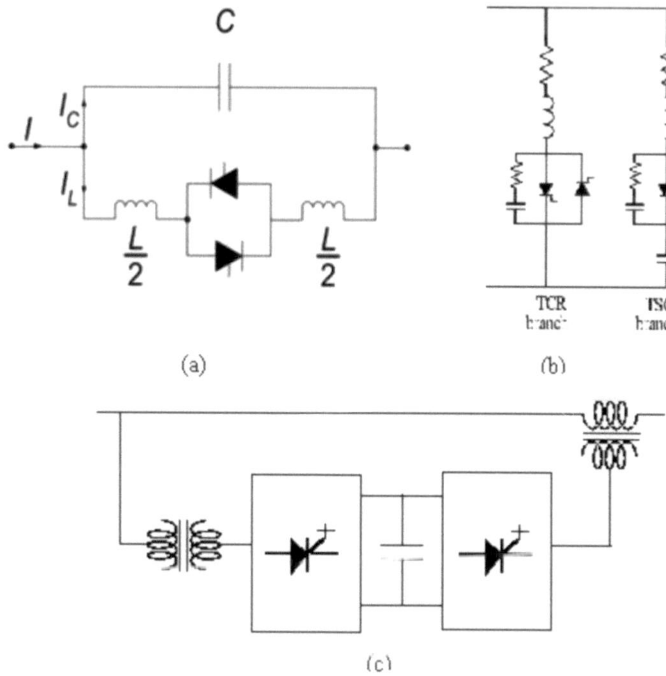

Figure 1. Models of FACTS Devices (a) TCSC, (b) SVC and (c) UPFC

$$X_{ij} = X_{line} + X_{TCSC} \tag{2}$$

$$X_{TCSC} = r_{TCSC} \cdot X_{line} \tag{3}$$

where, X_{line} is the reactance of the transmission line, and r_{TCSC} is the compensation factor of TCSC. The rating of TCSC depends on transmission line. To prevent overcompensation, we choose TCSC reactance between $-0.7X_{line}$ to $0.2X_{line}$ [26-27].

2.2.2. SVC

SVC can be used for both inductive and capacitive compensation. In this chapter, SVC is modeled as an ideal reactive power injection at bus i:

$$\Delta Q_i = Q_{SVC} \tag{4}$$

2.2.3. UPFC

Two types of UPFC models have been studied in the literature; one is the coupled model [28], and the other the decoupled type[29-31]. In the first, UPFC is modeled with series combination of a voltage source and impedance in the transmission line. In the decoupled model, UPFC is modeled with two separated buses. The first model is more complex than the second one because the modification of the Jacobian matrix is inevitable. In conventional power flow algorithms, we can easily implement the decoupled model. In this chapter, the decoupled model has been used to model the UPFC as in Fig. 2.

UPFC controls power flow of the transmission lines. To present UPFC in load flow studies, the variables P_{u1}, Q_{u1}, P_{u2}, and Q_{u2} are used. Assuming a lossless UPFC, real power flow from bus i to bus j can be expressed as follows:

$$P_{ij} = P_{u1} \tag{5}$$

Although UPFC can control the power flow but, it cannot generate the real power. Therefore, we have:

$$P_{u1} + P_{u2} = 0 \tag{6}$$

Reactive power output of UPFC, Q_{u1} , and Q_{u2}, can be set to an arbitrary value depending on the rating of UPFC to maintain bus voltage.

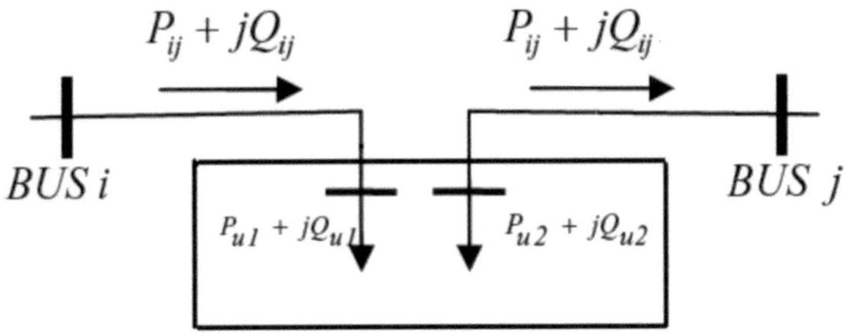

Figure 2. Decoupled model for UPFC

3. Security index

The security index for contingency analysis of power systems can be expressed as in the following [32-33]:

$$J_V = \sum_i w_i \left| V_i - V_{ref,i} \right|^2 \tag{7}$$

$$J_P = \sum_j w_j \left(\frac{S_j}{S_{j,amx}}\right)^2 \tag{8}$$

Here we have:

V_i, w_i as voltage amplitude and associated weighting factor for the i_{th} bus, respectively, S_j, w_j as apparent power and associated weighting factor for the j_{th} line, respectively, $V_{ref,i}$ as nominal voltage magnitude, which is assumed to be $1pu$ for all load buses (i.e., PQ buses), and to be equal to specified value for generation buses (i.e., PV buses), and $S_{j,\max}$ as nominal apparent power of the j_{th} line or transformer. J_P is the security index for the even distribution of the total active flow, and J_V is the security index for the closeness of bus voltage to the reference voltage. If the number of overloaded lines decreases, the value of J_P reduces too. Similarly, when the bus voltage have a value close to the desired level, J_V becomes a small value. Minimization of both J_P and J_V means the maximization of security margins.

4. The proposed algorithm

4.1. Harmony search algorithm

Harmony Search Algorithm (HSA) has recently been developed in an analogy with music improvisation process, where music players improvise the pitches of their instruments to obtain better harmony [34]. The steps in the procedure of harmony search are as follows [35]:

Step 1: Initialize the problem and algorithm parameters

Step 2: Initialize the harmony memory

Step 3: Improvise a new harmony

Step 4: Update the harmony memory

Step 5: Check the stopping criterion

The next following five subsections describe these steps.

a. Initialize the problem and algorithm parameters

In step 1, the optimization problem is specified as follows:

$\min\{f(x) \mid x \in X\}$ subject to $g(x) \geq 0$ and $h(x) = 0$

where, $f(x)$ is the objective function, $g(x)$ the inequality constraint function, and $h(x)$ the equality constraint function. x is the set of each decision variable x_i, and X is the set of the possible range of values for each decision variable; that is $X_{i,\min} \leq X_i \leq X_{i,\max}$ where, $X_{i,\min}$ and $X_{i,\max}$ are the lower and upper bounds of each decision variable. The HS algorithm parameters are also specified in this step. These are the harmony memory size (HMS), or the number of solution vectors in the harmony memory, harmony memory considering rate (HMCR), pitch adjusting rate (PAR), the number of decision variables (N), and the number of improvisations (NI), or stopping criterion. The harmony memory (HM) is a memory location where all the solution vectors (sets of decision variables) are stored. This HM is similar to the genetic pool in the GA [36]. Here, HMCR and PAR are parameters that are used to improve the solution vector and are defined in step 3.

b. Initialize the harmony memory

In step 2, the HM matrix is filled with as many randomly generated solution vectors as the HMS in the following:

$$HM = \begin{bmatrix} x_1^1 & x_2^1 & \cdots & x_{N-1}^1 & x_N^1 \\ x_1^2 & x_2^2 & \cdots & x_{N-1}^2 & x_N^2 \\ \vdots & \vdots & \vdots & \vdots & \vdots \\ x_1^{HMS-1} & x_2^{HMS-1} & \cdots & x_{N-1}^{HMS-1} & x_N^{HMS-1} \\ x_1^{HMS} & x_2^{HMS} & \cdots & x_{N-1}^{HMS} & x_N^{HMS} \end{bmatrix} \tag{9}$$

c. Improvise a new harmony

A new harmony vector, $x'_i=(x'_1, x'_2, ..., x'_N)$, is generated based on three rules; (1) memory consideration, (2) pitch adjustment, and (3) random selection. Generating a new harmony is called 'improvisation' [36]. In the memory consideration, the value of the first decision variable x'_1 for the new vector is chosen from any value in the specified HM range ($x_1^1 - x_1{}^{HMS}$). Values of the other decision variables ($x'_2, x'_3, ..., x'_N$), are chosen in the same manner. The HMCR, which varies between zero and one, is the rate of choosing one value from the historical values stored in the HM, while (1-HMCR) is the rate of randomly selecting one value from the possible range of values.

$$x'_i \leftarrow \begin{cases} x'_i \in \left\{x_i^1, x_i^2, ..., x_i{}^{HMS}\right\} & with\,probability\,HMCR \\ x'_i \in X_i & with\,probability\,(1-HMCR) \end{cases} \tag{10}$$

For example, an HMCR of 0.85 indicates that the HS algorithm will choose the decision variable value from historically stored values in the HM with 85% probability or from the entire possible range with (100–85) % probability. Every component obtained by the memory consideration is examined to determine whether it should be pitch-adjusted. This operation uses the PAR parameter, which is the rate of pitch adjustment as follows:

http://dx.doi.org/10.5772/55509

x'_i

$$\leftarrow \begin{cases} Yes & with\,probability\,PAR \\ No & with\,probability\,(1-PAR) \end{cases} \tag{11}$$

The value of (1-PAR) sets the rate of doing nothing. If the pitch adjustment decision for x'_i is "Yes", x'_i will be replaced as follows:

$x'_i \leftarrow x'_i \pm rand() * b_w$

where, b_w is an arbitrary distance bandwidth and rand () is a random number between 0 and 1.

In step 3, HM consideration, pitch adjustment or random selection in turn is applied to each variable of the new harmony vector.

d. Update harmony memory

If the new harmony vector $x'_i=(x'_1, x'_2, \ldots, x'_N)$ is better than the worst harmony in the HM, judged in terms of the objective function value, the new harmony is included in the HM, and the existing worst harmony is excluded from the HM.

e. Check stopping criterion

If the stopping criterion (maximum number of improvisations) is satisfied, the computation terminates. Otherwise, steps 3, and 4 are repeated.

4.2. Cost of FACTS devices

Using database of [32], cost function for SVC, TCSC, and UPFC shown in Fig. 3 are modeled as follows:

For TCSC:

$$C_{TCSC} = 0.0015s^2 - 0.713s + 153.75 \tag{12}$$

For SVC:

$$C_{SVC} = 0.0003s^2 - 0.3015s + 127.38 \tag{13}$$

For UPFC:

$$C_{UPFC} = 0.0003s^2 - 0.2691s + 188.22 \tag{14}$$

Here, s is the operating range of the FACTS devices in MVAR, and C_{TCSC}, C_{SVC}, and C_{UPFC} are in $\$US / kVAR$.

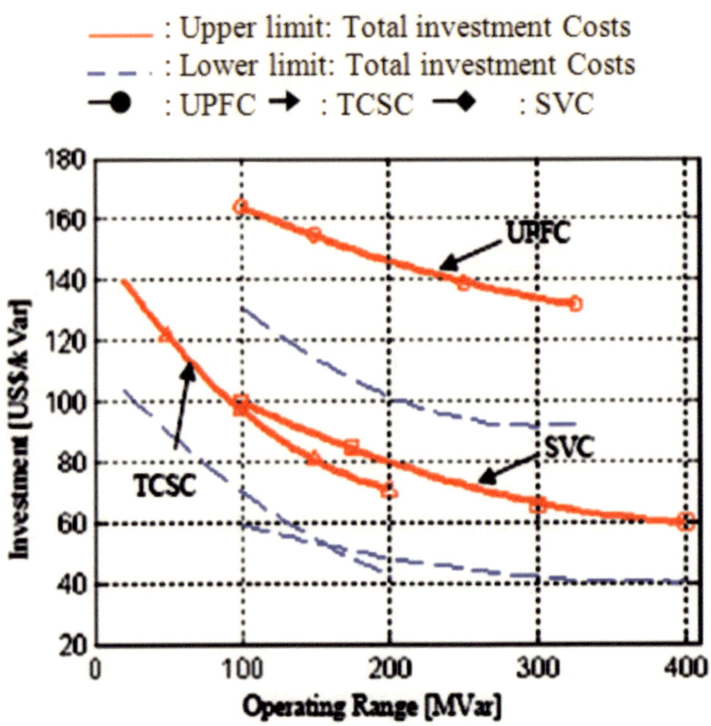

Figure 3. Cost Function of the FACTS devices: SVC, TCSC, and UPFC

4.3. Objective function

The goal of the optimization algorithm is to place FACTS devices in order to enhance power system security-level considering cost function of FACTS devices. These devices should be placed to prevent congestion in transmission lines and transformers, and to maintain bus voltages close to their reference values. Security index introduced in section III has been used in objective function considering cost function of FACTS devices and power system losses. Fitness function FF is expressed by the following equation:

$$FF = a_1.J_p + a_2.J_v + a_3(Total\ Investment\ Cost) + a_4(Losses) \tag{15}$$

The coefficient a_1 to a_4 have been selected by trial and error, which are 0.2665, 0.5714, 0.1421, and 0.02, respectively. These values a_1 to a_4 give better optimization results for different runs of the algorithm. In the equation above, the third term is sum of cost functions of TCSC, UPFC and SVC, described in equations (12) to (14). In addition, the forth term is total active power loss of the power system.

5. Simulation results

Simulation studies are carried out for different scenarios in the IEEE 30-bus power system. Five different cases have been considered:

- Case 1: power system normal operation (without installation of FACTS devices),
- Case 2: one TCSC is installed,
- Case 3: one SVC is installed,
- Case 4: one UPFC is installed, and
- Case 5: Multi-type (TCSC, SVC, and UPFC) FACTS devices are installed.

The first case is the normal operation of network without using any FACTS device. In the second, third, and fourth cases, installation of only one device has been considered. Each device is placed in an optimal location obtained by HSA. Multi-type FACTS devices installation is considered in the 5th scenario. In this case, three different kinds of FACTS devices have been considered to be placed in optimal locations to enhance power system security.

The performance index evolutions of implemented methods are shown in Fig.4, and Fig.5. The average, and maximum performance indices are shown in Fig.4. Tables 1, and 2 show optimal locations of devices for different cases. These results illustrate that the installation of one device in the network could not lead to improved security of power system and reduction in power system losses simultaneously, and that multi-type FACTS devices should be placed in optimal locations to improve security margins and reduce losses in the network.

Device Type	UPFC		TCSC		SVC	
Size/Location	Size (MVA)	Location (Bus No-Bus No)	Size (MVA)	Location (Bus No-Bus No)	Size (MVA)	Location (Bus No.)
TCSC	-	-	90.6	1-2	-	-
SVC	-	-	-	-	39	1
UPFC	48.3	12-15	-	-	-	
Multi-type	75.9	12-15	73.1	2-5	66.7	1

Table 1. The results for FACTS allocations and sizes

Scenario	J_P	J_V	($) Cost	Losses (MVA)
1	3.45	24.2	-	28.36
2	3.39	19.6	9197500	22.51
3	3.33	19.1	4521600	26.3
4	3.21	16.4	8501600	23.09
5	3.09	10.5	27897000	20.49

Table 2. Simulation results for different cases

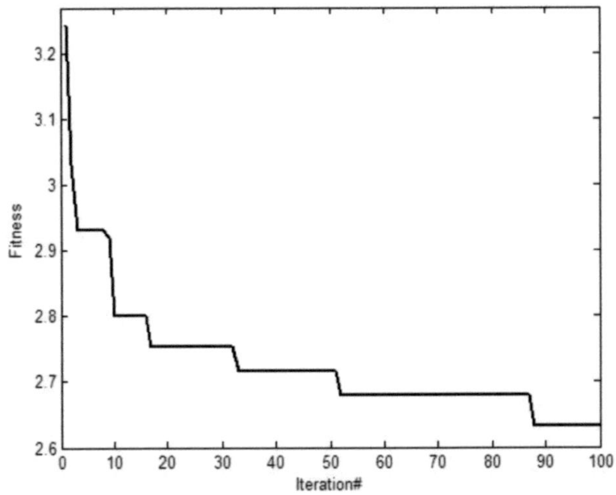

Figure 4. Performance index evolution (average of fitnesses in every iteration)

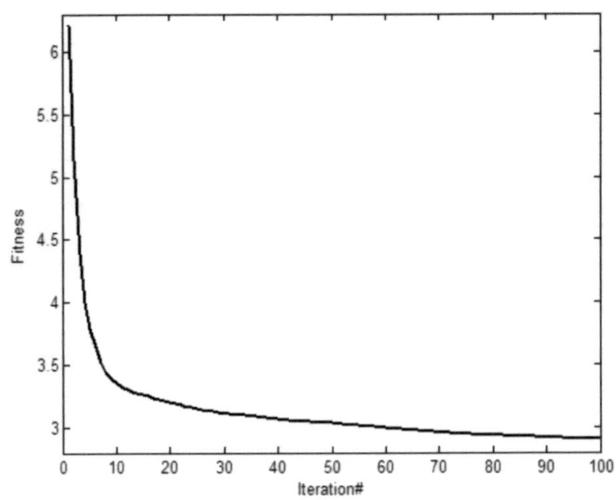

Figure 5. Performance index evolution (maximum of fitnesses in every iteration)

6. Harmonic optimization in multi-level inverters using harmony search algorithm

6.1. Introduction

Nowadays, dc-to-ac inverters are widely used in industry. All applications are mainly divided into two general groups; 1- Electric drives for all ac motors when dc supply is used, and 2- in systems including high voltage direct current (HVDC) transmission systems, custom power and flexible ac transmission systems (FACTS) devices, flexible distributed generation (FDG), and interconnection of distributed generation (DG) units to a grid. Several switching algorithm such as pulse width modulation (PWM), sinusoidal pulse width modulation (SPWM), space-vector modulation (SVM), selective harmonic eliminated pulse width modulation (SHEPWM), or programmed-waveform pulse width modulation (PWPWM) are applied extensively to control and determine switching angles to achieve the desired output voltage. In the recent decade, a new kind of inverter named multi-level inverter has been introduced. In various publications, this inverter has been used in place of the common inverters to indicate its advantages in different applications. Being multi-level, it can be used in high-power and high-voltage applications. In order to reach the desired fundamental component of voltage, all of various switching methods produce harmonics and hence, it is of interest to select the best method to achieve minimum harmonics and total harmonic distortion (THD). It is suggested to use optimized harmonic stepped waveform (OHSW) to eliminate low order harmonics by determining proper angles, and then removing the rest of the harmonics via filters. In addition, this technique lowers switching frequency down to the fundamental frequency and consequently, power losses and cost are reduced.

Traditionally, there are two states for DC sources in multi-level inverters: 1- Equal DC sources, 2- Non-equal DC sources. Several algorithms have been suggested for the above purposes. In [37] Newton-Raphson method has been used to solve equations. Newton-Raphson method is fast and exact for those modulation indices (M) that can satisfy equations, but it cannot obtain the best answer for other indices. Also, [38] has used the mathematical theory of resultants to find the switching angles such that all corresponding low-order harmonics are completely canceled out sequentially for both equal and non-equal DC sources separately. However, by increasing levels of multi-level converters, equation set tends to a high-order polynomial, which narrows its feasible solution space. In addition, this method cannot suggest any answer to minimize harmonics of some particular modulation indices where there is no acceptable solution for the equation set. Genetic algorithm (GA) method has been presented in [39] to solve the same problem with any number of levels for both eliminating and minimizing the harmonics, but it is not fast and exact enough. This method has also been used in [40] to eliminate the mentioned harmonics for non-equal DC sources. Moreover, all optimal solutions have used main equations in fitness function. This means that the fundamental component cannot be satisfied exactly

Here, a harmony search (HS) algorithm approach will be presented that can solve the problem with a simpler formulation and with any number of levels without extensive derivation of analytical expressions. It is also faster and more precise than GA.

7. Cascade H-bridges

The cascaded multi-level inverter is one of the several multi-level configurations. It is formed by connecting several single-phase, H-bridge converters in series as shown in Fig. 1a for a 13-level inverter. Each converter generates a square-wave voltage waveform with different duty ratios. Together, these form the output voltage waveform, as shown in Fig. 1b. A three-phase configuration can be obtained by connecting three of these converters in Y, or Δ. For harmonic optimization, the switching angles θ_1, θ_2,... and θ_6 (for a 13-level inverter) shown in Fig. 1b have to be selected, so that certain order harmonics are eliminated.

8. Problem statement

Fig. 6b shows a 13-level inverter, where θ_1, θ_2,...and θ_6 are variables and should be determined. Each full-bridge inverter produces a three level waveform $+V_{dc}$, $-V_{dc}$ and 0, and each angle θ_i is related to the ith inverter $i=1, 2, ..., S$. S is the number of DC sources that is equal to the number of switching angles (in this study S=6). The number of levels L, is calculated as $L=2S+1$. Considering equal amplitude of all dc sources, the Fourier series expansion of the output voltage waveform is as follows:

$$V(t)=\sum_{n=1}^{\infty} V_n \sin(n\omega t) \tag{16}$$

where, V_n is the amplitude of harmonics. The angles are limited between 0 and 90 ($0<\theta_i<\pi/2$). Because of odd quarter-wave symmetric characteristic, the harmonics with even orders become zero. Consequently, V_n will be as follows:

$$V_n=\begin{cases} \frac{4V_{dc}}{n\pi}\sum_{i=1}^{k}\cos(n\theta_i) & \textit{for odd}\, n \\ 0 & \textit{for even}\ n \end{cases} \tag{17}$$

There are two approaches to adjust the switching angles:

1. Minimizing the THD that is not common, because some low order harmonics may remain.
2. Canceling the lower order harmonics and removing the remained harmonics with a filter.

The second approach is preferred. For motor drive applications, it is necessary to eliminate low order harmonics from 5 to 17. Hence, in this section, a 13-level inverter is chosen to eliminate low-order harmonics from 5 to 17. It is not needed to delete triple harmonics because they will be eliminated in three-phase circuits. Thus, for a 13-level inverter, Eq. (17) changes into (18).

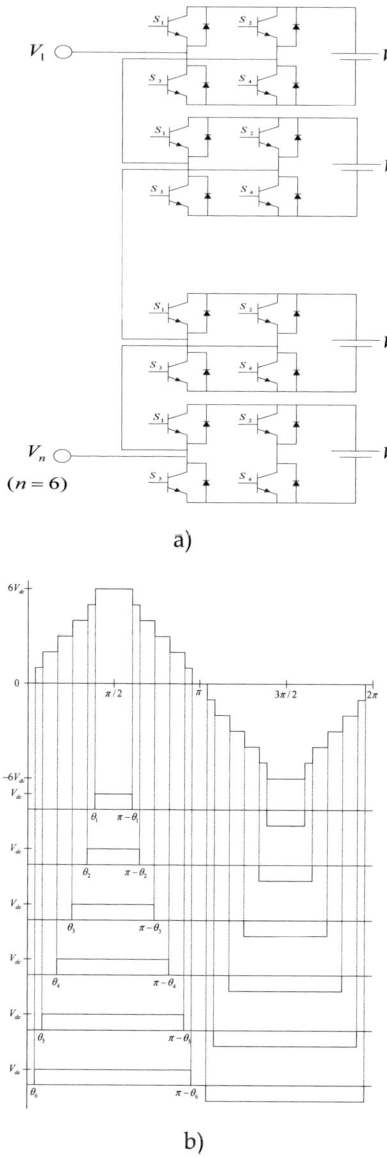

Figure 6. a) Multi-Level Inverter b) Multi-Level waveform generation

$$\begin{aligned} M &= \cos(\theta_1) + \cos(\theta_2) + \ldots + \cos(\theta_6) \\ 0 &= \cos(5\theta_1) + \cos(5\theta_2) + \ldots + \cos(5\theta_6) \\ &\vdots \\ 0 &= \cos(17\theta_1) + \cos(17\theta_2) + \ldots + \cos(17\theta_6) \end{aligned} \tag{18}$$

Here, M is the modulation index and defined as:

$$M = \frac{V_1}{4V_{dc}/\pi} \quad (0 < M \leq 6) \tag{19}$$

It is necessary to determine six switching angles, namely θ_1, θ_2,…and θ_6 so that equation set (18) is satisfied. These equations are nonlinear and different methods can be applied to solve them.

9. Genetic algorithm

In order to optimize the THD, genetic algorithm (GA) that is based on natural evolution and population is implemented. This algorithm is usually applied to reach a near global optimum solution. In each iteration of GA (referred as generation), a new set of strings (i.e. chromosomes) with improved fitness is produced using genetic operators (i.e. selection, crossover and mutation).

4.75	13.02	30.26	43.55	87.36	89.82

Table 3. A typical chromosome

a. Chromosome's structure

Chromosome structure of a GA is shown in table 1 that involves θ_i as parameter of the inverter.

b. Selection

The method of tournament selection is used for selections in a GA [41-42]. This method chooses each parent by choosing n_t (Tournament size) players randomly, and choosing the best individual out of that set to be a parent. In this section, n_t is chosen as 4.

c. Cross Over

Crossover allows the genes from different parents to be combined in children by exchanging materials between two parents. Crossover function randomly selects a gene at the same coordinate from one of the two parents and assigns it to the child. For each chromosome, a random number is selected. If this number is between 0.01 and 0.3 [42], the two parents are combined; else chromosome is transferred with no crossover.

d. Mutation

GA creates mutation-children by randomly changing the genes of individual parents. In this section, GA adds a random vector from a Gaussian distribution to the parents. For each chromosome, random number is selected. If this number is between 0.01 and 0.1 [42], mutation process is applied; else chromosome is transferred with no mutation.

10. Harmony search algorithm

Harmony Search Algorithm (HSA) has been implemented based on the algorithm described in section 4 of the first part of this chapter [34-35].

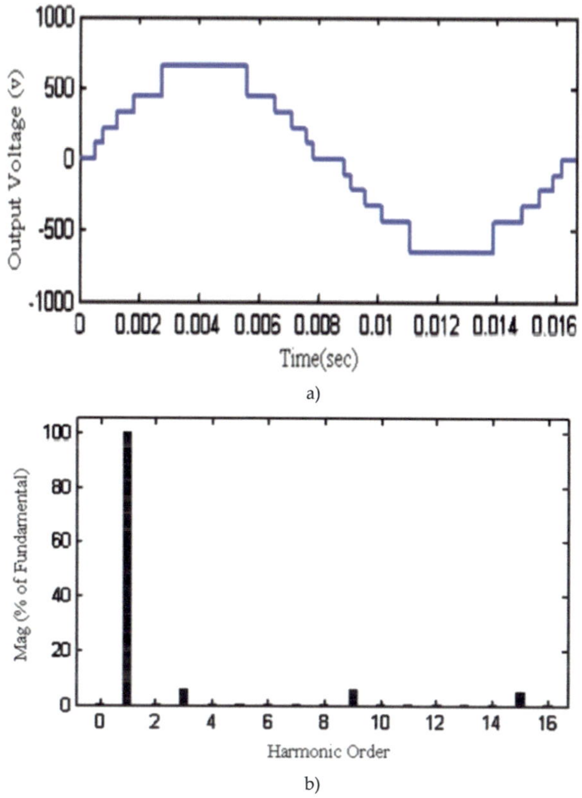

Figure 7. a) output voltage waveform b) harmonic spectrum

11. Simulation results

Harmony Search algorithm has been used to solve the optimization problem. The objective function has been chosen as follows:

$$f = \left\{ \left(100 \frac{V_1^* - V_1}{V_1^*} \right)^4 + \sum_{i=2}^{6} \frac{1}{h} \left(50 \frac{V_i}{V_1} \right) \right\} \tag{20}$$

where, V_1^* is the desired fundamental harmonic, h_1=1, h_2=5 … and h_6=17, are orders of the first six viable harmonics at the output of a three-phase multi-level inverter, respectively. The parameters of the harmony search algorithm have been chosen as: *HMS*=10, *HMCR*=0.9, *PAR*=0.6, and b_w =0.01. The optimal solution vector is obtained after 1000 iterations as: [10.757, 16.35, 26.973, 39.068, 59.409, 59.409]. With these switching angles, the output voltage waveform and its spectrum will be obtained as shown in Fig. 2. The values of the objective function and the total harmonic distortion (THD) has been obtained as: *THD*=4.73%, and f =4.8e−8. Simulation has also been performed by GA and results obtained as: *THD*=7.11%, and f =0.05. It is obvious that the harmony search algorithm performed much better than GA approach.

12. Conclusion

In the first part of the presented chapter, we presented a novel approach for optimal placement of multi-type FACTS devices based on harmony search algorithm. Simulations of IEEE 30-bus test system for different scenarios demonstrate that the placement of multi-type FACTS devices leads to improvement in security, and reduction in losses of power systems.

In the second part, the harmony search algorithm was proposed for harmonic optimization in multi-level inverters. Harmony search algorithm has more flexibility than conventional methods. This method can obtain optimum switching angles for a wide range of modulation indices. This advantage is of importance, especially when the number of switching angles goes up, where equation set may not have any solution, or when it is solvable only for a short range of modulation indices. Moreover, the implementation of the harmony search algorithm is very straightforward compared to the conventional methods like Newton-Raphson, where it is necessary to calculate the Jacobean matrix. In addition, one of the most attractive features of intelligent algorithms is their independency from case studies. Actually, intelligent algorithm can be imposed to a variety of different problems without any need for extensive manipulations. For example, the harmony search algorithm and GA algorithms are able to find optimum switching angles in order to cancel out low-order harmonics, and if it is not possible to completely remove them, they can suggest optimum switching angles so that, low-order harmonics will be reduced as much as possible. Furthermore, with a little manipulation in the defined objective function, one can use HSA and GA as a tool for THD optimization. Also, the results indicate that, harmony search algorithm has many benefits over GA such as simplicity in the implementation, precision, and speed in global convergence.

Author details

H. R. Baghaee*, M. Mirsalim and G. B. Gharehpetian

*Address all correspondence to: hrbaghaee@aut.ac.ir

Department of Electrical Engineering, Amirkabir University of Technology, Tehran, Iran

References

[1] Hingurani, N. G, & Gyugyi, L. (2000). Understanding FACTS: Concepts and Technology of Flexible AC Transmission Systems", USA, IEEE Press, New York.

[2] Alabduljabbar, A. A, & Milanovic, J. V. Assessment of techno-economic contribution of FACTS devices to power system operation", Electric Power Systems Research, Octber (2010)., 80, 1247-1125.

[3] Nagalakshmi, S, & Kamaraj, N. Comparison of computational intelligence algorithms for loadability enhancement of restructured power system with FACTS devices", Comparison of computational intelligence algorithms for loadability enhancement of restructured power system with FACTS devices", Swarm and Evolutionary Computation, August (2012)., 5, 17-27.

[4] Roy Billinton, Mahmud Fotuhi-Firuzabad, Sherif Omar Faried, Saleh Aboreshaid "Impact of Unified Power Flow Controllers on Power System Reliability", IEEE Tran. on Power Systems, (2000)., 15, 410-415.

[5] Narmatha, R. Banu, D. Devaraj, "Multi-objective GA with fuzzy decision making for security enhancement in power system", Applied Soft Computing, September (2012)., 12(9), 2756-2764.

[6] Sung-Hwan Song; Jung-Uk Lim; Seung-Il MoonFACTS Operation Scheme for Enhancement of Power System Security", IEEE Power Tech Conference Proceedings, (2003). Bologna,, 3, 36-41.

[7] Sun-Ho Kim; Jung-Uk Lim; Seung-Il Moon "Enhancement of Power System Security Level through The Power Flow Control Of UPFC" IEEE Power Engineering Society Summer Meeting, (2000)., 1, 38-43.

[8] Kazemi, A, Shayanfar, H. A, Rabiee, A, & Aghaie, J. Power System Security Improvement using the Unified Power Flow Controller (UPFC)", IEEE Power India Conference, (2006)., 1, 1-5.

[9] Singh, J. G, Singh, S. N, & Srivastava, S. C. Placement of FACTS Controllers for Enhancing Power System Loadability" IEEE Power India Conference, (2006)., 1, 1-7.

[10] Jurado, F, & Rodriguez, J. A. Optimal location of SVC based on system loadability and contingency analysis", Proceeding of Emerging Technologies and Automation, (1999)., 2, 1193-1199.

[11] Sing, S. N, & David, A. K. A New Approach for Placement of FACTS Devices in Open Power Markets", IEEE Power Engineering Review, (2001) , 21, 58-60.

[12] Galiana, F. D, Almeida, K, Toussaint, M, Griffin, J, & Atanackovic, D. Assessment and Control of the Impact of FACTS Devices on Power System Performance", IEEE Trans. Power Systems, (2001)., 11, 1931-1936.

[13] Thukaram, D, Jenkins, L, & Visakha, K. Improvement Of System Security With Unified-Power-Flow Controller At Suitable Locations Under Network Contingencies Of Interconnected Systems", IEE Proc. Generation, Transmission and Distribution, (2005)., 152, 682-690.

[14] Verma, K. S, Singh, S. N, & Gupta, H. O. Location of Unified Power Flow Controller for Congestion Management", Electric Power System Research,, 58, 89-96.

[15] Gerbex, S, Chekaoui, R, & Germond, A. J. Optimal Location of Multi-type FACTS Devices in a Power System be Means of Genetic Algorithm", IEEE Trans. on Power Systems, (2001)., 16, 537-544.

[16] Gerbex, S, Cherkaoui, R, & Germond, A. J. Optimal Location of FACTS Devices to Enhance Power System Security", IEEE Power Tech Conference Proceedings, (2003). Bologna, Italy,, 3, 1-7.

[17] Cai, L. J. and Erlich, "Optimal choice and allocation of FACTS devices in deregulated electricity market using genetic algorithms", IEEE PES Power Systems Conference and Exposition, (2004)., 1, 201-207.

[18] Baskaran, J, & Palanisamy, V. Genetic Algorithm Applied to Optimal Location of FACTS Device in a Power System Network Considering Economic Saving Cost", (2005). Academic Open Internet Journal,, 15, 1-8.

[19] Chung-Fu Chang, Ji-Jen Wong, Ji-Pyng Chiouand Ching-Tzong Su "Robust Searching Hybrid Differential Evolution Method for Optimal Reactive Power Planning in Large-Scale Distribution Systems", Electric Power System Research, (2007). , 77, 430-437.

[20] Ji-Pyng Chiou "Variable Scaling Hybrid Differential Evolution for Large-Scale Economic Dispatch Problems"Electric Power System Research", (2007)., 77, 212-218.

[21] Figueroa, N. G, & Cedeño, J. R. A Differential Evolution Solution Approach for Power System State Estimation", Power and Energy Systems, (2004)., 1, 448-456.

[22] Gerbex, R, & Cherkaoui, A. J. Germond "Optimal Location Of Multi-Type FACTS Devices In A Power System By Means Of Genetic Algorithm", IEEE Trans. On Power System, (2001)., 16, 537-544.

[23] Cai, J, & Erlich, I. Optimal choice and allocation of FACTS devices using genetic algorithms", IEEE Trans. on Power System, (2004)., 1, 201-207.

[24] Verma, K. S, Singh, S. N, & Gupda, H. O. Location of Unified Power Flow Controller for Congestion Management", Electric Power System Research, (2001)., 58, 89-96.

[25] Leung, H. C, & Chung, T. S. Optimal power flow with a versatile FACTS controller by Genetic algorithm approach", Proceeding of the 5th international conference on advances in power system control, operation and management, APSCOM, (2000)., 1, 178-183.

[26] Lie, T. T, & Deng, W. Optimal flexible AC transmission systems (FACTS) devices allocations", Electrical power and energy system, (1997). , 19, 125-134.

[27] Baskaran, J, & Palanisamy, V. Optimal location of FACTS device in a power system network considering power loss using genetic algorithm", EE-Pub on line journal www.eepub.com.

[28] Baghaee, H. R, Mirsalim, M, & Kashefi, A. Kaviani, G.B. Gharehpetian, "Security/cost-based optimal allocation of multi-type FACTS devices using multi-objective particle swarm optimization", Simulation: Transactions of the Society for Modeling and Simulation International, (2012). , 1(1), 1-12.

[29] Kim, S. H, Lim, J. U, & Moon, S. I. I. Enhancement of power system security level through the power control of UPFC", IEEE Power Engineering Society Summer Meeting, (2000). , 1, 38-43.

[30] Kim, T. H, Ch, J, Seo, J. U, Lim, S, Moon, J. K, & Park, B. M. Han," A Decoupled unified power flow controller model for power flow considering limit resolution", IEEE Power Engineering Society Winter Meeting, (2000). , 1, 1190-1195.

[31] Lim, J. U, & Moon, S. I. An operation scheme of UPFC's considering operation objective and states", IEEE Power Engineering Society Winter Meeting, (2002). , 1, 610-615.

[32] Atif, S. Debs, " Modem Power Systems Control and Operation" USA, (2008). Kluwer Academic Publishers.

[33] Allen, J. Wood, Bruce F. Wollenberg, "Power Generation, Operation, and Control", USA, (1996). John Wiley & Sons.

[34] Alireza Askarzadeh, Alireza Rezazadeh "Parameter identification for solar cell models using harmony search-based algorithms", Solar Energy, (2012). , 86, 3241-3249.

[35] Sergio Gil-López, Javier Del Ser, Sancho Salcedo-Sanz, Ángel M. Pérez-Bellido, José Marı´a Cabero, José A. Portilla-Figueras "A hybrid harmony search algorithm for the spread spectrum radar polyphase codes design problem", Expert Systems with Applications, (2012). , 39, 11089-11093.

[36] Lee, K. S, & Geem, Z. W. A New Structural Optimization Method Based On The Harmony Search Algorithm", Computer Structure, (2005).

[37] Jason LaiChair Dusan Borojevic, and Alex Q. Huang "Optimized Harmonic Stepped Waveform from Multi-Level Inverters", M.S thesis Virginia Polytechnic Institute and State University, (1999).

[38] Chiasson, J, & Tolbert, L. Keith McKenzie, and Zhong Du, "Eliminating Harmonics in a Multi-level Converter using Resultant Theory", IEEE Trans. on Control Systems Technology, (2005). , 13, 216-223.

[39] Ozpineci, B, Tolbert, L. M, & Chiasson, J. N. Harmonic Optimization of Multi-level Converters using Genetic Algorithms", Proceeding on 35th IEEE Power Electronics Specialists Conference, ((2005)., 5, 3911-3916.

[40] Hosseini, M. G, & Aghdam, S. H. Fathi, and A. Ghasemi, "Modeling and Simulation of Three-Phase OHSW Multi-level Voltage-Source Inverter by Means of Switching Functions", IEEE International Conference on Power Electronics and Drives Systems, (2005)., 1, 633-641.

[41] Miller, B. L, & Goldberg, D. E. Genetic Algorithms, Tournament Selection, and the Effects of Noise", Complex Systems, (1995)., 1, 193-212.

[42] The Math Works IncMATLAB Genetic Algorithm and Direct Search Toolbox User's Guide, Mathwork Inc., (2012).

Chapter 9

Heuristic Search Applied to Fuzzy Cognitive Maps Learning

Bruno Augusto Angélico, Márcio Mendonça,
Lúcia Valéria R. de Arruda and Taufik Abrão

Additional information is available at the end of the chapter

http://dx.doi.org/10.5772/52918

1. Introduction

Fuzzy Cognitive Maps were initially proposed by Kosko [1–3], as an extension of cognitive maps proposed by Axelrod [4]. FCM is a graph used for representing causal relationships among concepts that stand for the states and variables of the system, emulating the cognitive knowledge of experts on a specific area. FCM can be interpreted as a combination of Fuzzy Logic and Neural Networks, because it combines the sense rules of Fuzzy Logic with the learning of the Neural Networks. A FCM describes the behavior of a knowledge based system in terms of concepts, where each concept represents an entity, a state, a variable, or a characteristic of the system. The human knowledge and experience about the system determines the type and the number of the nodes as well as the initial conditions of the FCM.

FCM has been considered with great research interest in many scientific fields, such as political decision [5], medical decision [6, 7], industrial process control [8, 9], artificial life [10], social systems [11], corporative decision, policy analysis [12], among others.

The knowledge of the experts firstly defines the influence of a concept on the other, determining the causality relationships. Then, the concept values are qualitatively obtained by linguistic terms, such as strong, weak, null, and so on. These linguistic variables are transformed in numerical values using a *defuzzification* method, for instance, the center of gravity scheme described in [2].

Hence, in general, experts develop a FCM identifying key concepts, defining the causal relationships among the concepts, and estimating the strength of these relationships. However, when the experts are not able to express the causal relationships or they substantially diverge in opinion about it, data driven methods for learning FCMs may be necessary.

Particularly, this Chapter focuses on the FCM learning using three different population based metaheuristics: particle swarm optimization (PSO), genetic algorithm (GA) and differential evolution (DE). Two process control problems described in [8] and [9] are considered in this work. A complete convergence analysis of the PSO, GA and DE is carried out considering 10000 realizations of each algorithm in every scenario of the studied processes.

The rest of the Chapter has the following organization: Section 2 briefly describes the FCM modeling and the processes to be controlled. Section 3 considers the PSO, GA and the DE approaching for FCM learning, while Section 4 shows the simulation results. Lastly, Section 5 points out the main conclusions.

2. FCM modeling in control processes

In FCMs, concepts (nodes) are utilized to represent different aspects and behavior of the system. The system dynamics are simulated by the interaction of concepts. The concept C_i, $i = 1, 2, \ldots, N$ is characterized by a value $A_i \in [0, 1]$.

Concepts are interconnected concerning the underlying causal relationships amongst factors, characteristics, and components that constitute the system. Each interconnection between two concepts, C_i and C_j, has a weight, $W_{i,j}$, which is numerically represented by the strength of the causal relationships between C_i and C_j. The sign of $W_{i,j}$ indicates whether the concept C_i causes the concept C_j or vice versa. Hence, if:

$$\begin{cases} W_{i,j} > 0, & \text{positive causality} \\ W_{i,j} < 0, & \text{negative causality} \\ W_{i,j} = 0, & \text{no relation} \end{cases}$$

The number of concepts and the initial weights of the FCM are determined by human knowledge and experience. The numerical values, A_i, of each concept is a transformation of the fuzzy values assigned by the experts. The FCM converges to a steady state (limit cycle) according to the scheme proposed in [3]:

$$A_i(k+1) = f\left(A_i(k) + \sum_{\substack{j=1 \\ j \neq i}}^{N} W_{ji} A_j(k) \right), \tag{1}$$

where k is the interaction index and $f(\cdot)$ is the sigmoid function

$$f(x) = \frac{1}{1 + e^{-\lambda x}}, \tag{2}$$

that guarantees the values $A_i \in [0, 1]$. $\lambda > 0$ is a parameter representing the learning memory. In this work, $\lambda = 1$ has been adopted.

2.1. First control system (PROC1)

A simple chemical process frequently considered in literature [8, 13, 14], is initially selected for illustrating the need of a FCM learning technique. Figure 1 represents the process (PROC1) consisting of one tank and three valves that control the liquid level in the tank. Valves V_1 and V_2 fill the tank with different liquids. A chemical reaction takes place into the tank producing a new liquid that leaves the recipient by valve V_3. A sensor (gauger) measures the specific gravity of the resultant mixture.

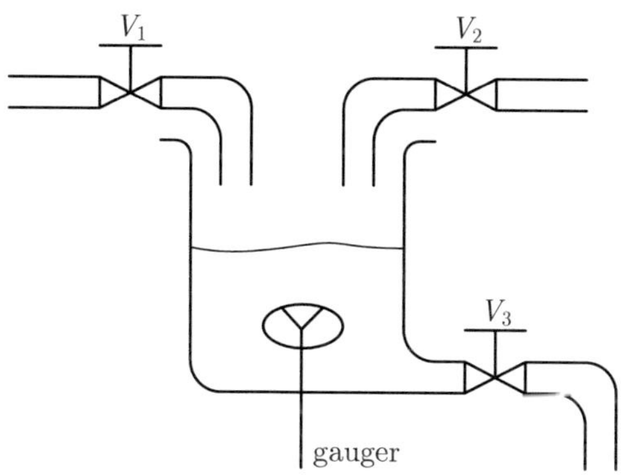

Figure 1. PROC1: a chemical process control problem described in [8].

When the value of the specific gravity, G, is in the range $[G_{min}, G_{max}]$, the desired liquid has been produced. The height of the liquid inside, H, must lie in the range $[H_{min}, H_{max}]$. The controller has to keep G and H within their bounds, i.e.,

$$H_{min} \leq H \leq H_{max}; \tag{3}$$

$$G_{min} \leq G \leq G_{max}. \tag{4}$$

The group of experts defined a list of five concepts, C_i, $i = 1, 2, \ldots, 5$, related to the main physical quantities of the process [8]:

- Concept C_1: volume of liquid inside the tank (depends on V_1, V_2, and, V_3);
- Concept C_2: state of V_1 (closed, open or partially open);
- Concept C_3: state of V_2 (closed, open or partially open);
- Concept C_4: state of V_3 (closed, open or partially open);
- Concept C_5: specific gravity of the produced mixture.

For this process, the fuzzy cognitive map in Figure 2 can be abstracted [8].

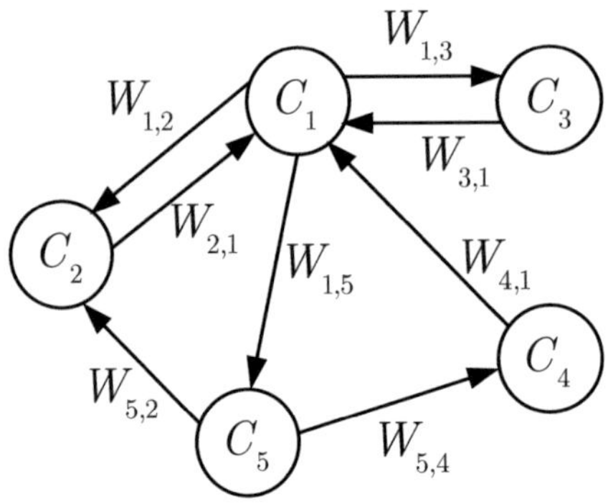

Figure 2. Fuzzy Cognitive Map proposed in [8] for the chemical process control problem.

The experts also had a consensus regarding the range of the weights between concepts, as presented in Equations (5a) to (5h).

$$-0.50 \leq W_{1,2} \leq -0.30; \tag{5a}$$
$$-0.40 \leq W_{1,3} \leq -0.20; \tag{5b}$$
$$0.20 \leq W_{1,5} \leq 0.40; \tag{5c}$$
$$0.30 \leq W_{2,1} \leq 0.40; \tag{5d}$$
$$0.40 \leq W_{3,1} \leq 0.50; \tag{5e}$$
$$-1.0 \leq W_{4,1} \leq -0.80; \tag{5f}$$
$$0.50 \leq W_{5,2} \leq 0.70; \tag{5g}$$
$$0.20 \leq W_{5,4} \leq 0.40. \tag{5h}$$

For this problem the following weight matrix is obtained:

$$\mathbf{W} = \begin{bmatrix} 0 & W_{1,2} & W_{1,3} & 0 & W_{1,5} \\ W_{2,1} & 0 & 0 & 0 & 0 \\ W_{3,1} & 0 & 0 & 0 & 0 \\ W_{4,1} & 0 & 0 & 0 & 0 \\ 0 & W_{5,2} & 0 & W_{5,4} & 0 \end{bmatrix}. \tag{6}$$

According to [8], all the experts agreed on the range of values for $W_{2,1}$, $W_{3,1}$, and $W_{4,1}$, and most of them agreed on the same range for $W_{1,2}$ and $W_{1,3}$. However, regarding the weights $W_{1,5}$, $W_{5,2}$, and $W_{5,4}$, their opinions varied significantly.

Finally, the group of experts determined that the values output concepts, C_1 and C_5, which are crucial for the system operation, must lie, respectively, in the following regions:

$$0.68 \leq A_1 \leq 0.70; \tag{7a}$$
$$0.78 \leq A_5 \leq 0.85. \tag{7b}$$

2.2. Second control system (PROC2)

In [9] it is considered a system consisting of two identical tanks with one input and one output valve each one, with the output valve of the first tank being the input valve of the second (PROC2), as illustrated in Figure 3.

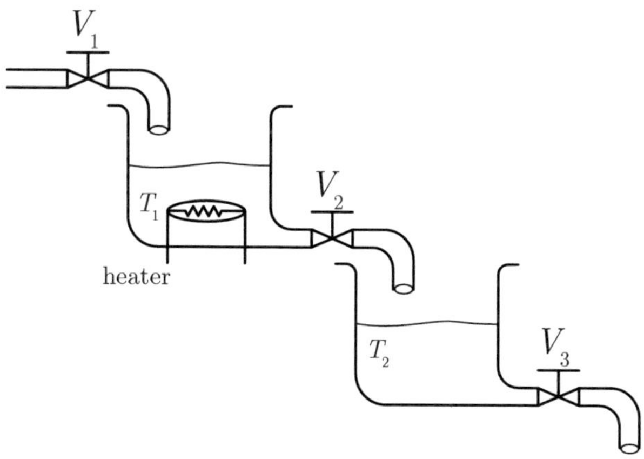

Figure 3. PROC2: a chemical process control problem described in [9].

The objective is to control the volume of liquid within the limits determined by the height H_{min} and H_{max} and the temperature of the liquid in both tanks within the limits T_{min} and T_{max}, such that

$$T^1_{min} \leq T^1 \leq T^1_{max}; \tag{8a}$$
$$T^2_{min} \leq T^2 \leq T^2_{max}; \tag{8b}$$
$$H^1_{min} \leq H^1 \leq H^1_{max}; \tag{8c}$$
$$H^2_{min} \leq H^2 \leq H^2_{max}. \tag{8d}$$

The temperature of the liquid in tank 1 is increased by a heater. A temperature sensor continuously monitors the temperature in tank 1, turning the heater on or off. There is also a temperature sensor in tank 2. When T_2 decreases, the valve V_2 is open and hot liquid comes into tank 2.

Based on this process, a FCM is constructed with eight concepts:

- Concept C_1: volume of liquid inside the tank 1 (depends on V_1 and V_2);
- Concept C_2: volume of liquid inside the tank 2 (depends on V_1 and V_2);
- Concept C_3: state of V_1 (closed, open or partially open);
- Concept C_4: state of V_2 (closed, open or partially open);
- Concept C_5: state of V_3 (closed, open or partially open);
- Concept C_6: Temperature of the liquid in tank 1;

- Concept C_7: Temperature of the liquid in tank 2;
- Concept C_8: Operation of the heater.

According to [9], the fuzzy cognitive map in Figure 4 can be constructed.

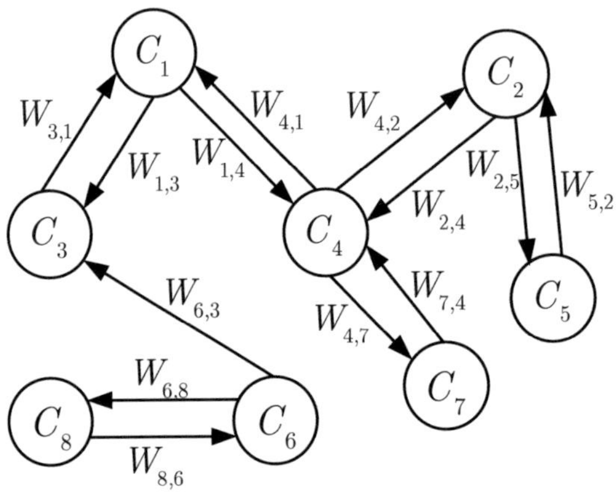

Figure 4. Fuzzy Cognitive Map proposed in [9] for the chemical process control problem.

It is assumed for PROC2 in this Chapter only causal constraints in the weights between concepts, where concepts $W_{4,1}$ and $W_{5,2}$ are $\in (-1, 0]$ and the others have positive causality. The weight matrix for PROC2 is given by

$$\mathbf{W} = \begin{bmatrix} 0 & 0 & W_{1,3} & W_{1,4} & 0 & 0 & 0 & 0 \\ 0 & 0 & 0 & W_{2,4} & W_{2,5} & 0 & 0 & 0 \\ W_{3,1} & 0 & 0 & 0 & 0 & 0 & 0 & 0 \\ W_{4,1} & W_{4,2} & 0 & 0 & 0 & 0 & W_{4,7} & 0 \\ 0 & W_{5,2} & 0 & 0 & 0 & 0 & 0 & 0 \\ 0 & 0 & W_{6,3} & 0 & 0 & 0 & 0 & W_{6,8} \\ 0 & 0 & 0 & W_{7,4} & 0 & 0 & 0 & 0 \\ 0 & 0 & 0 & 0 & 0 & W_{8,6} & 0 & 0 \end{bmatrix}. \tag{9}$$

Finally, the values output concepts, C_1, C_2, C_6 and C_7, which are crucial for the system operation, must lie, respectively, in the following regions:

$$0.64 \leq A_1 \leq 0.69; \tag{10a}$$
$$0.48 \leq A_2 \leq 0.52; \tag{10b}$$
$$0.63 \leq A_6 \leq 0.67; \tag{10c}$$
$$0.63 \leq A_7 \leq 0.67. \tag{10d}$$

Two significant weaknesses of FCMs are its critical dependence on the experts opinions and its potential convergence to undesired states. In order to handle these impairments, learning procedures can be incorporated, increasing the efficiency of FCMs. In this sense, heuristic optimization approach has been deployed as an effective learning method in FCMs [15].

3. Heuristic FCM learning

A FCM construction can be done in the following manner:

- Identification of concepts and its interconnections determining the nature (positive, negative or null) of the causal relationships between concepts.
- Initial data acquisition by the expert opinions and/or by an equation analysis when the mathematical system model is known.
- Submitting the data from the expert opinions to a fuzzy system which output represents the weights of the FCM.
- Weight adaptation and optimization of the initially proposed FCM, adjusting its response to the desired output.
- Validation of the adjusted FCM.

This section focuses on the weight adaptation (FCM learning). In [16] a very interesting survey on FCM learning is provided. The FCM weights optimization (FCM learning) can be classified into three different methods.

1. Hebbian learning based algorithm;
2. Heuristic optimization techniques, including genetic algorithm, particle swarm optimization, differential evolution, simulated annealing, etc;
3. Hybrid approaches.

In the Hebbian based methodologies, the FCM weights are iteratively adapted based on a law which depends on the concepts behavior [10], [17]. These algorithms require the experts' knowledge for initial weight values. The differential Hebbian learning (DHL) algorithm proposed by Dickerson and Kosko is a classic example [10]. On the other hand, heuristic (metaheuristic) techniques tries to find a proper **W** matrix by minimizing a cost function based on the error among the desired values of the output concepts and the current output concepts' values (13). The experts' knowledge is not totally necessary, except for the causality constraints, due to the physical restrictions[1]. These techniques are optimization tools and generally are computationally complex. Examples of hybrid approaching considering Hebbian learning and Heuristic optimization techniques can be found in [18], [19].

There are several works in the literature dealing with heuristic optimization learning. Most of them are population-based algorithms. For instance, in [8] the PSO algorithm with constriction factor is adopted; in [20] it is presented a FCM learning based on a Tabu Search (TS) and GA combination; in [21] a variation of GA named RCGA (real codec-G.A.) is proposed; in [22] a comparison between GA and Simulated Annealing (SA) is done; in [13] the authors presented a GA based algorithm named Extended Great Deluge Algorithm.

The purpose of the learning is to determine the values of the FCM weights that will produce a desired behavior of the system, which are characterized by M output concept values that

[1] For instance, a valve cannot be negatively open.

lie within desired bounds determined by the experts. Hence, the main goal is to obtain a connection (or weight) matrix

$$\mathbf{W} = \left[W_{i,j}\right], \quad i,j = 1, 2, \ldots, N, \tag{11}$$

that leads the FCM to a steady state with output concept values within the specified region. Note that, with this notation, and defining $\mathbf{A} = [A_1 \ \cdots \ A_N]^\top$, and $\underline{\mathbf{W}} = \mathbf{W}^\top + \mathbf{I}$, with $\{\cdot\}^\top$ meaning transposition and $\mathbf{I}$ identity matrix, Equation (1) can be compactly written as

$$\mathbf{A}(k+1) = f\left(\underline{\mathbf{W}} \cdot \mathbf{A}(k)\right). \tag{12}$$

After the updating procedure in (12), the following cost function is considered for obtaining the optimum weight matrix $\mathbf{W}$ [8]:

$$\begin{aligned} F(\mathbf{W}) = & \sum_{i=1}^{M} H\left(\min\left(A_{\text{out}}^{i}\right) - A_{\text{out}}^{i}\right)\left|\min\left(A_{\text{out}}^{i}\right) - A_{\text{out}}^{i}\right| \\ & + \sum_{i=1}^{M} H\left(A_{\text{out}}^{i} - \max\left(A_{\text{out}}^{i}\right)\right)\left|\max\left(A_{\text{out}}^{i}\right) - A_{\text{out}}^{i}\right|, \end{aligned} \tag{13}$$

where $H(\cdot)$ is the Heaviside function, and A_{out}^{i}, $i = 1, \ldots, M$, represents the value of the ith output concept.

3.1. Particle Swarm Optimization

The PSO is a meta-heuristic based on the movement of a population (swarm) of individuals (particles) randomly distributed in the search space, each one with its own position and velocity. The position of a particle is modified by the application of velocity in order to reach a better performance [23, 24]. In PSO, each particle is treated as a point in a $\mathcal{W}$-dimensional space[2] and represents a candidate vector. The ith particle position at instant t is represented as

$$\mathbf{x}_i(t) = \left[\, x_{i,1}(t) \ x_{i,2}(t) \ \cdots \ x_{i,\mathcal{W}}(t) \,\right]. \tag{14}$$

In this Chapter, each $x_{i,1}(t)$ represents one of the $W_{i,j}$ in the tth iteration. Each particle retains a memory of the best position it ever encountered. The best position among all particles until the tth iteration (best global position) is represented by $\mathbf{x}_g^{\text{best}}$, while the best position of the ith

[2] $\mathcal{W}$ is the number of FCM connections (relationships).

particle is represented as $\mathbf{x}_i^{\text{best}}$. As proposed in [25], the particles are manipulated according to the following velocity and position equations:

$$\mathbf{v}_i(t+1) = \omega \cdot \mathbf{v}_i(t+1) + \phi_1 \cdot \mathbf{U}_{1i}\left(\mathbf{x}_g^{\text{best}}(t) - \mathbf{x}_i(t)\right) + \phi_2 \cdot \mathbf{U}_{2i}\left(\mathbf{x}_i^{\text{best}}(t) - \mathbf{x}_i(t)\right) \quad (15)$$

$$\mathbf{x}_i(t+1) = \mathbf{x}_i(t) + \mathbf{v}_i(t+1), \quad (16)$$

where ϕ_1 and ϕ_2 are two positive constants representing the individual and global acceleration coefficients, respectively, $\mathbf{U}_{1i}$ and $\mathbf{U}_{2i}$ are diagonal matrices whose elements are random variables uniformly distributed (u.d.) in the interval $[0, 1]$, and ω is the inertial weight that plays the role of balancing the global search (higher ω) and the local search (smaller ω).

A typical value for ϕ_1 and ϕ_2 is $\phi_1 = \phi_2 = 2$ [24]. Regarding the inertia weight, experimental results suggest that it is preferable to initialize ω to a large value, and gradually decrease it.

The population size $\mathcal{P}$ is kept constant in all iterations. In order to obtain further diversification for the search universe, a factor $V_{\max}$ is added to the PSO model, which is responsible for limiting the velocity in the range $[\pm V_{\max}]$, allowing the algorithm to escape from a possible local solution.

Regarding the FCM, the ith candidate vector $\mathbf{x}_i$ is represented by a vector formed by $\mathcal{W}$ FCM weights. It is important to point out that after each particle update, restrictions must be imposed on $W_{i,j}$ according to the experts opinion, before the cost function evaluation.

3.2. Genetic Algorithm

Genetic Algorithm is an optimization and search technique based on selection mechanism and natural evolution, following Darwin's theory of species' evolution, which explains the history of life through the action of physical processes and genetic operators in populations or species. GA allows a population composed of many individuals to evolve under specified selection rules to a state that maximizes the "fitness" (maximizes or minimizes a cost function). Such an algorithm became popular through the work of John Holland in the early 1970s, and particularly his book Adaptation in Natural and Artificial Systems (1975). The algorithm can be implemented in a binary form or in a continuous (real-valued) form. This Chapter considers the latter case.

Initially, a set of $\mathcal{P}$ chromosomes (individuals) is randomly (uniformly distributed) defined, where each chromosome, $\mathbf{x}_i$, $i = 1, 2, \cdots, \mathcal{P}$ consists of a vector of variables to be optimized, which, in this case, is formed by FCM weights, respecting the constraints. Each variable is represented by a continuous floating-point number. The $\mathcal{P}$ chromosomes are evaluated through a cost function.

T strongest chromosomes are selected for mating, generating the mating pool, using the roulette wheel method, where the probability of choosing a given chromosome is proportional to its fitness value. In this work, each pairing generates two offspring with crossover. The weakest T chromosomes are changed by the T offspring from $T/2$ pairing.

The crossover procedure is similar to the one presented in [26]. It begins by randomly selecting a variable in the first pair of parents to be the crossover point

$$\alpha = \lceil u \cdot \mathcal{W} \rceil, \tag{17}$$

where u is an u.d. random variable (r.v.) in the interval $[0, 1]$, and $\lceil \cdot \rceil$ is the upper integer operator. The jth pair of parents, $j = 1, 2, \cdots, T/2$ is defined as

$$\begin{aligned} \text{dad}_j &= \left[x^d_{j,1} \; x^d_{j,2} \cdots x^d_{j,\alpha} \cdots x^d_{j,\mathcal{W}} \right] \\ \text{mom}_j &= \left[x^m_{j,1} \; x^m_{j,2} \cdots x^m_{j,\alpha} \cdots x^m_{j,\mathcal{W}} \right] \end{aligned}. \tag{18}$$

Then the selected variables are combined to form new variables that will appear in the offspring

$$\begin{aligned} x^o_{j,1} &= x^m_{j,\alpha} - \beta \left[x^m_{j,\alpha} - x^d_{j,\alpha} \right]; \\ x^o_{j,2} &= x^d_{j,\alpha} + \beta \left[x^m_{j,\alpha} - x^d_{j,\alpha} \right]. \end{aligned} \tag{19}$$

where β is also a r.v. u.d. in the interval $[0, 1]$. Finally,

$$\begin{aligned} \text{offspring}_1 &= \left[x^d_{j,1} \; x^d_{j,2} \; x^d_{j,\alpha-1} \cdots x^o_{j,1} \; x^m_{j,\alpha+1} \cdots x^m_{j,\mathcal{W}} \right] \\ \text{offspring}_2 &= \left[x^m_{j,1} \; x^m_{j,2} \; x^m_{j,\alpha-1} \cdots x^o_{j,2} \; x^d_{j,\alpha+1} \cdots x^d_{j,\mathcal{W}} \right]. \end{aligned} \tag{20}$$

In order to allow escaping from possible local minima, a mutation operation is introduced in the resultant population, except for the strongest one (elitism). It is assumed in this work a Gaussian mutation. If the rate of mutations is given by P_m, there will be $N_m = \lceil P_m \cdot (\mathcal{P} - 1) \cdot \mathcal{W} \rceil$ mutations uniformly chosen among $(\mathcal{P} - 1) \cdot \mathcal{W}$ variables. If $x_{i,w}$ is chosen, with $w = 1, 2, \cdots, \mathcal{W}$, than, after Gaussian mutation, it is substituted by

$$x'_{i,w} = x_{i,w} + \mathcal{N}\left(0, \sigma^2_m\right), \tag{21}$$

where $\mathcal{N}\left(0, \sigma^2_m\right)$ represents a normal r.v. with zero mean and variance σ^2_m.

After mutation, restrictions must be imposed on $W_{i,j}$ according to the experts opinion, before the cost function evaluation.

3.3. Differential Evolution

The Differential Evolution (DE) search has been introduced by Ken Price and Rainer Storn [27, 28]. DE is a parallel direct search method. As in GA, a population with $\mathcal{P}$ elements is randomly defined, where each $\mathcal{W}$-dimension element consists of a **x** vector of variables to be optimized (FCM weights in this case) respecting the constraints.

In the classical DE, a perturbation is created by using a difference vector based mutation,

$$\mathbf{y}_i = \mathbf{x}_{r0} + F_e \cdot (\mathbf{x}_{r1} - \mathbf{x}_{r2}), \qquad i = 1, 2, \cdots, \mathcal{P}, \tag{22}$$

where the real and constant factor F_e (typically $\in [0.5, 1.0]$) controls the gain of the differential variation. The indexes $r0$, $r1$ and $r2$ are randomly chosen and mutually exclusive. In this work, an alternative perturbation procedure named DE/current-to-best/1/bin is considered [28, 29], such that

$$\mathbf{y}_i = \mathbf{x}_i + F_e \cdot (\mathbf{x}_{\text{best}} - \mathbf{x}_i) + F_e \cdot (\mathbf{x}_{r1} - \mathbf{x}_{r2}), \qquad i = 1, 2, \cdots, \mathcal{P}, \tag{23}$$

There are also other variants of the perturbation procedure [28, 29]. A uniform crossover operation is applied in order to have diversity enhancement, which mixes parameters of the mutation vector $\mathbf{y}_i$ and $\mathbf{x}_i$, for generating the trial vector $\mathbf{u}_i$:

$$\mathbf{u}_i = \begin{cases} \mathbf{y}_i & \text{if } (r \leq \chi) \\ \mathbf{x}_i & \text{otherwise} \end{cases} \qquad i = 1, 2, \cdots, \mathcal{P}, \tag{24}$$

where χ is the crossover constant typically $\in [0.8, 1.0]$ and r is a random variable u.d. in the interval $[0, 1)$. In order to prevent the case $\mathbf{u}_i = \mathbf{x}_i$, at least one component is taken from the mutation vector $\mathbf{y}_i$.

For selecting, the algorithm uses a simple method where the trial vector $\mathbf{u}_i$ competes against the target vector $\mathbf{x}_i$, such that,

$$\mathbf{x}_i(t+1) = \begin{cases} \mathbf{u}_i(t) & \text{if } f(\mathbf{u}_i(t)) \leq f(\mathbf{x}_i(t)) \\ \mathbf{x}_i(t) & \text{otherwise} \end{cases} \tag{25}$$

4. Simulation results

All the simulations were taken considering 10^4 trials for PROC1 and PROC2. It is worth noting that PSO, GA and DE input parameters were chosen previously after exhaustive simulation tests. As a result, optimal or quasi-optimal input parameters for the PSO, GA and DE heuristic algorithms have been obtained.

4.1. Process PROC1

The adopted values for the input parameters of PSO, GA and DE are summarized on Table 1. DE has less parameter inputs than the other two methods, which is a relative advantage. Four different scenarios for the process PROC1 were analyzed. The main performance results for each scenario is described in the next subsections.

4.1.1. *Scenario 1*

This scenario considers all the constraints on the FCM weights shown in Equations (5a) to (5h). As mentioned in [8] and also verified here, there is no solution in this case.

PSO	
Population	$\mathcal{P} = 10, 20, 30$
Acceleration Coefficients	$\phi_1 = 2, \phi_2 = 0.2$
Inertial Weight	$\omega = 1.2$
Initial Velocity	$\mathbf{v}_i(0) = 0.1$
Maximum Velocity	$V_{\max} = 2$
GA	
Population	$\mathcal{P} = 10$
Mating Pool	$T = 2 \cdot \text{round}(\mathcal{P}/4)$
Rate of Mutation	$P_m = 0.2$
Mutation Std. Deviation	$\sigma_m = 0.3$
DE	
Population	$\mathcal{P} = 10, 20, 30$
Crossover Constant	$\chi = 0.9$
DE Gain Variation Factor	$F_e = 0.9$

Table 1. PSO, GA and DE input parameters values for PROC1.

4.1.2. Scenario 2

In this scenario, the constraints on the FCM weights $W_{1,5}$, $W_{5,2}$, and $W_{5,4}$ have been relaxed, since the experts' opinions have varied significantly. In this case the values of these weights were allowed to assume any value in the interval [0, 1] in order to keep the causality of relationships. Tables 2, 3 and 4 present the obtained simulation results for PSO, GA and DE, respectively.

As can be observed, in the Scenario 2, GA has a better performance than DE and PSO, achieving convergence without failure with $\mathcal{P} = 10$. PSO has the second best performance, presenting no convergence errors in 10^4 independent experiments when $\mathcal{P} = 20$. DE does not achieve 100% of performance success, even for $\mathcal{P} = 30$, when 63 errors occurred in 10^4 independent experiments. Figures 5 and 6 present the mean FCM concepts convergence in the Scenario 2 under 10^4 trials. Considering the best case simulated for each algorithm (GA with $\mathcal{P} = 10$, PSO with $\mathcal{P} = 20$ and DE with $\mathcal{P} = 30$), DE and PSO presented similar average convergence, being faster than GA.

PSO with $\mathcal{P} = 10$					
$A_{\max}$	0.6882	0.8058	0.6203	0.8389	0.8186
$A_{\min}$	0.6477	0.7297	0.5749	0.6590	0.7800
A_{average}	0.6840	0.7911	0.6178	0.6713	0.8148
Number of Failures: 49					
Probability of Success: 0.9951					
PSO with $\mathcal{P} = 20$					
$A_{\max}$	0.6882	0.8051	0.6183	0.6967	0.8186
$A_{\min}$	0.6800	0.7297	0.5750	0.6590	0.7801
A_{average}	0.6838	0.7926	0.6178	0.6730	0.8139
Number of Failures: 0					
Probability of Success: 1.0					

Table 2. PSO simulation results for Scenario 2, PROC1.

http://dx.doi.org/10.5772/52918

GA with $\mathcal{P} = 10$					
A_{max}	0.6882	0.8051	0.6183	0.6964	0.8186
A_{min}	0.6800	0.7297	0.5749	0.6590	0.7800
$A_{average}$	0.6833	0.7726	0.6123	0.6605	0.8059
Number of Failures: 0					
Probability of Success: 1.0					

Table 3. GA simulation results for Scenario 2, PROC1.

DE with $\mathcal{P} = 10$					
A_{max}	0.6882	0.8062	0.6217	0.8389	0.8186
A_{min}	0.6243	0.5563	0.5749	0.6590	0.6590
$A_{average}$	0.6746	0.7685	0.6074	0.6620	0.8078
Number of Failures: 4372					
Probability of Success: 0.5628					
DE with $\mathcal{P} = 20$					
A_{max}	0.6882	0.8058	0.6203	0.8389	0.8186
A_{min}	0.6477	0.5563	0.5749	0.6590	0.7776
$A_{average}$	0.6810	0.7845	0.6070	0.6616	0.8089
Number of Failures: 537					
Probability of Success: 0.9463					
DE with $\mathcal{P} = 30$					
A_{max}	0.6882	0.8054	0.6192	0.6965	0.8186
A_{min}	0.6653	0.5984	0.5749	0.6590	0.7800
$A_{average}$	0.6819	0.7861	0.6072	0.6614	0.8089
Number of Failures: 63					
Probability of Success: 0.9937					

Table 4. DE simulation results for Scenario 2, PROC1.

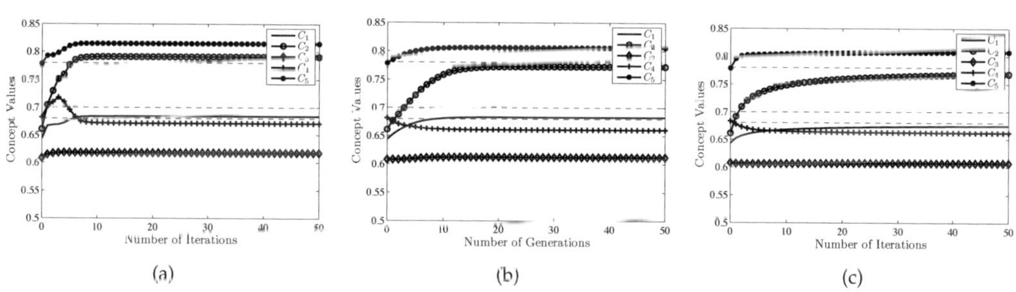

Figure 5. Mean convergence for (a) PSO, (b) GA and (c) DE in PROC1, Scenario 2 with $\mathcal{P} = 10$.

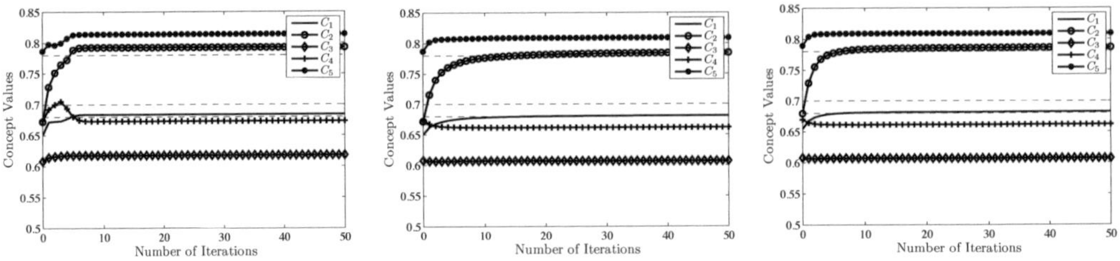

(a) (b) (c)

Figure 6. Mean convergence in PROC1, Scenario 2 for (a) PSO with $\mathcal{P} = 20$, (b) DE with $\mathcal{P} = 20$ and (c) DE with $\mathcal{P} = 30$.

4.1.3. Scenario 3

In this Scenario, all the weights constrains were relaxed, but the causalities were kept, i.e., the value of the weights were fixed in the interval $[0, 1]$ or in the interval $[-1, 0)$, according to the causality determined by the experts. Table 5 presents the obtained results. As can be seen, $\mathcal{P} = 10$ was enough for achieving 100% of convergence in GA and PSO. With $\mathcal{P} = 10$, DE was not able to find the proper solution in 36 trials, resulting in a probability of success equal to 0.9964, but when $\mathcal{P} = 20$, DE obtained 100% of success. Figure 7 presents the mean convergence of the concepts in 10^4 independent experiments. In this scenario, DE presented the fastest average convergence.

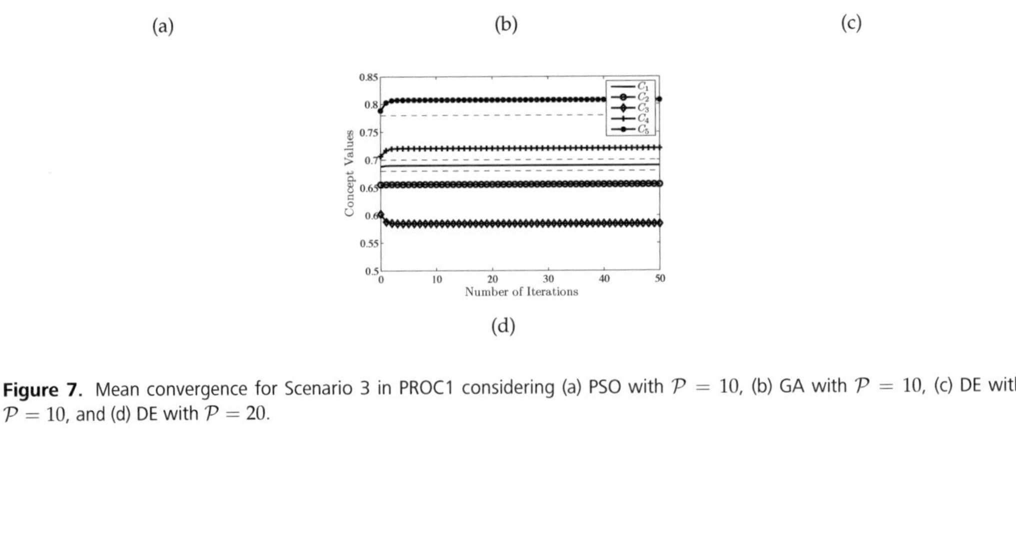

(a) (b) (c)

(d)

Figure 7. Mean convergence for Scenario 3 in PROC1 considering (a) PSO with $\mathcal{P} = 10$, (b) GA with $\mathcal{P} = 10$, (c) DE with $\mathcal{P} = 10$, and (d) DE with $\mathcal{P} = 20$.

PSO with $\mathcal{P} = 10$					
A_{max}	0.7000	0.8404	0.6590	0.8404	0.8206
A_{min}	0.6800	0.4339	0.4339	0.6590	0.7800
$A_{average}$	0.6907	0.6886	0.6112	0.7333	0.8080
Number of Failures: 0					
Probability of Success: 1.0					
GA with $\mathcal{P} = 10$					
A_{max}	0.7000	0.8404	0.6590	0.8404	0.8206
A_{min}	0.6800	0.4339	0.4339	0.6590	0.7800
$A_{average}$	0.6906	0.6496	0.5914	0.7139	0.8027
Number of Failures: 0					
Probability of Success: 1.0					
DE with $\mathcal{P} = 10$					
A_{max}	0.7189	0.8404	0.6590	0.8404	0.8206
A_{min}	0.6590	0.4294	0.4277	0.6590	0.6590
$A_{average}$	0.6896	0.6560	0.5839	0.7209	0.8060
Number of Failures: 36					
Probability of Success: 0.9964					
DE with $\mathcal{P} = 20$					
A_{max}	0.7000	0.8404	0.6590	0.8404	0.8206
A_{min}	0.6800	0.4339	0.4339	0.6590	0.7800
$A_{average}$	0.6897	0.6557	0.5848	0.7204	0.8074
Number of Failures: 0					
Probability of Success: 1.0					

Table 5. GA, PSO and DE simulation results for Scenario 3, PROC1.

4.1.4. Scenario 4

In this situation, the causality and the strength of the causality were totally relaxed. PSO and GA algorithms were able to determine proper weights for the connection matrix with $\mathcal{P} = 10$, and DE did not have success in only 4 experiments, resulting in a probability of success equal to 0.9996. When $\mathcal{P} = 20$, DE did not presented convergence errors. Table 6 presents the obtained results, while Figure 8 presents the mean convergence of the concepts in 10^4 independent experiments. As in Scenario 3, DE presented the fastest average convergence, while GA and PSO presented similar average convergence speed.

4.2. Process PROC2

For this process the value of the weights were fixed in the interval [0, 1] or in the interval [−1, 0), according to the causality determined by the experts. The adopted input simulation parameters for PSO, GA and DE algorithms were the same considered in PROC1, except for the population size in PSO that was assumed $\mathcal{P} = 10$ and $\mathcal{P} = 20$.

Table 7 summarizes the simulation results, while Figures 9 and 10 present the mean convergence of the concepts value considering 10^4 independent experiments. As can be seen, $\mathcal{P} = 10$ was enough for achieving convergence rate of 100% in GA. With $\mathcal{P} = 10$, PSO presented 30 failures and probability of success equal to 0.9970, while DE presented 897 failures and probability of success equal to 0.9103. However, with $\mathcal{P} = 20$, both PSO and DE were able to achieve 100% of success. For this process, DE has a faster average convergence than PSO and GA, being PSO the second fastest.

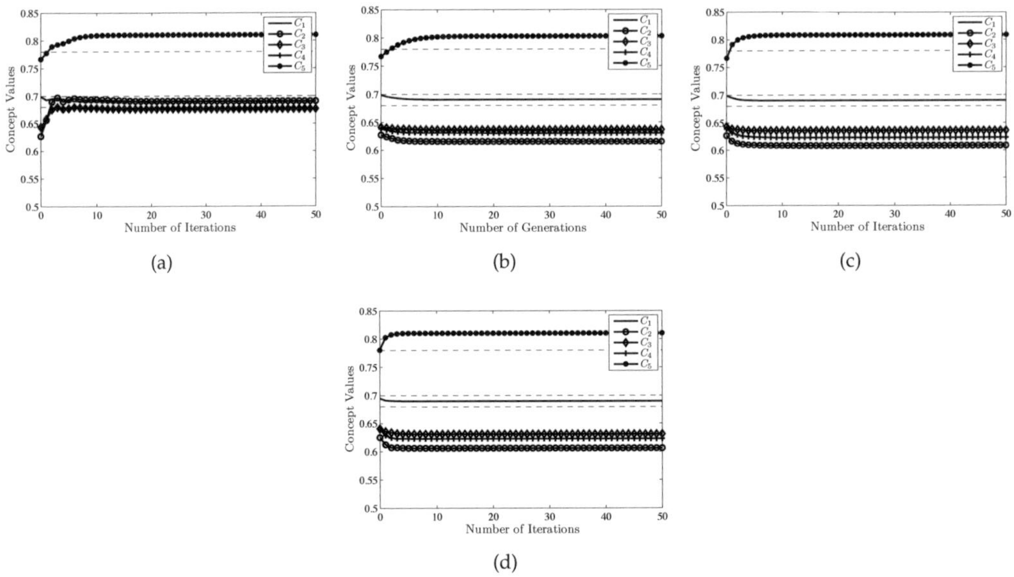

Figure 8. Mean convergence for Scenario 4 in PROC1 considering (a) PSO with $\mathcal{P} = 10$, (b) GA with $\mathcal{P} = 10$, (c) DE with $\mathcal{P} = 10$, and (d) DE with $\mathcal{P} = 20$.

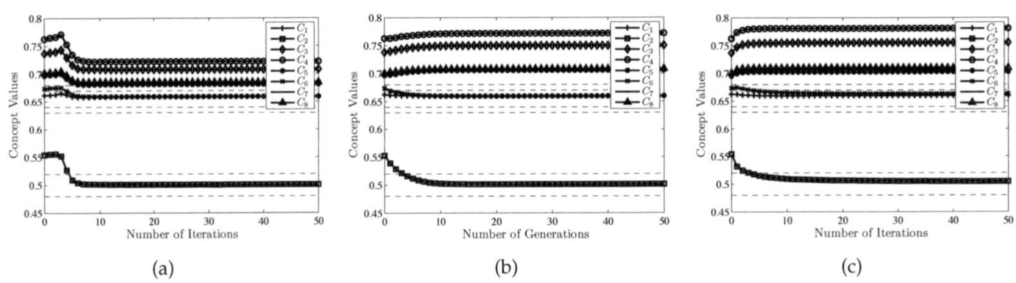

Figure 9. Mean convergence for (a) PSO, (b) GA and (c) DE in PROC2 with $\mathcal{P} = 10$.

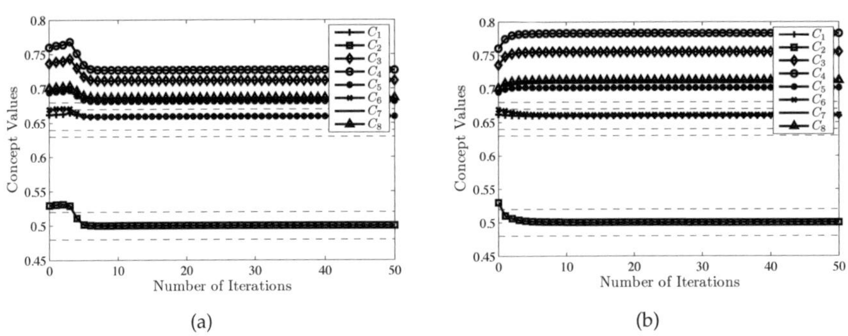

Figure 10. Mean convergence for (a) PSO and (b) DE in PROC2 with $\mathcal{P} = 20$.

http://dx.doi.org/10.5772/52918

PSO with $\mathcal{P} = 10$					
A_{max}	0.7000	0.9199	0.8206	0.8404	0.8206
A_{min}	0.6800	0.2129	0.4339	0.3952	0.7800
$A_{average}$	0.6901	0.6908	0.6767	0.6777	0.8101
Number of Failures: 0					
Probability of Success: 1.0					
GA with $\mathcal{P} = 10$					
A_{max}	0.7000	0.9192	0.8206	0.8404	0.8206
A_{min}	0.6800	0.2130	0.4339	0.3952	0.7800
$A_{average}$	0.6905	0.6154	0.6371	0.6306	0.8030
Number of Failures: 0					
Probability of Success: 1.0					
DE with $\mathcal{P} = 10$					
A_{max}	0.7166	0.9199	0.8206	0.8404	0.8229
A_{min}	0.6800	0.2112	0.4285	0.3948	0.7628
$A_{average}$	0.6901	0.6087	0.6358	0.6232	0.8084
Number of Failures: 4					
Probability of Success: 0.9996					
DE with $\mathcal{P} = 20$					
A_{max}	0.7000	0.9199	0.8206	0.8404	0.8206
A_{min}	0.6800	0.2129	0.4338	0.3952	0.7800
$A_{average}$	0.6902	0.6061	0.6311	0.6227	0.8102
Number of Failures: 0					
Probability of Success: 1.0					

Table 6. GA, PSO and DE simulation results for Scenario 4, PROC1.

5. Conclusions

There are in the literature, several models based on fuzzy cognitive maps developed and adapted to a large range of applications. The use of learning algorithms may be necessary for obtaining proper values for the weight matrix and reducing the dependence on the experts' knowledge.

Within this context, this Chapter presented a comparison of three heuristic search approaches, PSO, GA and DE, applied to FCM weight optimization in two processes control. In all the considered cases, GA presented a performance in terms of probability of success better or equal to the other two schemes, being PSO the second best technique in terms of probability of success in the two considered processes.

Specially in scenario 2 of PROC 1, when there are several weight constraints, GA achieved 100% of success in 10^4 independent experiments with a population of 10 chromosomes. PSO needed $\mathcal{P} = 20$ particles in the population in order to reach 100% of success. DE was not able to achieve 100% of success even for $\mathcal{P} = 30$.

PSO with $\mathcal{P} = 10$								
A_{max}	0.7040	0.6590	0.9029	0.9407	0.8134	0.7234	0.6700	0.8153
A_{min}	0.6400	0.4130	0.6590	0.6590	0.6590	0.6590	0.6590	0.6590
$A_{average}$	0.6592	0.5006	0.7079	0.7221	0.6809	0.6593	0.6592	0.6850
Number of Failures: 30								
Probability of Success: 0.9970								
GA with $\mathcal{P} = 10$								
A_{max}	0.6800	0.5200	0.9038	0.9409	0.7870	0.6700	0.6700	0.8153
A_{min}	0.6400	0.4800	0.6590	0.6590	0.6590	0.6590	0.6590	0.6590
$A_{average}$	0.6596	0.5014	0.7500	0.7717	0.7055	0.6596	0.6596	0.7082
Number of Failures: 0								
Probability of Success: 1.0								
DE with $\mathcal{P} = 10$								
A_{max}	0.7523	0.6591	0.9120	0.9477	0.8134	0.7608	0.7530	0.8271
A_{min}	0.5747	0.4472	0.6590	0.6590	0.6590	0.6590	0.6590	0.6590
$A_{average}$	0.6597	0.5046	0.7553	0.7808	0.7026	0.6632	0.6632	0.7097
Number of Failures: 897								
Probability of Success: 0.9103								
PSO with $\mathcal{P} = 20$								
A_{max}	0.6800	0.5200	0.9040	0.9411	0.7869	0.6700	0.6700	0.8154
A_{min}	0.6400	0.4800	0.6590	0.6590	0.6590	0.6590	0.6590	0.6590
$A_{average}$	0.6594	0.5002	0.7116	0.7267	0.6813	0.6594	0.6593	0.6865
Number of Failures: 0								
Probability of Success: 1.0								
DE with $\mathcal{P} = 20$								
A_{max}	0.6800	0.5200	0.9043	0.9427	0.7870	0.6700	0.6700	0.8154
A_{min}	0.6400	0.4800	0.6590	0.6590	0.6590	0.6590	0.6590	0.6590
$A_{average}$	0.6595	0.5003	0.7548	0.7827	0.7015	0.6608	0.6608	0.7124
Number of Failures: 0								
Probability of Success: 1.0								

Table 7. GA, PSO and DE simulation results for PROC2.

In most scenarios, DE has presented a faster convergence in comparison to PSO and GA, but the population size needed in DE was higher than the other two methods.

Author details

Bruno Augusto Angélico[1], Márcio Mendonça[1],
Lúcia Valéria R. de Arruda[2], Taufik Abrão[3]

1 Federal University of Technology - Paraná (UTFPR), Campus Cornélio Procópio, PR, Brazil
2 Federal University of Technology - Paraná (UTFPR), Campus Curitiba, PR, Brazil
3 State University of Londrina - Paraná (UEL), Londrina, PR, Brazil

References

[1] Kosko B. Fuzzy cognitive maps. *International Journal of Man-Machine Studies*, 24:65–75, January 1986.

[2] Kosko B. *Neural networks and fuzzy systems : a dynamical systems approach to machine intelligence*. Prentice Hall, Upper Saddle River, NJ, 1992.

[3] Kosko B. *Fuzzy Engineering*. Prentice-Hall, Inc., Upper Saddle River, NJ, 1997.

[4] Axelrod R.M. *Structure of Decision : The Cognitive Maps of Political Elites*. Princeton University Press, Princeton, NJ, 1976.

[5] Andreou A.S., Mateou N.H., and Zombanakis G.A. Soft computing for crisis management and political decision making: the use of genetically evolved fuzzy cognitive maps. *Soft Computing*, 9:194–210, March 2005.

[6] Stylios C.D., Georgopoulos V.C., Malandraki G.A., and Chouliara S. Fuzzy cognitive map architectures for medical decision support systems. *Applied Soft Computing*, 8:1243–1251, June 2008.

[7] Papageorgiou E.I., Stylios C.D., and Groumpos P.P. An integrated two-level hierarchical system for decision making in radiation therapy based on fuzzy cognitive maps. *IEEE Transactions on Biomedical Engineering*, 50(12):1326 –1339, dec. 2003.

[8] Papageorgiou E.I., Parsopoulos K.E., Stylios C.S., Groumpos P.P., and Vrahatis M.N. Fuzzy cognitive maps learning using particle swarm optimization. *Journal of Intelligent Information Systems*, 25:95–121, July 2005.

[9] Stylios C.D., Georgopoulos V.C., and Groumpos P.P. The use of fuzzy cognitive maps in modeling systems. In *Proceeding of 5th IEEE Mediterranean Conference on Control and Systems, Paphos*, 1997.

[10] Dickerson J. A. and Kosko B. Virtual worlds as fuzzy cognitive maps. In *IEEE Virtual Reality Annual International Symposium*, volume 3, pages 471 – 477, 1993.

[11] Taber R. Knowledge Processing With Fuzzy Cognitive Maps. *Expert Systems With Applications*, 1(2):83–87, 1991.

[12] Perusich K. Fuzzy cognitive maps for policy analysis. In *International Symposium on Technology and Society Technical Expertise and Public Decisions*, pages 369 373, jun 1996.

[13] Baykasoglu A., Durmusoglu Z.D.U., and K. Vahit. Training fuzzy cognitive maps via extended great deluge algorithm with applications. *Computers in Industry*, 62(2):187 – 195, 2011.

[14] Glykas M. *Fuzzy Cognitive Maps: Advances in Theory, Methodologies, Tools and Applications*. Springer Publishing Company, Incorporated, 1st edition, 2010.

[15] Parsopoulos K.E. and Groumpos P.P. A first study of fuzzy cognitive maps learning using particle swarm optimization. In *IEEE 2003 Congress on Evolutionary Computation*, pages 1440–1447, 2003.

[16] Papageorgiou E.I. Learning algorithms for fuzzy cognitive maps - a review study. *IEEE Transactions on Systems, Man, and Cybernetics - Part C*, 42(2):150 –163, march 2012.

[17] Stylios C.D. Papageorgiou E.I. and P.P. Groumpos. Active hebbian learning algorithm to train fuzzy cognitive maps. *International Journal of Approximate Reasoning*, 37(3):219–249, 2004.

[18] Papageorgiou E.I. and Groumpos P.P. A new hybrid learning algorithm for fuzzy cognitive maps learning. *Applied Soft Computing*, 5:409–431, 2005.

[19] Zhu Y. and Zhang W. An integrated framework for learning fuzzy cognitive map using rcga and nhl algorithm. In *Int. Conf. Wireless Commun., Netw. Mobile Comput., Dalian, China*, 2008.

[20] Alizadeh S., Ghazanfari M., Jafari M., and Hooshmand S. Learning fcm by tabu search. *International Journal of Computer Science*, 3:142–149, 2007.

[21] Stach W., Kurgan L.A., Pedrycz W., and Reformat M. Genetic learning of fuzzy cognitive maps. *Fuzzy Sets and Systems*, 153(3):371–401, 2005.

[22] Ghazanfari M., Alizadeh S., Fathian M., and Koulouriotis D.E. Comparing simulated annealing and genetic algorithm in learning fcm. *Applied Mathematics and Computation*, 192(1):56–68, 2007.

[23] Kennedy J. and Eberhart R. Particle swarm optimization. In *IEEE International Conference on Neural Networks*, pages 1942–1948, 1995.

[24] Kennedy J. and Eberhart R.C. *Swarm Intelligence*. Morgan Kaufmann, first edition, March 2001.

[25] Shi Y. and Eberhart R.C. A modified particle swarm optimizer. In *IEEE International Conference on Evolutionary Computation*, pages 69–73, 1998.

[26] Haupt R.L. and Haupt S.E. *Practical Genetic Algorithms*. John Wiley & Sons, Hoboken, NJ, USA, 2nd edition, 2004.

[27] Storn R. and Price K. Differential evolution - a simple and efficient heuristic for global optimization over continuous spaces. *J. of Global Optimization*, 11(4):341–359, December 1997.

[28] Storn R.M. Price K.V. and Lampinen J.A. *Differential Evolution A Practical Approach to Global Optimization*. Natural Computing Series. Springer-Verlag, Berlin, Germany, 2005.

[29] Das S. and Suganthan P.N. Differential evolution: A survey of the state-of-the-art. *IEEE Transactions on Evolutionary Computation*, 15(1):4 –31, feb. 2011.

Chapter 10

Optimal Allocation of Reliability in Series Parallel Production System

Rami Abdelkader, Zeblah Abdelkader,
Rahli Mustapha and Massim Yamani

Additional information is available at the end of the chapter

http://dx.doi.org/10.5772/55725

1. Introduction

One of the most important problems in many industrial applications is the redundancy optimization problem. This latter is well known combinatorial optimization problem where the design goal is achieved by discrete choices made from elements available on the market. The natural objective function is to find the minimal cost configuration of a series-parallel system under availability constraints. The system is considered to have a range of performance levels from perfect working to total failure. In this case the system is called a *multi-state* system (MSS). Let consider a multi-state system containing n components C_i ($i = 1, 2, \ldots, n$) in series arrangement. For each component C_i there are various versions, which are proposed by the suppliers on the market. Elements are characterized by their cost, performance and availability according to their version. For example, these elements can represent machines in a manufacturing system to accomplish a task on product in our case they represent the whole of electrical power system (generating units, transformers and electric carrying lines devices). Each component C_i contains a number of elements connected in parallel. Different versions of elements may be chosen for any given system component. Each component can contain elements of *different* versions as sketched in figure 1.

A limitation can be undesirable or even unacceptable, where only identical elements are used in parallel (i.e. homogeneous system) for two reasons. First, by allowing different versions of the devices to be allocated in the same system, one can obtain a solution that provides the desired availability or reliability level with a lower cost than in the solution with identical parallel devices. Second, in practice the designer often has to include additional devices in the existing system. It may be necessary, for example, to modernize a production line system according to a new demand levels from customers or according to new reliability requirements.

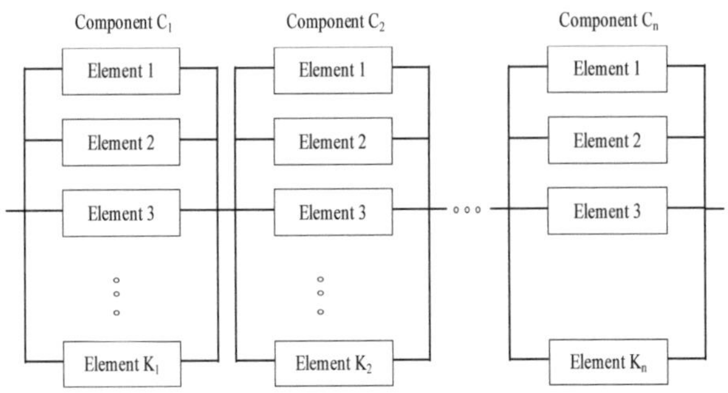

Figure 1. Series Parallel Production System

1.1. Literature review

The vast majority of classical reliability or availability analysis and optimization assume that components and system are in either of *two* states (i.e., complete working state and total failure state). However, in many real life situations we are actually able to distinguish among various levels of performance for both system and components. For such situation, the existing dichotomous model is a gross oversimplification and so models assuming multi-state (degradable) systems and components are preferable since they are closer to reliability. Recently much works treat the more sophisticated and more realistic models in which systems and components may assume many states ranging from perfect functioning to complete failure. In this case, it is important to develop MSS reliability theory. In this paper, an MSS reliability theory will be used, where the binary state system theory is extending to the multi-state case. As is addresses in recent review of the literature for example in (Ushakov, Levitin and Lisnianski, 2002) or (Levitin and Lisnianski, 2001). Generally, the methods of MSS reliability assessment are based on four different approaches:

i. The structure function approach.

ii. The stochastic process (mainly Markov) approach.

iii. The Monte-Carlo simulation technique.

iv. The universal moment generating function (UMGF) approach.

In (Ushakov, Levitin and Lisnianski, 2002), a comparison between these four approaches highlights that the UGF approach is fast enough to be used in the optimization problems where the search space is sizeable.

The problem of total investment-cost minimization, subject to reliability or availability constraints, is well known as the redundancy optimization problem (ROP). The ROP is studied in many different forms as summarized in (Tillman, Hwang and Kuo, 1977), and more recently in (Kuo and Prasad, 2000). The ROP for the multi-state reliability was introduced in (Ushakov, 1987). In (Lisnianski, Levitin, Ben-Haim and Elmakis, 1996) and (Levitin, Lisnianski, Ben-Haim and Elmakis, 1997), genetic algorithms were used to find the optimal or nearly optimal power system structure.

This work uses an *ant colony* optimization approach to solve the ROP for multi-state system. The idea of employing a colony of cooperating agents to solve combinatorial optimization problems was recently proposed in (Dorigo, Maniezzo and Colorni, 1996). The ant colony approach has been successfully applied to the classical traveling salesman problem (Dorigo and Gambardella, 1997), and to the quadratic assignment problem (Maniezzo and Colorni, 1999). Ant colony shows very good results in each applied area. It has been recently adapted for the reliability design of binary state systems (Liang and Smith, 2001). The ant colony has also been adapted with success to other combinatorial optimization problems such as the vehicle routing problem (Bullnheimer, Hartl and Strauss, 1997). The ant colony method has been used to solving the redundancy allocation problem (Nahas N., Nourelfath M., Aït-Kadi Daoud, 2006).

In this paper, we extend the work of other researchers by proposing ant colony system algorithm to solve the ROP characterised in the problem of optimization of the structure of power system where redundant elements are included in order to provide a desired level of reliability through optimal allocation of elements with different parameters (optimal structure with series-parallel elements) in continuous production system.

The use of this algorithm is within a general framework for the comparative and structural study of metaheuristics. In a first step the application of ant colonies in its primal form is necessary and thereafter in perspective the study will be completed.

1.2. Approach and outlines

The problem formulated in this chapter lead to a complicated combinatorial optimization problem. The total number of different solution to be examined is very large, even for rather small problems. An exhaustive examination of all possible solutions is not feasible given reasonable time limitations. Because of this, the ant colony optimization (or simply ACO) approach is adapted to find optimal or nearly optimal solutions to be obtained in a short time. The newer developed meta-heuristic method has the advantage to solve the ROP for MSS *without* the limitation on the diversity of versions of elements in parallel. Ant colony optimization is inspired by the behavior of real ant colonies that exhibit the highly structured behavior. Ants lay down in some quantity an aromatic substance, known as *pheromone*, in their way to food. An ant chooses a specific path in correlation with the intensity of the pheromone. The pheromone trail evaporates over time if no more pheromone in laid down by others ants, therefore the best path has more intensive pheromone and higher probability to be chosen.

During the optimization process, artificial ants will have to evaluate the availability of a given selected structure of the series-parallel system (electrical network). To do this, a fast procedure of availability estimation is developed. This procedure is based on a modern mathematical technique: the z-transform or UMGF which was introduced in (Ushakov, 1986). It was proven to be very effective for high dimension combinatorial problems: see e.g. (Ushakov, 2002), (Levitin, 2001). The universal moment generating function is an extension of the ordinary moment generating function (UGF) (Ross, 1993). The method developed in this chapter allows the availability function of reparable series-parallel MSS to be obtained using a straightforward numerical procedure.

2. Formulation of redundancy optimization problem

2.1. Series-parallel system with different redundant elements

Let consider a series-parallel system containing n subcomponents C_i ($i = 1, 2, \ldots, n$) in series as represented in figure 1. Every component C_i contains a number of different elements connected in parallel. For each component i, there are a number of element versions available in the market. For any given system component, different versions and number of elements may be chosen. For each subcomponent i, elements are characterized according to their version v by their cost (C_{iv}), availability (A_{iv}) and performance (Σ_{iv}). The structure of system component i can be defined by the numbers of parallel elements (of each version) k_{iv} for $1 \le v \le V_i$, where V_i is a number of versions available for element of type i. Figure 2 illustrates these notations for a given component i. The entire system structure is defined by the vectors $\boldsymbol{k}_i = \{k_{iv_i}\}$ ($1 \le i \le n$, $1 \le v \le V_i$). For a given set of vectors $\boldsymbol{k}_1, \boldsymbol{k}_2, \ldots, \boldsymbol{k}_n$ the total cost of the system can be calculated as:

$$C = \sum_{i=1}^{n} \sum_{v=1}^{V_i} k_{iv} C_{iv} \tag{1}$$

2.2. Availability of reparable multi-state systems

The series-parallel system is composed of a number of failure prone elements, such that the failure of some elements leads only to a degradation of the system performance. This system is considered to have a range of performance levels from perfect working to complete failure. In fact, the system failure can lead to decreased capability to accomplish a given task, but not to complete failure. An important MSS measure is related to the ability of the system to satisfy a given demand.

In electric power systems, reliability is considered as a measure of the ability of the system to meet the load demand (D), i.e., to provide an adequate supply of electrical energy (Σ). This definition of the reliability index is widely used in power systems: see e.g., (Ross, 1993), (Murchland, 1975), (Levitin, Lisnianski, Ben-Haim and Elmakis, 1998), (Lisnianski, Levitin, Ben-Haim and Elmakis, 1996), (Levitin, Lisnianski, and Elmakis, 1997). The Loss of Load Probability index (LOLP) is usually used to estimate the reliability index (Billinton and Allan, 1990). This index is the overall probability that the load demand will not be met. Thus, we can write $R = Probab(\Sigma \ge D)$ or $R = 1$-LOLP with LOLP $= Probab(\Sigma < D)$. This reliability index depends on consumer demand D.

For reparable MSS, a multi-state steady-state availability E is used as $Probab(\Sigma \ge D)$ after enough time has passed for this probability to become constant (Levitin, Lisnianski, Ben-Haim and Elmakis, 1998). In the steady-state the distribution of states probabilities is given by equation (2), while the multi-state stationary availability is formulated by equation (3):

$$P_j = \lim_{t\to\infty}[\Pr obab(\Sigma(t) = \Sigma_j)] \tag{2}$$

$$E = \sum_{\Sigma_j \geq D} P_j \tag{3}$$

If the operation period T is divided into M intervals (with durations $T_1, T_2, \ldots, T_M$) and each interval has a required demand level ($D_1, D_2, \ldots, D_M$, respectively), then the generalized MSS availability index A is:

$$A = \frac{1}{\sum_{j=1}^{M} T_j} \sum_{j=1}^{M} \Pr obab(\Sigma \geq D_j)\, T_j \tag{4}$$

We denote by $\boldsymbol{D}$ and $\boldsymbol{T}$ the vectors $\{D_j\}$ and $\{T_j\}$ $(1 \leq j \leq M)$, respectively. As the availability A is a function of $\boldsymbol{k_1}, \boldsymbol{k_2}, \ldots, \boldsymbol{k_n}$, $\boldsymbol{D}$ and $\boldsymbol{T}$, it will be written $A(\boldsymbol{k_1}, \boldsymbol{k_2}, \ldots, \boldsymbol{k_n}, \boldsymbol{D}, \boldsymbol{T})$. In the case of a power system, the vectors $\boldsymbol{D}$ and $\boldsymbol{T}$ define the cumulative load curve (consumer demand). In reality the load curves varies randomly; an approximation is used from random curve to discrete curve see (Wood and Ringlee, 1970). In general, this curve is known for every power system.

2.3. Optimal design problem formulation

The multi-state system redundancy optimization problem of electrical power system can be formulated as follows: find the minimal cost system configuration $\boldsymbol{k_1}, \boldsymbol{k_2}, \ldots, \boldsymbol{k_n}$, such that the corresponding availability exceeds or equal the specified availability A_0. That is,

$$\text{Minimize } C = \sum_{i=1}^{n} \sum_{v=1}^{V_i} k_{iv} C_{iv} \tag{5}$$

$$\text{subject to } A(k_1, k_2, \ldots, k_n, D, T) \geq A_0 \tag{6}$$

The input of this problem is the specified availability and the outputs are the minimal investment-cost and the corresponding configuration determined. To solve this combinatorial optimization problem, it is important to have an effective and fast procedure to evaluate the availability index for a series-parallel system of elements. Thus, a method is developed in the next section to estimate the value of $A(\boldsymbol{k_1}, \boldsymbol{k_2}, \ldots, \boldsymbol{k_n}, \boldsymbol{D}, \boldsymbol{T})$.

3. Multi-state system availability estimation

The procedure used in this chapter is based on the universal z-transform, which is a modern mathematical technique introduced in (Ushakov, 1986). This method, convenient for numerical implementation, is proved to be very effective for high dimension combinatorial problems. In the literature, the universal z-transform is also called universal moment generating function (UMGF) or simply u-function or u-transform. In this chapter, we mainly use the acronym UMGF. The UMGF extends the widely known ordinary moment generating function (Ross, 1993).

3.1. Definition and properties

The UMGF of a discrete random variable Σ is defined as a polynomial:

$$u(z) = \sum_{j=1}^{J} P_j z^{\Sigma_j} \tag{7}$$

where the variable Σ has J possible values and P_j is the probability that Σ is equal to Σ_j.

The probabilistic characteristics of the random variable Σ can be found using the function $u(z)$. In particular, if the discrete random variable Σ is the MSS stationary output performance, the availability E is given by the probability $Probab(\Sigma \geq D)$ which can be defined as follows:

$$\text{Probab}(\Sigma \geq D) = \Psi\left(u(z) z^{-D}\right) \tag{8}$$

where Ψ is a distributive operator defined by expressions (9) and (10):

$$\Psi(Pz^{\sigma-D}) = \begin{cases} P, \text{ if } \sigma \geq D \\ 0, \text{ if } \sigma < D \end{cases} \tag{9}$$

$$\Psi\left(\sum_{j=1}^{J} P_j z^{\Sigma_j - D}\right) = \sum_{j=1}^{J} \Psi\left(P_j z^{\Sigma_j - D}\right) \tag{10}$$

It can be easily shown that equations (7) – (10) meet condition $Probab(\Sigma \geq D) = \sum_{\Sigma_j \geq D} P_j$. By using the operator Ψ, the coefficients of polynomial $u(z)$ are summed for every term with $\Sigma_j \geq D$, and the probability that Σ is not less than some arbitrary value D is systematically obtained.

Consider single elements with total failures and each element i has nominal performance Σ_i and availability A_i. Then, $Probab(\Sigma = \Sigma_i) = A_i$ and $Probab(\Sigma = 0) = 1 - A_i$. The UMGF of such an element has only two terms and can be defined as:

$$u_i = (1 - A_i)z^0 + A_i z^{\Sigma_i} = (1 - A_i) + A_i z^{\Sigma_i} \tag{11}$$

To evaluate the MSS availability of a series-parallel system, two basic composition operators are introduced. These operators determine the polynomial $u(z)$ for a group of elements.

3.2. Composition operators

3.2.1. Properties of the operators

The essential property of the UMGF is that it allows the total UMGF for a system of elements connected in parallel or in series to be obtained using simple algebraic operations on the individual UMGF of elements. These operations may be defined according to the physical nature of the elements and their interactions. The only limitation on such an arbitrary operation is that its operator ϕ should satisfy the following Ushakov's conditions (Ushakov, 1986):

$\varphi(p_1 z^{g_1}, \; p_2 z^{g_2}) = p_1 p_2 z^{\varphi(g_1, g_2)}$,

$\varphi(g) = g$,

$\varphi(g_1, \; \dots, \; g_n) = \varphi(\varphi(g_1, \; \dots, \; g_k), \; \varphi(g_{k+1}, \; \dots, \; g_n))$,

$\varphi(g_1, \; \dots, \; g_k, \; g_{k+1}, \; \dots, \; g_n) = \varphi(g_1, \dots, g_{k+1}, g_k, \dots, g_n)$ for any k.

3.2.2. Parallel elements

Let consider a system component m containing J_m elements connected in parallel. As the performance measure is related to the system productivity, the total performance of the parallel system is the *sum* of performances of all its elements. In power systems engineering, the term capacity is usually used to indicate the quantitative performance measure of an element (Lisnianski, Levitin, Ben-Haim and Elmakis, 1996). It may have different physical nature. Examples of elements capacities are: generating capacity for a generator, pipe capacity for a water circulator, carrying capacity for an electric transmission line, etc. The capacity of an element can be measured as a percentage of nominal total system capacity. In a manufacturing system, elements are machines. Therefore, the total performance of the parallel machine is the sum of performances (Dallery and Gershwin, 1992).

The u-function of MSS component m containing J_m parallel elements can be calculated by using the Γ operator:

$u_p(z) = \Gamma(u_1(z), \; u_2(z), \; \dots, \; u_n(z))$, where $\Gamma(g_1, \; g_2, \; \dots, \; g_n) = \sum_{i=1}^{n} g_i$.

Therefore for a pair of elements connected in parallel:

$$\Gamma(u_1(z),\ u_2(z))=\Gamma(\sum_{i=1}^{n} P_i z^{a_i},\ \sum_{j=1}^{m} Q_j z^{b_j})=\sum_{i=1}^{n}\sum_{j=1}^{m} P_i Q_j z^{a_i+b_j}.$$

Parameters a_i and b_j are physically interpreted as the respective performances of the two elements. n and m are numbers of possible performance levels for these elements. P_i and Q_j are steady-state probabilities of possible performance levels for elements.

One can see that the Γ operator is simply a product of the individual u-functions. Thus, the component UMGF is:

$$u_p(z)=\prod_{j=1}^{J_m} u_j(z).$$

Given the individual UMGF of elements defined in equation (11), we have:

$$u_p(z)=\prod_{j=1}^{J_m}(1-A_j+A_j z^{\Sigma_i}).$$

3.2.3. Series elements

When the elements are connected in series, the element with the least performance becomes the bottleneck of the system. This element therefore defines the total system productivity. To calculate the u-function for system containing n elements connected in series, the operator η should be used: $u_s(z)=\eta(u_1(z),\ u_2(z),\ \ldots,\ u_m(z))$, where $\eta(g_1,\ g_2,\ \ldots,\ g_m)=\min\{g_1,\ g_2,\ \ldots,\ g_m\}$

so that

$$\eta(u_1(z),\ u_2(z))=\eta\left(\sum_{i=1}^{n} P_i z^{a_i},\ \sum_{j=1}^{m} Q_j z^{b_j}\right)=\sum_{i=1}^{n}\sum_{j=1}^{m} P_i Q_j z^{\min\{a_i,\ b_j\}}$$

Applying composition operators Γ and η consecutively, one can obtain the UMGF of the entire series-parallel system.

4. The ant colony optimization approach

The problem formulated in this chapter is a complicated combinatorial optimization problem. The total number of different solutions to be examined is very large, even for rather small problems. An exhaustive examination of the enormous number of possible solutions is not feasible given reasonable time limitations. Thus, because of the search space size of the ROP for MSS, a new meta-heuristic is developed in this section. This meta-heuristic consists in an adaptation of the ant colony optimization method.

4.1. The ACO principle

Recently, (Dorigo, Maniezzo and Colorni, 1996) introduced a new approach to optimization problems derived from the study of any colonies, called "Ant System". Their system inspired by the work of real ant colonies that exhibit the highly structured behavior. Ants lay down in some quantity an aromatic substance, known as pheromone, in their way to food. An ant chooses a specific path in correlation with the intensity of the pheromone. The pheromone trail evaporates over time if no more pheromone in laid down by others ants, therefore the best paths have more intensive pheromone and higher probability to be chosen. This simple behavior explains why ants are able to adjust to changes in the environment, such as new obstacles interrupting the currently shortest path.

Artificial ants used in ant system are agents with very simple basic capabilities mimic the behavior of real ants to some extent. This approach provides algorithms called ant algorithms. The Ant System approach associates pheromone trails to features of the solutions of a combinatorial problem, which can be seen as a kind of adaptive memory of the previous solutions. Solutions are iteratively constructed in a randomized heuristic fashion biased by the pheromone trails, left by the previous ants. The pheromone trails, τ_{ij}, are updated after the construction of a solution, enforcing that the best features will have a more intensive pheromone. An Ant algorithm presents the following characteristics. It is a natural algorithm since it is based on the behavior of ants in establishing paths from their colony to feeding sources and back. It is parallel and distributed since it concerns a population of agents moving simultaneously, independently and without supervisor. It is cooperative since each agent chooses a path on the basis of the information, pheromone trails, laid by the other agents with have previously selected the same path. It is versatile that can be applied to similar versions the same problem. It is robust that it can be applied with minimal changes to other combinatorial optimization problems. The solution of the travelling salesman problem (TSP) was one of the first applications of ACO.

Various extensions to the basic TSP algorithm were proposed, notably by (Dorigo and Gambardella, 1997a). The improvements include three main aspects: the state transition rule provides a direct way to balance between exploration of new edges and exploitation of a priori and accumulated knowledge about the problem, the global updating rule is applied only to edges which belong to the best ant tour and while ants construct solution, a local pheromone updating rule is applied. These extensions have been included in the algorithm proposed in this paper.

4.2. ACO-based solution approach

In our reliability optimization problem, we have to select the best combination of parts to minimize the total cost given a reliability constraint. The parts can be chosen in any combination from the available components. Components are characterized by their reliability, capacity and cost. This problem can be represented by a graph (figure 2) in which the set of nodes comprises the set of subsystems and the set of available components (i.e. max (M_j), $j = 1..n$) with a set of connections partially connect the graph (i.e. each subsystem is connected only to its available components). An additional node (blank node) is connected to each subsystem.

In figure 2, a series-parallel system is illustrated where the first and the second subsystem are connected respectively to their 3 and 2 available components. The nodes cp_{i3} and $cp_{i4,}$ represent

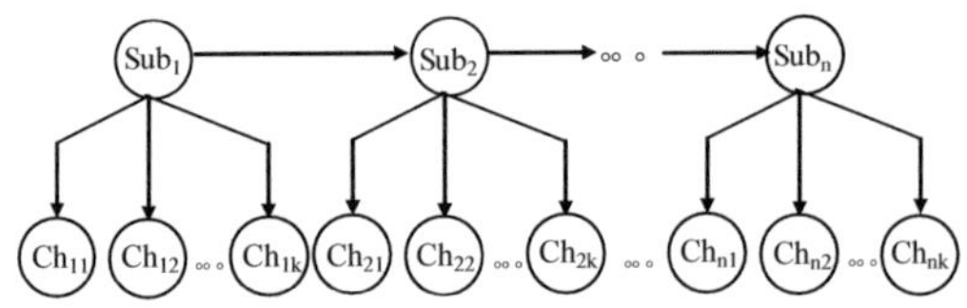

Figure 2. Definition of a series-parallel system with tree subsystems into a graph

the blank components of the two subsystems. At each step of the construction process, an ant uses problem-specific heuristic information, denoted by η_{ij} to choose the optimal number of components in each subsystem. An imaginary heuristic information is associated to each blank node. These new factors allow us to limit the search surfaces (i.e. tuning factors). An ant positioned on subsystem i chooses a component j by applying the rule given by:

$$j = \begin{cases} \arg \max\limits_{m \in AC_i} ([\tau_{im}]^{\alpha}[\eta_{im}]^{\beta}) & \text{if } q \le q_o \\ J & \text{if } q > q_o \end{cases} \tag{12}$$

and J is chosen according to the probability:

$$p_{ij} = \begin{cases} \dfrac{[\tau_{ij}]^{\alpha}[\eta_{ij}]^{\beta}}{\sum\limits_{m \in AC_i} [\tau_{im}]^{\alpha}[\eta_{im}]^{\beta}} & \text{if } j \in AC_i \\ 0 & \text{otherwise} \end{cases} \tag{13}$$

α : The relative importance of the trail.

β : The relative importance of the heuristic information η_{ij}.

AC_i: The set of available components choices for subsystem i.

q: Random number uniformly generated between 0 and 1.

The heuristic information used is : $\eta_{ij} = 1/(1+c_{ij})$ where c_{ij} represents the associated cost of component j for subsystem i. A "tuning" factor $t_i = \eta_{ij} = 1/(1+c_{i(Mi+1)})$ is associated to blank component (M_i+1) of subsystem i. The parameter q_o determines the relative importance of exploitation versus exploration: every time an ant in subsystem i have to choose a component j, it samples a random number $0 \le q \le 1$. If $q \le q_o$ then the best edge, is chosen (exploitation), otherwise an edge is chosen according to (12) (biased exploration).

The pheromone update consists of two phases: local and global updating. While building a solution of the problem, ants choose components and change the pheromone level on subsystem-component edges. This local trail update is introduced to avoid premature convergence and effects a temporary reduction in the quantity of pheromone for a given subsystem-component edge so as to discourage the next ant from choosing the same component during the same cycle. The local updating is given by:

$$\tau_{ij}^{new} = (1-\rho)\tau_{ij}^{old} + \rho\tau_o \tag{14}$$

where ρ is a coefficient such that $(1-\rho)$ represents the evaporation of trail and τ_o is an initial value of trail intensity. It is initialized to the value $(n.TC_{nn})^{-1}$ with n is the size of the problem (i.e. number of subsystem and total number of available components) and TC_{nn} is the result of a solution obtained through some simple heuristic.

After all ants have constructed a complete system, the pheromone trail is then updated at the end of a cycle (i.e. global updating), but only for the best solution found. This choice, together with the use of the pseudo-random-proportional rule, is intended to make the search more directed: ants search in a neighbourhood of the best solution found up to the current iteration of the algorithm. The pheromone level is updated by applying the following global updating rule:

$$\tau_{ij}^{new} = (1-\rho)\tau_{ij}^{old} + \rho\Delta\tau_{ij}$$

$$\Delta\tau_{ij} = \begin{cases} \dfrac{1}{TC_{best}} & if\,(i,\ j) \in best \quad tour \\ 0 & otherwise \end{cases}$$

4.3. The algorithm

An ant-cycle algorithm is stated as follows. At time zero an initialization phase takes place during wish *NbAnt* ants select components in each subsystem according to the Pseudo-random-proportional transition rule. When an ant selects a component, a local update is made to the trail for that subsystem-component edge according to equation (13). In this equation, ρ is a parameter that determines the rate of reduction of the pheromone level. The pheromone reduction is small but sufficient to lower the attractiveness of precedent subsystem-component edge. At the end of a cycle, for each ant k, the value of the system's reliability R_k and the total cost TC_k are computed. The best feasible solution found by ants (i.e. total cost and assignments) is saved. The pheromone trail is then updated for the best solution obtained according to (13). This process is iterated until the tour counter reaches the maximum number of cycles NC_{max} or all ants make the same tour (stagnation behavior).

The followings are formal description of the algorithm.

1. Set NC:=0 (NC: cycle counter)

   For every edge (i,j) set an initial value $\tau_{ij}(0)= \tau_o$

2. For k=1 to NbAnt do

For i=1 to NbSubSystem do

For j=1 to MaxComponents do

Choose a component, including blanks, according to (1) and (2).

Local update of pheromone trail for chosen subsystem- component edge (i,j) :

$$\tau_{ij}^{new} = (1-\rho)\tau_{ij}^{old} + \rho\tau_o$$

End For

End For

3. Calculate R_k (system reliability for each ant)

   Calculate the total cost for each ant TC_k

   Update the best found feasible solution

4. Global update of pheromone trail:

   For each edge (i,j)∈ best feasible solution, update the pheromone trail according to:

$$\tau_{ij}^{new} = (1-\rho)\tau_{ij}^{old} + \rho\Delta\tau_{ij}$$

$$\Delta\tau_{ij} = \begin{cases} \dfrac{1}{TC_{best}} & if\ (i,\ j) \in best\ \ tour \\ 0 & otherwise \end{cases}$$

   End For

5. cycle=cycle +1

6. if ($NC < NC_{max}$) and (not stagnation behavior)

   Then

   Goto step 2

   Else

   Print the best feasible solution and components selection.

   Stop.

5. Illustrative example

Description of the system to be optimized

The power station coal transportation system which supplies the boilers is designed with five basic components as depicted in figure.3.

The process of coal transportation is: The coal is loaded from the bin to the primary conveyor (Conveyor 1) by the primary feeder (Feeder 1). Then the coal is transported through the conveyor 1 to the Stacker-reclaimer, when it is left up to the burner level. The secondary feeder (Feeder 2) loads the secondary conveyor (Conveyor 2) which supplies the burner feeding system of the boiler. Each element of the system is considered as unit with total failures.

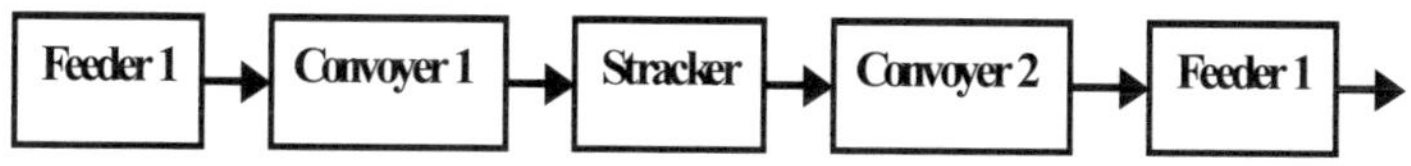

Figure 3. Synoptic of the detailed power station coal transportation

Comp#	Vers#	Availability A	Cost C	Capacity Ξ
	1	0.980	0.590	120
	2	0.977	0.535	100
	3	0.982	0.470	85
1	4	0.978	0.420	85
	5	0.983	0.400	48
	6	0.920	0.180	31
	7	0.984	0.220	26
	1	0.995	0.205	100
	2	0.996	0.189	92
2	3	0.997	0.091	53
	4	0.997	0.056	28
	5	0.998	0.042	21
	1	0.971	7.525	100
3	2	0.973	4.720	60
	3	0.971	3.590	40
	4	0.976	2.420	20
	1	0.977	0.180	115
	2	0.978	0.160	100
	3	0.978	0.150	91
	4	0.983	0.121	72
4	5	0.981	0.102	72
	6	0.971	0.096	72
	7	0.983	0.071	55
	8	0.982	0.049	25
	9	0.977	0.044	25
	1	0.984	0.986	100
5	2	0.983	0.825	60
	3	0.987	0.490	40
	4	0.981	0.475	20

Acronyms:

Comp #: System component number.

Vers #: System version number.

Table 1. Characteristics of available system components on the market

Demand level (%)	100	80	50	20
Duration (h)	4203	788	1228	2536
Probability	0.479	0.089	0.140	0.289

Table 2. Parameters of the cumulative demand curve

A0	A_{01}	Optimal Structure	Computed Availability A	Computed Cost C
0.975	0.9760	1: Components **3 – 6 – 5 -7** 2: Components **2 – 3 – 4 - 4** 3: Components **1- 4** 4: Components **2 - 5 - 7 - 8** 5: Components **3 - 3 - 4**	0.9773	13.4440
0.980	0.9826	1: Components **2 - 2** 2: Components **3 – 3 -5** 3: Components **2 - 3 - 3** 4: Components **5 - 6 - 7** 5: Components **3 - 3 - 4**	0.9812	14.9180
0.990	0.9931	1: Components **2 - 1** 2: Components **3 - 3** 3: Components **2 - 2 - 3** 4: Components **5 - 5 - 6** 5: Components **2 - 2**	0.9936	**16.2870**

Table 3. Optimal Solution Obtained By Ant Algorithm

Optimal availabilities obtained by Ant Algorithm were compared to availabilities given by genetic algorithm (presented by symbol A_0 in table 3) in the reference (Levitin et al., 1997), and to those obtained by harmony search (presented by symbol A_{01} in table 3) given in (Rami et al., 2009).

For this type of problem, we define the minimal cost system configuration which provides the desired reliability level $A \geq A_0$ (where A_0 is given in (Levitin et al, 1997) taken as reference).

We will clearly remark the improvement of the reliability of the system at price equal compared to the two other methods.

We gave more importance to the reliability of the system compared to its cost what justifies the increase in the cost compared to the reference.

The compromise of the cost/reliability was treated successfully in this work.

The objective is to select the optimal combination of elements used in series-parallel structure of power system. This has to correspond to the minimal total cost with regard to the selected level of the system availability. The ACO allows each subsystem to contain elements with different technologies. The ACO algorithm proved very efficient in solving the ROP and better quality results in terms of structure costs and reliability levels have been achieved compared to GA (Levitin et al., 1997).

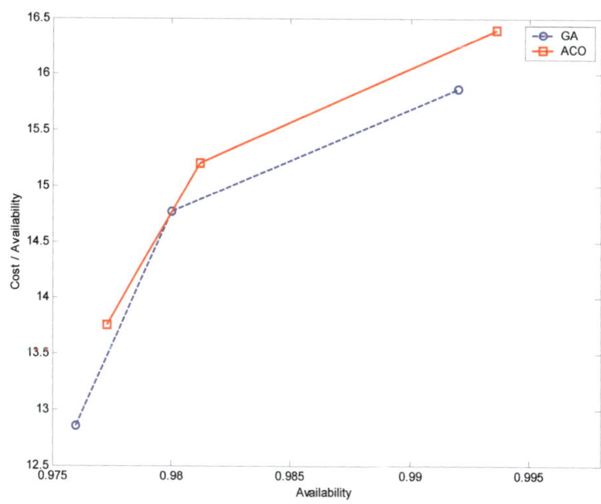

Figure 4. Cost-availability rate of GA and ACO algorithm versus availability

A0	% of A_{GA}	% of A_{ACO}	% of C/A
0.975	0.1	0.23	58.5
0.980	0.0	0.12	13.3
0.990	0.2	0.36	39.4

Table 4. Comparison of Optimal Solutions Obtained by ACO and Genetic Algorithms For Different Availability Requirements

From figure 4 and the table, one can observe:

ACO achieved better quality results in terms of structure cost and reliability in different reliability levels (figure 4). We remark in all case, GA performed better by achieving a less expensive configuration, however ACO algorithm achieved a near optimal configuration with a slightly higher reliability level (table 4).

We take, for example, for reference reliability level ($A_0 = 0.975$, table 4), GA prove an augmentation of 0.1 percent compared to 0.23 percent given by ACO this for a difference in rate Cost-reliability of 58.3%. It is noticed, according to figure 4, that ACO tends, at equal price, to increase the reliability of the system.

6. Conclusion

A new algorithm for choosing an optimal series-parallel power structure configuration is proposed which minimizes total investment cost subject to availability constraints. This algorithm seeks and selects devices among a list of available products according to their availability, nominal capacity (performance) and cost. Also defines the number and the kind of parallel machines in each sub-system. The proposed method allows a practical way to solve wide instances of structure optimization problem of multi-state power systems without limitation on the diversity of versions of machines put in parallel. A combination is used in this algorithm is based on the universal moment generating function and an ant colony optimization algorithm.

Author details

Rami Abdelkader[1], Zeblah Abdelkader[1*], Rahli Mustapha[2] and Massim Yamani[1]

*Address all correspondence to: azeblah@yahoo.fr

1 University Of Sidi Bel Abbes, Engineering Faculty, Algeria

2 University Of Oran, Engineering Faculty, Algeria

References

[1] Ait-Kadi and Nourelfath(2001). Availability optimization of fault-tolerant systems. *International Conference on Industrial Engineering and Production Management (IEPM'2001)*, Québec , 20-23.

[2] BauerBullnheimer, Hartl and Strauss, ((2000). Minimizing Total Tardiness on a Single machine using Ant Colony Optimization. Central European Journal of Operations research, , 8(2)

[3] BullnheimerHartl and Strauss, ((1997). Applying the Ant System to the vehicle Routing problem. *2nd Metaheuristics International Conference (MIC-97)*, Sophia-Antipolis, France, , 21-24.

[4] Billinton and Allan(1990). Reliability evaluation of power systems. *Pitman.*

[5] Chern(1992). On the Computational Complexity of Reliability redundancy Allocation in a Series System. Operations Research Letters, , 11

[6] Costa and Hertz(1997). Ants Can Colour Graphs. Journal of the Operational Research Society, , 48

[7] Dallery and Gershwin(1992). Manufacturing Flow Line Systems: A Review of Models and Analytical Results. *Queueing Systems theory and Applications,* Special Issue on Queueing Models of Manufacturing Systems, , 12(1-2), 3-94.

[8] Den BestoStützle and Dorigo, ((2000). Ant Colony Optimization fort he Total Weighted tardiness Problem. Proceeding of the 6th International Conference on parallel problem Solving from nature (PPSNVI), LNCS 1917, Berlin, , 611-620.

[9] Di Caro and Dorigo(1998). Mobile Agents for Adaptive Routing. *Proceedings for the 31st Hawaii International Conference On System Sciences,* Big Island of Hawaii, , 74-83.

[10] DorigoManiezzo and Colorni, ((1996). The Ant System: Optimization by a colony of cooperating agents. IEEE Transactions on Systems, Man and Cybernetics- Part B, , 26(1), 1-13.

[11] Dorigo and Gambardella(1997a). Ant Colony System: A Cooperative Learning Approach to the Traveling Salesman Problem", IEEE Transactions on Evolutionary computation, , 1(1), 53-66.

[12] Dorigo and Gambardella(1997b). Dorigo, M. and L. M. Gambardella. Ant Colonies for the Travelling Salesman Problem. Bio Systems, , 43

[13] Kuo and Prasad(2000). An Annotated Overview of System-reliability Optimization. IEEE Transactions on Reliability, , 49(2)

[14] Levitin and Lisnianski(2001). A new approach to solving problems of multi-state system reliability optimization. *Quality and Reliability Engineering International,* , 47(2), 93-104.

[15] LevitinLisnianski, Ben-Haim and Elmakis, ((1997). Structure optimization of power system with different redundant elements. *Electric Power Systems Research,* , 43(1), 19-27.

[16] (LevitinLisnianski, Ben-Haim and Elmakis, ((1998). Redundancy optimization for series-parallel multi-state systems. *IEEE Transactions on Reliability* , 47(2), 165-172.

[17] Liang and Smith(2001). An Ant Colony Approach to Redundancy Allocation. Submitted to *IEEE Transactions on Reliability.*

[18] LisnianskiLevitin, Ben-Haim and Elmakis, ((1996). Power system structure optimization subject to reliability constraints. *Electric Power Systems Research,* , 39(2), 145-152.

[19] Maniezzo and Colorni(1999). The Ant System Applied to the Quadratic Assignment Problem. IEEE Transactions on Knowledge and data Engineering, , 11(5)

[20] Murchland(1975). Fundamental concepts and relations for reliability analysis of multi-state systems. *Reliability and Fault Tree Analysis,* ed. R. Barlow, J. Fussell, N. Singpurwalla. SIAM, Philadelphia.

[21] Nahas, N, & Nourelfath, M. Aït-Kadi Daoud, Efficiently solving the redundancy allocation problem by using ant colony optimization and the extended great deluge algo-

rithm, International Conference on Probabilistic Safety Assessment and Management (PSAM) and ESREL, New Orleans, USA, May 14_19, (2006).

[22] NourefathAit-Kadi and Soro, ((2002). Optimal design of reconfigurable manufacturing systems. *IEEE International Conference on Systems, Man and Cybernetics (SMC'02),* Tunisia.

[23] RamiZeblah and Massim, ((2009). Cost optimization of power system structure subject to reliability constraints using harmony search", PRZEGLĄD ELEKTROTECHNICZNY (Electrical Review), R. 85 NR 4/2009.

[24] Ross(1993). Introduction to probability models. *Academic press.*

[25] Schoofs and Naudts(2000). Schoofs, L. and B. Naudts, "Ant Colonies are Good at Solving Contraint Satisfaction Problem," Proceeding of the 2000 Congress on Evolutionary Computation, San Diego, CA, July , 2000, 1190-1195.

[26] TillmanHwang and Kuo, ((1977). Tillman, F. A., C. L. Hwang, and W. Kuo, "Optimization Techniques for System Reliability with Redundancy- A review," IEEE Transactions on Reliability, , R-26(3)

[27] UshakovLevitin and Lisnianski, ((2002). Multi-state system reliability:from theory to practice. *Proc. of 3 Int. Conf. on mathematical methods in reliability, MMR 2002,* Trondheim, Norway, , 635-638.

[28] Ushakov(1987). optimal standby problems and a universal generating function. *Sov. J. Computing System Science,* N 4, , 25, 79-82.

[29] Ushakov(1986). Universal generating function. *Sov. J. Computing System Science,* N 5, , 24, 118-129.

[30] Wagner and Bruckstein(1999). Wagner, I. A. and A. M. Bruckstein. Hamiltonian(t)-An Ant inspired heuristic for Recognizing Hamiltonian Graphs. Proceeding of the 1999 Congress on Evolutionary Compuation, Washington, D.C., , 1465-1469.

[31] Wood, A. J. & R. J Ringlee ((1970). Frequency and duration methods for power reliability calculations ', Part II, ' Demand capacity reserve model ',IEEE Trans. On PAS, , 94, 375-388.

Section 4

Grover-Type Quantum Search

Chapter 11

Geometry and Dynamics of a Quantum Search Algorithm for an Ordered Tuple of Multi-Qubits

Yoshio Uwano

Additional information is available at the end of the chapter

http://dx.doi.org/10.5772/53187

1. Introduction

1.1. Geometry and dynamics viewpoints to quantum search algorithms

Quantum computation has been one of the hottest interdisciplinary research areas over some decades, where informatics, physics and mathematics are crossing with (see [1] including an excellent historical overview and [2–4] as later publications for general references). In the middle of 1990's, two great discoveries are made by Shor [5] in 1994 and by Grover [6] in 1996 that roused bubbling enthusiasm to quantum computation. As one of those, Grover found in 1996 the quantum search algorithm for the linear search through unsorted lists [6, 7], whose efficiency exceeds the theoretical bound of the linear search in classical computing: For an unsorted list of N data, the Grover search algorithm needs only $O(\sqrt{N})$ trials to find the target with high probability, while the linear search in classical computing needs $O(N)$ trials. Throughout this paper, the term *classical computing* means the computation theory based on the conventional binary-code operations. The adjective *classical* here is used as an antonym of *quantum*; like quantum mechanics vs classical mechanics.

Though the classical linear search is not of high complexity, the speedup by Grover's algorithm is exciting due to its wide applicability to other search-based problems; G-BBHT algorithm, the quantum counting problem, the minimum value search, the collision problem and the SAT problem, for example [8, 9]. A number of variations and extensions of the Grover algorithm have been made (see [10–13], for example): As far as the author made a search for academic articles in 2012 with keywords 'Grover', 'quantum' and 'search' by Google scholar (accessed 5 September 2012), more than five hundred 'hits' are available. Many of those can be traced from the preprint archive [14].

Among numbers of studies concerning Grover's quantum search algorithm, a pioneering geometric study on the algorithm is made by Miyake and Wadati in 2001 [15]: The sequence of quantum states generated by the Grover algorithm in 2^n data is shown to be on a geodesic in $(2^{n+1}-1)$-dimensional sphere. Further, the reduced search sequence is given rise to the complex projective space $\mathbf{C}P^{2^n-1}$ through a geometric reduction, which is also shown to be on a geodesic in $\mathbf{C}P^{2^n-1}$. Roughly speaking, the reduction in [15] is made through the phase-factor elimination from quantum states, so that $\mathbf{C}P^{2^n-1}$ is thought of as the space of rays. Note that the geodesics above are associated with the standard metric on $(2^{n+1}-1)$-dimensional sphere and with the Fubini-Study metric on $\mathbf{C}P^{2^n-1}$, respectively. The Fubini-Study metric on $\mathbf{C}P^{2^n-1}$ is utilized also in [15] to measure the minimum distance from each state involved in the search sequence to the submanifold consisting of non-entangled states, which characterizes the entanglement of the states along the search.

As expected benefits of geometric and dynamical views on quantum algorithmic studies like [15], the following would be worth listed;

1 By revealing underlying geometry of quantum algorithms (not necessarily universal), numbers of results in geometry are expected to be applied to make advances in quantum computation and information.

2 On looked upon the iterations made in algorithms as (discrete) time-evolutions of states, numbers of results in dynamical systems are expected to be applied to make advances in quantum computation and information.

3 In view of a close connection between geometry and dynamical systems, geometric and dynamical-systems studies on quantum algorithms may provide interesting examples of dynamical systems.

It would be worth noting here that there exists another approach to quantum searches using adiabatic evolution [16–18]. That approach, however, is outside the scope of this chapter since the search dealt with in this chapter is organized on the so-called amplitude magnification technique [8] which differs from the adiabatic evolution.

1.2. Quantum search for an ordered tuple of multi-qubits – a brief history –

Motivated by the work [15], the author studied in [19] a Grover-type search algorithm for an ordered tuple of multi-qubits together with a geometric reduction other than the reduction made in [15]. While the search algorithm is organized as a natural extension of Grover's original algorithm, the reduction of the search space made in [19] provides a nontrivial result: On denoting the degree of multi-qubits by n and the number of multi-qubits enclosed in each ordered tuple by ℓ, the space of $2^n \times \ell$ complex matrices with unit norm denoted by $\mathrm{M}_1(2^n,\ell)$ is taken as the extended space of ordered tuples of multi-qubits (ESOT), where the collection of all the ordered tuples denoted by $\mathrm{M}_1^{OT}(2^n,\ell)$ is included. The reduction is applied to the regular part, $\dot{\mathrm{M}}_1(2^n,\ell)$, of the ESOT, $\mathrm{M}_1(2^n,\ell)$, to give rise to the space denoted by $\dot{\mathrm{P}}_\ell$ of regular density matrices of degree ℓ which plays a key role in quantum information theory. Roughly speaking, the reduction applied in [19] is made by the elimination of 'complex rotations' leaving the relative configuration of multi-qubit states placed in each ordered tuples, so that the reduction is understood to be a very natural geometric projection of $\dot{\mathrm{M}}_1(2^n,\ell)$ to the space, $\dot{\mathrm{P}}_\ell$.

A significant result arising from the reduction is that the Riemannian metric on $\dot{\mathrm{P}}_\ell$ is shown to be derived 'consistently' from the standard metric on $\dot{\mathrm{M}}_1(2^n,\ell)$, which coincides with the SLD-Fisher metric on $\dot{\mathrm{P}}_\ell$ up to a constant multiple [19]. Namely, as a Riemannian manifold,

$M_1(2^n, \ell)$ is reduced to the space of regular density matrices of degree ℓ endowed with the SLD-Fisher metric, so that $\dot{P}_\ell$ is referred to as the quantum information space (QIS). Put another way, the reduction made in [19] reveals a direct nontrivial connection between the ESOT and the QIS. The former is a stage of quantum computation and the latter the stage of quantum information theory.

Due to the account given below, however, geometric studies were not made in [19] either on the search sequence in the ESOT generated by the Grover-type search algorithm or on the reduced search sequence in the QIS: Instead of geometric studies on the search sequences, it is the gradient dynamical system associated with the negative von Neumann entropy as the potential that is discussed in [19] on inspired by a series works of Nakamura [20–22] on complete integrablity of algorithms arising in applied mathematics. The result on the gradient system in [19] has drawn the author's interest to publish [23, 24] on the gradient systems on the QIS realizing the Karmarkar flow for linear programming and a Hebb-type learning equation for multivariate analysis, while geometric studies on the search sequences were left undone.

1.3. Chapter purpose, summary and organization

The purpose of this chapter is therefore to study the Grover-type search sequence for an ordered tuple of multi-qubits from geometric and dynamical viewpoints, which has been left since [19]. In particular, the reduced search sequence in the QIS is intensively studied from the viewpoint of quantum information geometry.

As an extension of [15] on the original search sequence, the Grover-type search sequence in the ESOT, $M_1(2^n, \ell)$, is shown to be on a geodesic. As a nontrivial result on the reduced search sequence in the QIS, the sequence is characterized in terms of an important geometric object in quantum information geometry:

Main Theorem *Through the reduction of the regular part, $\dot{M}_1(2^n, \ell)$, of the extended space of ordered tuples of multi-qubits (ESOT) to the quantum information space (QIS), $\dot{P}_\ell$, the reduced search sequence is on a geodesic in the QIS with respect to the m-parallel transport.*

Note that the m-parallel transport is the abbreviation of the *mixture* parallel transport [25–27], which is characteristic of the QIS.

To those who are not familiar to differential geometry, an important remark should be made on the term *geodesic* before the outline of chapter organization: One might hear that the geodesic between a pair of points is understood to be the shortest path connecting those points. This is true if geodesics are discussed on a Riemannian manifold endowed with the Levi-Civita (or Riemannian) parallel transport. As a reference accessible to potential readers, the book [28] is worth cited. Geodesics in the ESOT, $M_1(2^n, \ell)$, discussed in this chapter are the case. In general, however, geodesics are *not* characterized by shortest-path property *but* by autoparallel curves which have the shortest paths only in the case of the Levi-Civita parallel transport. What is needed to define geodesics is a parallel transport, while Riemannian metrics are not always necessary. The m-parallel transport of the QIS is the very example of parallel transport whose geodesics do not have the shortest-path property. As another crucial parallel transport in the QIS, the *exponential* parallel transport (the e-parallel transport) is well-known [25–27], whose geodesics do not have the shortest-path property either, though it is not dealt with in this chapter.

The organization of this chapter is outlined in what follows. Section 2 is for the quantum search for an ordered tuple of multi-qubits. The section starts with a brief review of the classical linear search in unsorted lists. The second subsection is for preliminaries to the quantum search: Mathematics for multi-qubits and ordered tuples of them is introduced. In the third subsection, the Grover-type quantum search algorithm is organized for an ordered tuple of multi-qubits along with the idea of Grover [6]. Dynamical behavior of the search sequence thus obtained is studied in the fourth subsection from the geometric viewpoint: The search sequence in the ESOT, $M_1(2^n, \ell)$, is shown to be on a geodesic in the ESOT. Section 3 is devoted to a study on the reduced search sequence in the QIS from geometric and dynamical points of view. The first subsection is a brief introduction of the QIS. The geometric reduction of the ESOT to the QIS is made in the second subsection: To be precise, our interest is focused on the reduction of the regular part, $\dot{M}_1(2^n, \ell)$, of the ESOT to simplify our geometric analysis. The third subsection starts with the standard parallel transport in the Euclidean space as a very familiar and intuitive example of the parallel transport. After the Euclidean case, the m-parallel transport in the QIS is introduced. It is shown that the reduced search sequence in the QIS is on a geodesic in the QIS with respect to the m-parallel transport. Section 4 is for concluding remarks, in which a significance of the main theorem (or Theorem 3.3) and some questions for future studies are included. A mathematical detail of Sec. 3 is consigned to Appendices following Sec. 4. Many symbols are introduced for geometric setting-up and analysis, which are listed in Appendix 1.

2. Quantum search for an ordered tuple of multi-qubits

2.1. Classical search: Review

The classical linear search in unsorted lists is outlined very briefly in what follows.

Let N be the number of unsorted data in a list, so that the data are labeled as $d_1, d_2, \cdots, d_N$. The N is assumed to be large enough. We start with a very figurative description of the search by taking the counter-consultation of a thick telephone book as an example; namely, the identification of the subscriber of a given telephone number. In the telephone book, the superscripts, $j = 1, 2, \cdots N$, of the data, $\{d_j\}_{j=1,2,\cdots,N}$, correspond to the names of subscribers sorted alphabetically and each d_j shows the telephone number of the j-th subscriber. Among the data, there assumed to be one telephone number, say d_M, that we wish to know its subscriber. The d_M is referred to as the target or the marked datum. A very naive way of finding d_M is to check the telephone number from d_1 in ascending order whether or not it is the same as the the target datum until we find d_M. The label M turns out to be the subscriber who we wish to identify. In average, this way requires $\frac{N}{2}$ trials of checking to find d_M.

The linear search is described in a smarter form than above in terms of the oracle function. In the same setting above, the oracle function, denoted by f, is defined to be the function of $\{1, 2, \cdots, N\}$ to $\{0, 1\}$ subject to

$$f(j) = \begin{cases} 1 & (j = M) \\ 0 & (j \neq M) \end{cases} \tag{1}$$

for $j = 1, 2, \cdots N$. Namely, $f(j) = 1$ means the 'hit' while $f(j) = 0$ a 'miss'. In theory, the evaluation of $f(j)$ is assumed to be done instantaneously, so that the evaluations does not affect the complexity of the problem. The search is therefore made by evaluating $f(j)$ from $j = 1$ in ascending order until we have $f(j) = 1$. The expected number of evaluations is $\frac{N}{2}$, linear in N, so that we say the classical search needs $O(N)$ evaluations. It is well known that the estimate $O(N)$ is the theoretical lowest bound of the classical linear search.

2.2. Quantum search: Preliminaries

2.2.1. Single-qubit

As known well, information necessary to classical computing is encoded into sequences of '0' and '1'. The minimum unit carrying '0' or '1' is said to be a *bit*. A quantum analogue of a *bit* is called a *qubit*, that takes 2-dimensional-vector form with complex-valued components. In particular, the basis vectors, $(1,0)^T$ and $(0,1)^T$, are taken to play the role of symbols '0' and '1' for classical computing, so that they are referred to as the *computational basis vectors*. We note here that the superscript ^{T} indicates the transpose operation to vectors and matrices henceforth. A significant difference between *qubit* and *bit* is that superposition of the computational basis vectors is allowed in qubit while it is not so of '0' and '1' in bit. Namely, superposition, $\alpha(1,0)^T + \beta(0,1)^T$ ($\alpha, \beta \in \mathbf{C}$), is allowed in qubit, so that we refer to the space of 2-dimensional column complex vectors, denoted by $\mathbf{C}^2$, as the single-qubit space. The $\mathbf{C}^2$ is endowed with the natural Hermitian inner product, say $\phi^\dagger\psi$ for $\phi, \psi \in \mathbf{C}^2$, where the superscript † indicates the Hermitian conjugate operation to vectors and matrices.

2.2.2. Multi-qubits

In order to express classical n-bit information in quantum computing, it is clearly necessary to prepare 2^n computational basis vectors, which span 2^n-dimensional complex vector space $\mathbf{C}^{2^n}$: For any integer x subject to $1 \leq x \leq 2^n$, let us denote by $e(x)$ the canonical basis vector in $\mathbf{C}^{2^n}$, whose x-th component equals 1 ($x = 1, 2, \cdots, 2^n$), while the others are naught. Then every $e(x)$ corresponds to the binary sequence $x_1 x_2 \cdots x_n$ ($x_j = 0, 1$, $j = 1, 2, \cdots, n$) with $x - 1 = \sum_{j=1}^n x_j 2^{n-j}$, so that the basis $\{e(x)\}_{x=1,2,\cdots,2^n}$ turns out to be the computational basis. To be precise mathematically, $\mathbf{C}^{2^n}$ should be understood as n-tensor product,

$$\mathbf{C}^{2^n} \cong (\mathbf{C}^2)^{\otimes n} = \overbrace{\mathbf{C}^2 \otimes \cdots \otimes \mathbf{C}^2}^{n}, \tag{2}$$

of the single-qubit spaces ($\mathbf{C}^2$s). The $\mathbf{C}^{2^n}$ ($\cong (\mathbf{C}^2)^{\otimes n}$) is called the n-qubit space (more generally, the multi-qubit space), which usually thought of as a Hilbert space for a combined quantum system consisting of n single-qubit systems. In the n-qubit space, any vectors with unit length are called *state vectors*;

$$\phi = \sum_{x=1}^{2^n} \alpha_x e(x) \quad (\alpha_x \in \mathbf{C},\ x = 1, 2, \cdots, n) \quad \text{with} \quad \sum_{x=1}^{2^n} |\alpha_x|^2 = 1. \tag{3}$$

It is worth noting here that, in a context of quantum computing or of quantum information, the n-qubit space, $(\mathbf{C}^2)^{\otimes n}$, is often assumed to be a 2^n-dimensional subspace of a complex Hilbert space (usually of infinite dimension) where a quantum dynamical system is described.

2.2.3. Ordered tuples of multi-qubits

We move on to introduce ordered tuples of multi-qubits: The degree of multi-qubits is set to be n and the number of multi-qubit data enclosed in any tuple to be ℓ henceforth. Let $\mathrm{M}(2^n, \ell)$ be the set of $2^n \times \ell$ matrices, which is made into a complex Hilbert space of dimension $2^n \times \ell$ endowed with the Hermitian inner product

$$\langle \Phi, \Phi' \rangle = \frac{1}{\ell} \text{trace } \Phi^\dagger \Phi' \quad (\Phi, \Phi' \in \mathrm{M}(2^n, \ell)). \tag{4}$$

The $\mathrm{M}(2^n, \ell)$ with $\langle \, , \, \rangle$ is the Hilbert space for our quantum search. The subset,

$$\mathrm{M}_1(2^n, \ell) = \{\Phi \in \mathrm{M}(2^n, \ell) \,|\, \langle \Phi, \Phi \rangle = 1\}, \tag{5}$$

of $\mathrm{M}(2^n, \ell)$ is what we are going to dealt with henceforth. An ordered tuple of multi-qubits is a matrix in $\mathrm{M}_1(2^n, \ell)$ of the form

$$\Phi = (\phi_1, \phi_2, \cdots, \phi_\ell) \quad \text{with} \quad \phi_j^\dagger \phi_j = 1 \quad (\phi_j \in \mathbf{C}^{2^n}, \, j = 1, 2, \cdots, \ell). \tag{6}$$

Namely, every column vector of an ordered tuple of multi-qubits stands for a n-qubit state vector. Then the subset of $\mathrm{M}_1(2^n, \ell)$ defined by

$$\mathrm{M}_1^{OT}(2^n, \ell) = \{\Phi = (\phi_1, \phi_2, \cdots, \phi_\ell) \in \mathrm{M}_1(2^n, \ell) \,|\, \phi_j^\dagger \phi_j = 1 \, (j = 1, 2, \cdots, \ell)\} \tag{7}$$

is the space of ordered tuples of multi-qubits, so that we refer to $\mathrm{M}_1(2^n, \ell)$ including $\mathrm{M}_1^{OT}(2^n, \ell)$ as the *extended* space of ordered tuples of multi-qubits (ESOT).

On closing this subsubsection, a remark on a vector-space structure of the ESOT, $\mathrm{M}_1(2^n, \ell)$, is made: As a vector space, the ESOT allows the following isomorphisms,

$$\mathrm{M}_1(2^n, \ell) \cong \mathbf{C}^{2^n} \otimes \mathbf{C}^\ell \cong \overbrace{\mathbf{C}^2 \otimes \cdots \otimes \mathbf{C}^2}^{n} \otimes \mathbf{C}^\ell, \tag{8}$$

which is usually lokked upon as a Hilbert space of the combined system consisting of n two-level particle systems (single-qubit systems) and an ℓ-level particle system. The structure (8) will be a clue to think about a physical realization of the present algorithm.

2.3. Quantum search for an ordered tuple

We are now in a position to present a Grover-type algorithm for an ordered tuple of multi-qubits. Our recipe traces, in principle, Grover's original scenario for the single-target state search [6]. We start with the initial state denoted by A and the target state W, which are defined to be

http://dx.doi.org/10.5772/53187

$$A = \frac{1}{\sqrt{2^n}} \begin{pmatrix} 1 & \cdots & 1 \\ \vdots & \cdots & \vdots \\ 1 & \cdots & 1 \end{pmatrix} \quad \text{and} \quad W = (e(\sigma_1), e(\sigma_2), \cdots, e(\sigma_\ell)), \tag{9}$$

where σ is an injection of $\{1, 2, \cdots, \ell\}$ into $\{1, 2, \cdots, 2^n\}$. On recalling that the state vector $e(x)$ corresponds to the binary sequence $x_1 x_2 \cdots x_n$, the target W corresponds to the ordered tuple of the binary sequences, $\sigma_{j,1}\sigma_{j,2} \cdots \sigma_{j,\ell}$, associated with $e(\sigma_j)$ $(j = 1, 2, \cdots, \ell)$. Note that σ is not necessarily injective in general, but, for simplicity in the succeeding section, it is required to be an injection. Through this chapter, we further assume that n is sufficiently larger than ℓ, so that, in W, the number ℓ of binary sequences is relatively quite smaller than the length n of each binary sequence.

Like in many literatures on quantum computation, we apply the description without the *oracle qubit* below. A treatment and a role of the oracle qubit can be seen, for example, in [1] .

The quantum search is proceeded by applying iteratively the unitary transformation

$$I_G = (-I_A) \circ I_W \tag{10}$$

of $\mathrm{M}(2^n, \ell)$ looked upon as a Hilbert space, where I_A and I_W are the unitary transformations defined to be

$$I_A : \Phi \in \mathrm{M}(2^n, \ell) \mapsto \Phi - 2\langle A, \Phi \rangle A \in \mathrm{M}(2^n, \ell), \tag{11}$$

$$I_W : \Phi \in \mathrm{M}(2^n, \ell) \mapsto \Phi - 2\langle W, \Phi \rangle W \in \mathrm{M}(2^n, \ell). \tag{12}$$

A very crucial remark is that, on implementation, I_W will be of course *not* realized with the target W (see [1] for example).

To express the action of I_G to the initial state A, it is convenient to introduce the $2^n \times \ell$ matrix,

$$R = \sqrt{\frac{2^n}{2^n - 1}} A - \sqrt{\frac{1}{2^n - 1}} W \in \mathrm{M}_1(2^n, \ell). \tag{13}$$

The pair $\{W, R\}$ forms an orthonormal basis of the subspace, denoted by $\mathrm{span}\{W, R\}$, of $\mathrm{M}(2^n, \ell)$ consisting of all the superpositions of the initial state A and the target W. The action of the operator I_G leaves the subspace, $\mathrm{span}\{W, R\}$, invariant; $I_G(\mathrm{span}\{W, R\}) = \mathrm{span}\{W, R\}$. The action of I_G can be therefore restricted on $\mathrm{span}\{W, R\}$ to be

$$I_G : (W, R) \mapsto (W, R) \begin{pmatrix} \cos\theta & -\sin\theta \\ \sin\theta & \cos\theta \end{pmatrix}, \tag{14}$$

where θ is defined by

$$\sin\frac{\theta}{2} = \sqrt{\frac{1}{2^n}}, \quad \cos\frac{\theta}{2} = \sqrt{\frac{2^n - 1}{2^n}}, \quad 0 < \theta < \pi. \tag{15}$$

On putting (10), (13) and (14) together, the k-times iteration I_G^k of I_G applied to A results in

$$I_G^k(A) = \left(\sin(k + \frac{1}{2})\theta\right) W + \left(\cos(k + \frac{1}{2})\theta\right) R \quad (k = 1, 2, 3, \cdots). \tag{16}$$

Hence $I_G^k(A)$ gets closed to the target W if $(k + \frac{1}{2})\theta$ does to $\frac{\pi}{2}$. Indeed, under the assumption $n >> 1$, Eq. (15) yields $\theta \simeq 2\sqrt{\frac{1}{2^n}}$, so that the probability of observing the state W from the state $I_G^k(A)$ gets the highest (closed to one) at the iteration number nearest to $\frac{\pi}{4}\sqrt{2^n} - \frac{1}{2}$. Namely, like Grover's original search algorithm, complexity of the quantum search presented above for an ordered tuple of multi-qubits is of the order of square root of 2^n, the length of binary sequences allowed to be expressed in multi-qubits. In the case of $\ell = 1$, our search of course becomes Grover's original ones, so that our search is thought of as a natural generalization of Grover's original one [6] based on the amplitude magnification technique (see [8], for example).

On closing this subsection, a remark should be made in what follows, which would be of importance to think of a physical implementation in future: We have organized the Grover-type search algorithm I_G as a unitary transformation of the ESOT, $M_1(2^n, \ell)$. Since physically acceptable tuples, however, are in the subset, $M_1^{OT}(2^n, \ell)$, of the ESOT, it is worth checking whether or not I_G leaves $M_1^{OT}(2^n, \ell)$ invariant. By a straightforward calculation with (9), (13) and (14), I_G indeed leaves $M_1^{OT}(2^n, \ell)$ invariant. Though this fact is very basic and simple, this supports, to an extent, a physical feasibility of the present algorithm.

2.4. Geodesic property of the search sequence

We show that the search sequence $\{I_G^k(A)\}$ generated by (16) is on the geodesic starting from the initial state A to the target state W, like in Wadati and Miyake [15] on Grover's original search.

2.4.1. Geometric setting-up

As is briefly mentioned of in Sec. 1, the term 'geodesics' can deal with a wider class of curves in differential geometry than that in usual sense. In usual sense especially among non-geometers, for example, one might have an experience of hearing a phrase like 'the shortest path between a pair of points is a geodesic'. In contrast with phrases like this, geodesics are defined to be autoparallel curves in differential geometry. Put in another way, we have to fix a parallel transport to discuss geodesics in the geometric framework. We have a variety of parallel transports, among which the Levi-Civita (or Riemannian) parallel transport can provide the shortest-path property. Note here that the Levi-Civita parallel transport is defined as the parallel transport that leaves the Riemannian metric endowed with the space. The geodesics to be mentioned of in this subsection can be understood as the familiar *shortest paths*.

Our discussion is made on the ESOT, $\mathrm{M}_1(2^n,\ell)$ defined by (5), with which the standard Riemannian metric is endowed in the following way. To those who are not familiar to geometry, it is recommended to think of the 2-dimensional unit-radius sphere, S^2 , in place of $\mathrm{M}_1(2^n,\ell)$, since $\mathrm{M}_1(2^n,\ell)$ is a $(2^{n+1}\ell-1)$-dimensional analogue of S^2. A Riemannian metric of $\mathrm{M}_1(2^n,\ell)$ has a role of an inner product in every tangent space,

$$T_\Phi \mathrm{M}_1(2^n,\ell) = \{X \in \mathrm{M}(2^n,\ell) \mid \Re(\mathrm{trace}\,\Phi^\dagger X) = 0\} \quad (\Phi \in \mathrm{M}_1(2^n,\ell)), \tag{17}$$

of $\mathrm{M}_1(2^n,\ell)$ at Φ as follows, where $\Re$ indicates the operation of taking the real part of complex numbers: On recalling the intuitive case of S^2, the tangent space at a point $p \in S^2 \subset \mathbf{R}^3$ is thought of as the collection of all the vectors normal to the radial vector p, which can be understood as all the velocity vectors from the dynamical viewpoint. The Riemannian metric, denoted by $((\cdot,\cdot))^{ESOT}$, of $\mathrm{M}_1(2^n,\ell)$ is defined to give the inner product

$$((X,X'))_\Phi^{ESOT} = \frac{1}{\ell}\Re(\mathrm{trace}\,X^\dagger X') \quad (X,X' \in T_\Phi \mathrm{M}_1(2^n,\ell),\ \Phi \in \mathrm{M}_1(2^n,\ell)) \tag{18}$$

in each tangent space $T_\Phi \mathrm{M}_1(2^n,\ell)$.

2.4.2. Geodesics

We are to give an explicit form of geodesics in a very intuitive manner as follows. Let us recall the 2-dimensional case, in which a geodesic with the initial position $p \in S^2 \subset \mathbf{R}^3$ is known well to be realized as a big circle passing through p. By the initial velocity, say $v \in \mathbf{R}^3$, always normal to p, the geodesic is uniquely determined as the intersection of S^2 and the plane spanned by the vectors p and v. The same story is valid for geodesics in $\mathrm{M}_1(2^n,\ell)$, so that we get an explicit form,

$$\Phi(s) = \big(\cos\sqrt{\ell}s\big)\Phi_0 + \big(\sin\sqrt{\ell}s\big)X_0 \quad (s \in \mathbf{R}), \tag{19}$$

of the geodesic with the initial position $\Phi_0 \in \mathrm{M}_1(2^n,\ell)$ and the initial vector $X_0 \in T_{\Phi_0}\mathrm{M}_1(2^n,\ell)$ of unit length tangent to the geodesic. In (19), s is taken to be the length parameter measured from the initial point Φ_0. To be precise from differential geometric viewpoint, the geodesics given by (19) are said to be associated with the Levi-Civita (or Riemannian) parallel transport in $\mathrm{M}_1(2^n,\ell)$.

We are to determine a geodesic which the search sequence $\{I_G^k(A)\}$ is placed on. From (13), (16) and (19), we can construct the geodesic from the big circle passing both W and R, so that we obtain

$$\begin{aligned}\Psi(s) &= \big(\cos\sqrt{\ell}s\big)\Big(\sqrt{\frac{1}{2^n}}W + \sqrt{\frac{2^n-1}{2^n}}R\Big) + \big(\sin\sqrt{\ell}s\big)\Big(\sqrt{\frac{2^n-1}{2^n}}W - \sqrt{\frac{1}{2^n}}R\Big)\\ &= \big(\cos(\sqrt{\ell}s + \frac{\theta}{2})\big)W + \big(\sin(\sqrt{\ell}s + \frac{\theta}{2})\big)R \quad (s \in \mathbf{R})\end{aligned} \tag{20}$$

as the desired geodesic, where s is the length parameter and θ is defined by (15). Setting the parameter sequence $\{s_k\}_{k=0,1,2,\cdots}$ to be $s_k = \frac{k}{\sqrt{\ell}}\theta$, Eq. (20) with $s = s_k$ indeed provides the search sequence $\{I_G^k(A)\}_{k=0,1,2,\cdots}$; $\Psi(s_k) = I_G^k(A)$ (see (16)). To summarize, we have the following.

Theorem 2.1. *The Grover-type search sequence $\{I_G^k(A)\}$ given by (16) for an ordered tuple of multi-qubits is on the geodesic curve $\Psi(s)$ given by (20) in the ESOT, $\mathrm{M}_1(2^n, \ell)$.*

As the closing remark of this section, it should be pointed out that in the case of $\ell = 1$, Theorem 2.1 reproduces the result of Miyake and Wadati on Grover's original search sequence on $S^{2^{n+1}-1}$ in [15].

3. Geometry and dynamics of the projected search sequence in the QIS

In this section, the reduced search sequence in the QIS is shown to be on a geodesic with respect to the m-parallel transport, one of the two significant parallel transports of the QIS. The reduced search sequence is derived from the Grover-type sequence $\{I_G^k(A)\}$ along with the reduction of the regular part of the ESOT to the QIS. The reduction method applied here is entirely different from that in Miyake and Wadati [15].

3.1. The QIS

This subsection is devoted to a brief introduction of the quantum information space (QIS), the space of regular density matrices endowed with the quantum SLD-Fisher metric (see also [19] for another brief introduction and [25, 26] for a detailed one).

Let us consider the space of $\ell \times \ell$ density matrices

$$\mathrm{P}_\ell = \{\rho \in \mathrm{M}(\ell, \ell) \,|\, \rho^\dagger = \rho,\ \mathrm{trace}\,\rho = 1,\ \rho : \text{positive semidefinite}\}, \tag{21}$$

and its regular part

$$\dot{\mathrm{P}}_\ell = \{\rho \in \mathrm{M}(\ell, \ell) \,|\, \rho^\dagger = \rho,\ \mathrm{trace}\,\rho = 1,\ \rho : \text{positive definite}\}, \tag{22}$$

where $\mathrm{M}(\ell, \ell)$ denotes the set of $\ell \times \ell$ complex matrices. The tangent space of $\dot{\mathrm{P}}_\ell$ at ρ can be described by

$$T_\rho \dot{\mathrm{P}}_\ell = \left\{\Xi \in \mathrm{M}(\ell, \ell) \,|\, \Xi^\dagger = \Xi,\ \mathrm{trace}\,\Xi = 0\right\}. \tag{23}$$

In this chapter, the regular part $\dot{\mathrm{P}}_\ell$ of P_ℓ plays a central role, while P_ℓ is usually dealt with as the quantum information space. A plausible account for taking $\dot{\mathrm{P}}_\ell$ is that we can be free from dealing with the boundary of P_ℓ which requires us an extra effort especially in differential calculus.

To any tangent vector $\Xi \in T_\rho \dot{\mathrm{P}}_\ell$, the symmetric logarithmic derivate (SLD) is defined to provide the Hermitian matrix $\mathcal{L}_\rho(\Xi) \in \mathrm{M}(\ell,\ell)$ subject to

$$\frac{1}{2}\left\{\rho\mathcal{L}_\rho(\Xi) + \mathcal{L}_\rho(\Xi)\,\rho\right\} = \Xi \qquad (\Xi \in T_\rho\dot{\mathrm{P}}_\ell)\,. \tag{24}$$

The quantum SLD-Fisher metric, denoted by $((\cdot,\cdot))^{QF}$, is then defined to be

$$((\Xi,\Xi'))_\rho^{QF} = \frac{1}{2}\mathrm{trace}\,\left[\rho\left(L_\rho(\Xi)L_\rho(\Xi') + L_\rho(\Xi')L_\rho(\Xi)\right)\right] \qquad (\Xi,\Xi' \in T_\rho\dot{\mathrm{P}}_\ell) \tag{25}$$

(see [25, 26]), which plays a central role in quantum information theory.

A more explicit expression of $((\cdot,\cdot))^{QF}$ is given in what follows. Let $\rho \in \dot{\mathrm{P}}_\ell$ be expressed as

$$\begin{aligned} &\rho = h\Theta h^\dagger, \quad h \in \mathrm{U}(l) \\ &\Theta = \mathrm{diag}(\theta_1,\cdots,\theta_\ell) \quad \text{with} \quad \mathrm{trace}\,\Theta = 1, \quad \theta_k > 0\ (k=1,2,\cdots,\ell), \end{aligned} \tag{26}$$

where $\mathrm{U}(l)$ denotes the group of $\ell\times\ell$ unitary matrices,

$$\mathrm{U}(l) = \{h \in \mathrm{M}(\ell,\ell)\,|\,h^\dagger h = I_\ell\}, \tag{27}$$

and I_ℓ the identity matrix of degree-ℓ. On expressing $\Xi \in T_\rho\dot{\mathrm{P}}_\ell$ as

$$\Xi = h\chi h^\dagger \tag{28}$$

with $h \in \mathrm{U}(l)$ in (26), the SLD $\mathcal{L}_\rho(\Xi)$ to $\Xi \in T_\rho\dot{\mathrm{P}}_\ell$ takes an explicit expression [19]

$$(h^\dagger\mathcal{L}_\rho(\Xi)h)_{jk} = \frac{2}{\theta_j+\theta_k}\chi_{jk} \quad (j,k=1,2,\cdots,\ell). \tag{29}$$

Putting (26)-(29) into (25), we have

$$((\Xi,\Xi'))_\rho^{QF} = 2\sum_{j,k=1}^{\ell}\frac{\overline{\chi}_{jk}\chi'_{jk}}{\theta_j+\theta_k} \tag{30}$$

[19], where $\Xi' \in T_\rho\dot{\mathrm{P}}_\ell$ is expressed as

$$\Xi' = h\chi' h^\dagger. \tag{31}$$

The space of $\ell\times\ell$ regular density matrices, $\dot{\mathrm{P}}_\ell$, endowed with the quantum SLD-Fisher metric $((\cdot,\cdot))^{QF}$ defined above is what we are referring to as the quantum information space (QIS) in the present chapter, which will be denoted also as the pair $(\dot{\mathrm{P}}_\ell, ((\cdot,\cdot))^{QF})$ henceforth.

3.2. Geometric reduction of the regular part of the ESOT to the QIS

We move to show how the regular part, denoted by $\dot{\mathrm{M}}_1(2^n,\ell)$, of the ESOT is reduced to the QIS through the geometric way, where $\dot{\mathrm{M}}_1(2^n,\ell)$ is defined to be

$$\dot{\mathrm{M}}_1(2^n,\ell) = \{\Phi \in \mathrm{M}_1(2^n,\ell) \,|\, \mathrm{rank}\,\Phi = \ell\}. \tag{32}$$

A key to the reduction is the $\mathrm{U}(2^n)$ action on $\mathrm{M}_1(2^n,\ell)$,

$$\alpha_g : \Phi \in \mathrm{M}_1(2^n,\ell) \mapsto g\Phi \in \mathrm{M}_1(2^n,\ell) \quad (g \in \mathrm{U}(2^n)), \tag{33}$$

where $\mathrm{U}(2^n)$ stands for the group of $2^n \times 2^n$ unitary matrices,

$$\mathrm{U}(2^n) = \{g \in \mathrm{M}(2^n,2^n) \,|\, g^\dagger g = I_{2^n}\}, \tag{34}$$

with $\mathrm{M}(2^n,2^n)$ denoting the set of $2^n \times 2^n$ complex matrices and I_{2^n} the identity matrix of degree-2^n. The $\mathrm{U}(2^n)$ action (33) is well-defined also on $\dot{\mathrm{M}}_1(2^n,\ell)$ since it leaves $\dot{\mathrm{M}}_1(2^n,\ell)$ invariant; $\alpha_g(\dot{\mathrm{M}}_1(2^n,\ell)) = \dot{\mathrm{M}}_1(2^n,\ell)$.

The $\mathrm{U}(2^n)$ action given above provides us with the equivalence relation $\sim$ both on $\mathrm{M}_1(2^n,\ell)$ and on $\dot{\mathrm{M}}_1(2^n,\ell)$;

$$\begin{aligned} \Phi \sim \Phi' \quad & \text{if and only if } {}^{\exists}g \in \mathrm{U}(2^n) \text{ s.t. } \alpha_g\Phi = \Phi' \\ & (\Phi, \Phi' \in \mathrm{M},\ \mathrm{M} = \mathrm{M}_1(2^n,\ell),\ \dot{\mathrm{M}}_1(2^n,\ell)). \end{aligned} \tag{35}$$

The subset of M defined by

$$[\Phi] = \{\Phi' \in \mathrm{M} \,|\, \Phi \sim \Phi'\} \quad (\mathrm{M} = \mathrm{M}_1(2^n,\ell), \dot{\mathrm{M}}_1(2^n,\ell)) \tag{36}$$

is called the equivalence class whose representative is $\Phi \in \mathrm{M}$ ($\mathrm{M} = \mathrm{M}_1(2^n,\ell), \dot{\mathrm{M}}_1(2^n,\ell)$). Note that $[\Phi] = [\Phi']$ holds true if and only if $\Phi \sim \Phi'$. The collection of the equivalence classes is called the quotient space, denoted by $\mathrm{M}/\sim$, of M by $\sim$ ($\mathrm{M} = \mathrm{M}_1(2^n,\ell), \dot{\mathrm{M}}_1(2^n,\ell)$).

To describe a geometric structure of the quotient spaces, $\mathrm{M}/\sim$ ($\mathrm{M} = \mathrm{M}_1(2^n,\ell), \dot{\mathrm{M}}_1(2^n,\ell)$), let us introduce the group of $(2^n-\ell) \times (2^n-\ell)$ unitary matrices,

$$\mathrm{U}(2^n-\ell) = \{\kappa \in \mathrm{M}(2^n-\ell, 2^n-\ell) \,|\, \kappa^\dagger\kappa = I_{2^n-\ell}\}, \tag{37}$$

with $\mathrm{M}(2^n-\ell,2^n-\ell)$ denoting the set of $(2^n-\ell) \times (2^n-\ell)$ complex matrices and $I_{2^n-\ell}$ the identity matrix of degree-$(2^n-\ell)$. We have the following lemma [19].

Lemma 3.1. *The quotient space* $\mathrm{M}_1(2^n,\ell)/\sim$ *is realized as* P_ℓ *defined by (21), where the projection of* $\mathrm{M}_1(2^n,\ell)$ *to* P_ℓ *is given by*

$$\pi^{(n,l)} : \Phi \in \mathrm{M}_1(2^n,\ell) \mapsto \frac{1}{\ell}\Phi^\dagger\Phi \in \mathrm{P}_\ell. \tag{38}$$

Similarly, the quotient space $\dot{\mathrm{M}}_1(2^n,\ell)/\sim$ *is realized as* $\dot{\mathrm{P}}_\ell$ *defined by (22). The projection is given by* $\pi^{(n,l)}$ *restricted to* $\dot{\mathrm{M}}_1(2^n,\ell)$*. The* $\dot{\mathrm{M}}_1(2^n,\ell)$ *admits the fibered manifold structure with the fiber* $\mathrm{U}(2^n)/\mathrm{U}(2^n-\ell)$*. Namely, the inverse image* $(\pi^{(n,l)})^{-1}(\rho) = \{\Phi \in \dot{\mathrm{M}}_1(2^n,\ell) \mid \pi^{(n,l)}(\Phi) = \rho\}$ *of any* $\rho \in \dot{\mathrm{P}}_\ell$ *is diffeomorphic to* $\mathrm{U}(2^n)/\mathrm{U}(2^n-\ell)$*.*

Note that the fibered manifold structure of $\dot{\mathrm{M}}_1(2^n,\ell)$ allows us to proceed differential calculus on $\dot{\mathrm{P}}_\ell \cong \dot{\mathrm{M}}_1(2^n,\ell)/\sim$ freely, while not on P_ℓ due to a collapse of the fibered structure on the boundary.

What is an intuitive interpretation of the quotient spaces, $\mathrm{M}_1(2^n,\ell)/\sim$ and $\dot{\mathrm{M}}_1(2^n,\ell)/\sim$? Let us consider any pair of points Φ and $\Phi' = {}^\exists g\Phi$ in M (M $= \mathrm{M}_1(2^n,\ell), \dot{\mathrm{M}}_1(2^n,\ell)$). Then since $g \in \mathrm{U}(2^n)$, the inner products between column vectors in Φ (see (6)) are kept invariant under α_g;

$$\langle \phi'_j, \phi'_k\rangle = \langle g\phi_j, g\phi_k\rangle = \langle \phi_j, \phi_k\rangle \quad (j,k = 1,2,\cdots,\ell). \tag{39}$$

This implies that the relative configuration of column vectors (namely multi-qubits) is kept invariant under the $\mathrm{U}(2^n)$ action. Hence each of the quotient spaces of $\mathrm{M}_1(2^n,\ell)$ and of $\dot{\mathrm{M}}_1(2^n,\ell)$, is understood to be a space of relative configurations of multi-qubits [19]. We wish to explain the relative configurations in more detail in a very simple setting-up with $n = 6$ and $\ell = 6$. Let us consider the set, $S = \{A, B, \cdots, Z, a, \cdots, z, 0, 1, \cdots, 9, \text{“,”}, \text{“.”}\}$, consisting of the capital Roman letters, the small ones, the arabic digits, a comma and a period. The correspondence of the 2^6-computational basis vectors, $e(x)$ $(x = 1,\cdots,2^6 = 64)$ (see subsubsec. 2.2.2), to the elements of S starts from $e(1) \mapsto A$ in ascending order. Then, under the equivalence relation $\sim$ defined by (35), the word ‘Search’ is identified with ‘Vhduk’ since the latter can be obtained from the former through α_g with the three-step shift matrix g. On choosing $g \in \mathrm{U}(2^n)$ to exchange the capital letters for the small ones, ‘Search’ is identified with ‘sEARCH’.

We are now in a position to show that the QIS $(\dot{\mathrm{P}}_\ell, ((\cdot,\cdot))^{QF})$ is a very natural outcome of the reduction of $(\dot{\mathrm{M}}_1(2^n,\ell), ((\cdot,\cdot))^{ESOT})$. Note here that the Riemannian metric $((\cdot,\cdot))^{ESOT}$ of $\mathrm{M}_1(2^n,\ell)$ naturally turns out to be a metric of $\dot{\mathrm{M}}_1(2^n,\ell)$ under the restriction $\Phi \in \dot{\mathrm{M}}_1(2^n,\ell)$, so that we apply the same symbol, $((\cdot,\cdot))^{ESOT}$, to the metric of $\dot{\mathrm{M}}_1(2^n,\ell)$. A crucial key is the direct-sum decomposition of the tangent space,

$$T_\Phi\dot{\mathrm{M}}_1(2^n,\ell) = \{X \in \mathrm{M}(2^n,\ell) \mid \Re(\mathrm{trace}\,\Phi^\dagger X) = 0\} \quad (\Phi \in \dot{\mathrm{M}}_1(2^n,\ell)), \tag{40}$$

of $\dot{\mathrm{M}}_1(2^n,\ell)$ at Φ, which is associated with the fibered-manifold structure of $\dot{\mathrm{M}}_1(2^n,\ell)$ mentioned of in Lemma 3.1. Note that $T_\Phi\dot{\mathrm{M}}_1(2^n,\ell)$ is identical with $T_\Phi\mathrm{M}_1(2^n,\ell)$ if $\Phi \in \dot{\mathrm{M}}_1(2^n,\ell)$.

Let us consider the pair of subspaces, $\mathrm{Ver}(\Phi)$ and $\mathrm{Hor}(\Phi)$, of $T_\Phi \dot{\mathrm{M}}_1(2^n,\ell)$, which are defined by

$$\mathrm{Ver}(\Phi) = \{X \in T_\Phi \dot{\mathrm{M}}_1(2^n,\ell) \mid X = \xi\Phi,\ \xi \in u(2^n)\} \tag{41}$$

and

$$\mathrm{Hor}(\Phi) = \{X \in T_\Phi \dot{\mathrm{M}}_1(2^n,\ell) \mid ((X',X))^{ESOT}_\Phi = 0,\ X' \in \mathrm{Ver}(\Phi)\}. \tag{42}$$

The $u(2^n)$ is the Lie algebra of $\mathrm{U}(2^n)$ consisting of all the $2^n \times 2^n$ anti-Hermitian matrices,

$$u(2^n) = \{\xi \in \mathrm{M}(2^n,2^n) \mid \xi^\dagger = -\xi\}. \tag{43}$$

The $\mathrm{Ver}(\Phi)$ and $\mathrm{Hor}(\Phi)$ are often called the vertical subspace and the horizontal subspace of $T_\Phi \dot{\mathrm{M}}_1(2^n,\ell)$, respectively. The $\mathrm{Ver}(\Phi)$ is understood to be the tangent space at Φ of the fiber space,

$$\mathrm{U}(2^n)\cdot\Phi = \{\Phi' \in \dot{\mathrm{M}}_1(2^n,\ell) \mid \Phi' = \alpha_g(\Phi),\ g \in \mathrm{U}(2^n)\}, \tag{44}$$

passing Φ, and $\mathrm{Hor}(\Phi)$ to be the subspace of $T_\Phi \dot{\mathrm{M}}_1(2^n,\ell)$ normal to $\mathrm{Ver}(\Phi)$ with respect to $((\cdot,\cdot))^{ESOT}_\Phi$. Thus the orthogonal direct-sum decomposition

$$T_\Phi \dot{\mathrm{M}}_1(2^n,\ell) = \mathrm{Ver}(\Phi) \oplus \mathrm{Hor}(\Phi) \quad (\Phi \in \dot{\mathrm{M}}_1(2^n,\ell)) \tag{45}$$

with respect to the inner product $((\cdot,\cdot))^{ESOT}_\Phi$ is allowed to the tangent space $T_\Phi \dot{\mathrm{M}}_1(2^n,\ell)$.

On using (45), the horizontal lift of any tangent vector of the QIS is given as follows: Let us fix $\rho \in \dot{\mathrm{P}}_\ell$ arbitrarily and any $\Phi \in \dot{\mathrm{M}}_1(2^n,\ell)$ subject to $\pi^{(n,l)}(\Phi) = \rho$. For any tangent vector $\Xi \in T_\rho \dot{\mathrm{P}}_\ell$ (see (23)), the horizontal lift of Ξ at Φ is the unique tangent vector, denoted by Ξ^*, in $T_\Phi \dot{\mathrm{M}}_1(2^n,\ell)$ that satisfies

$$(\pi^{(n,l)})_{*\Phi}(\Xi^*) = \Xi \quad \text{and} \quad \Xi^* \in \mathrm{Hor}(\Phi), \tag{46}$$

where $(\pi^{(n,l)})_{*\Phi} : T_\Phi \dot{\mathrm{M}}_1(2^n,\ell) \to T_{\pi^{(n,l)}(\Phi)} \dot{\mathrm{P}}_\ell = T_\rho \dot{\mathrm{P}}_\ell$ is the differential map of $\pi^{(n,l)}$ at Φ. For a detail of differential maps, see Appendix 2. Recalling, further, the orthogonal direct-sum decomposition (45), we can understand that the horizontal lift Ξ^* of $\Xi \in T_\rho \dot{\mathrm{P}}_\ell$ is of minimum length among vectors, say Xs, in $T_\Phi \dot{\mathrm{M}}_1(2^n,\ell)$ subject to $(\pi^{(n,l)})_{*\Phi}(X) = \Xi$.

Accordingly, the horizontal lift (46) and the Riemannian metric $((\cdot,\cdot))^{ESOT}$ are put together to give rise the Riemannian metric, denoted by $((\cdot,\cdot))^{RS}$, of $\dot{\mathrm{P}}_\ell$, which is defined to satisfy

$$((\Xi,\Xi'))^{RS}_\rho = ((\Xi^*,\Xi'^*))^{ESOT}_\Phi \quad \text{with} \quad \pi^{(n,l)}(\Phi) = \rho \quad (\Xi,\Xi' \in T_\rho \dot{\mathrm{P}}_\ell,\ \rho \in \dot{\mathrm{P}}_\ell). \tag{47}$$

http://dx.doi.org/10.5772/53187

The Ξ^* and Ξ'^* are the horizontal lift at Φ of Ξ and Ξ', respectively, and the superscript RS implies that $((\cdot,\cdot))^{RS}$ is the Riemannian metric of $\dot{\mathrm{P}}_\ell$ looked upon as the reduced space. Note here that the rhs of (47) is well-defined owing to the invariance,

$$(((\alpha_g)_{*\Phi}(X),(\alpha_g)_{*\Phi}(X')))^{ESOT}_{\alpha_g(\Phi)} = ((X,X'))^{ESOT}_{\Phi} \quad (X,X' \in T_\Phi \dot{\mathrm{M}}_1(2^n,\ell), \Phi \in \dot{\mathrm{M}}_1(2^n,\ell)), \tag{48}$$

of $((\cdot,\cdot))^{ESOT}$ and the equivariance,

$$\mathrm{Hor}(\alpha_g(\Phi)) = (\alpha_g)_{*\Phi}(\mathrm{Hor}(\Phi)) \quad (\Phi \in \dot{\mathrm{M}}_1(2^n,\ell)), \tag{49}$$

of $\mathrm{Hor}(\Phi)$ under the $\mathrm{U}(2^n)$ action, where $(\alpha_g)_{*\Phi}$ is the differential map of α_g at Φ (see (33) and Appendix 2). In view of (47), we say that the projection $\pi^{(n,l)} : \dot{\mathrm{M}}_1(2^n,\ell) \to \dot{\mathrm{P}}_\ell$ is a Riemannian submersion [29].

We have the following on the coincidence of $((\cdot,\cdot))^{RS}$ and $((\cdot,\cdot))^{QF}$ [19].

Theorem 3.2. *The Riemannian metric $((\cdot,\cdot))^{RS}$ defined by (47) to make $\pi^{(n,l)}$ the Riemannian submersion coincides with the SLD-Fisher metric defined by (25) up to the constant multiple 4; $4((\cdot,\cdot))^{RS} = ((\cdot,\cdot))^{QF}$.*

On closing this subsection, a comparison between the reduction here and the one by Miyake and Wadati is made. The reduction methods are essentially different since our reduction is made under 'left' $\mathrm{U}(2^n)$ action while 'right' $\mathrm{U}(1)$ action is dealt with in [15] The resultant spaces, namely the reduced spaces, are of course different mutually.

3.3. Geodesic property of the reduced search sequence

We are now in a position to show that the reduced search sequence $\{\pi^{(n,l)}(I_G^k(A))\}$ in the QIS is on an m-geodesic, a geodesic with respect to the m-parallel transport, of the QIS.

3.3.1. Intuitive example of parallel transports: The Euclidean case

Let us start with thinking of parallel transport in the 3-dimensional Euclidean space $\mathbf{R}^3$, the conventional model space not only for basic mathematics and physics but for our daily life. In $\mathbf{R}^3$, the notion of *parallel* seems to be a trivial one, which is usually not presented in a differential geometric framework to those who are not familiar to differential geometry. As minimum geometric knowledge necessary in this subsection, we introduce below a coordinate expression of tangent vectors. As known well, $\mathbf{R}^3$ is endowed with the Cartesian coordinates $y = (y_1, y_2, y_3)^T$ valid globally in $\mathbf{R}^3$. The tangent vectors at any point $p \in \mathbf{R}^3$ can be understood to be the infinitesimal limit of displacements from p. The tangent vector understood to be the displacement-limit $\lim_{\varepsilon\to 0}(p + \varepsilon e^{(j)})$ is then written as $\left(\frac{\partial}{\partial y_j}\right)_p$, where $e^{(j)}$s are the orthonormal vectors along the j-th axis ($j = 1,2,3$). The account for the expression $\left(\frac{\partial}{\partial y_j}\right)_p$ is that we have $\lim_{\varepsilon\to 0} F(p+\varepsilon e^{(j)}) = \left(\frac{\partial F}{\partial y_j}\right)(p)$ for any differentiable functions F. The parallel transport is defined to be a rule to transfer the tangent vectors at $p \in \mathbf{R}^3$ to those at another $p' \in \mathbf{R}^3$, which is of course have to be subject to several mathematical claims not gotten in detail here: The well-known parallel transport in the conventional Euclidean space, $\mathbf{R}^3$, is clearly expressed as

$$\sum_{j=1}^{3} v_j \left(\frac{\partial}{\partial y_j}\right)_p \in T_p\mathbf{R}^3 \mapsto \sum_{j=1}^{3} v_j \left(\frac{\partial}{\partial y_j}\right)_{p'} \in T_{p'}\mathbf{R}^3 \quad (v_j \in \mathbf{R}). \tag{50}$$

An important note is that parallel transports in general differentiable manifolds (including the familiar sphere S^2) are defined in terms of curves specifying the way of point-translations (see Appendix 3 for the case of S^2). The Euclidean case (50) is hence understood to be a curve-free case.

Once the parallel transport (50) is given to $\mathbf{R}^3$, geodesics in $\mathbf{R}^3$ are defined to be autoparallel curves: Let $\gamma(t)$ $(t \in {}^{\exists}[t_0, t_1] \in \mathbf{R})$ be a curve in $\mathbf{R}^3$, whose tangent vector at $t = \tau$ of the curve is given by $\frac{d\gamma}{dt}(\tau)$. The curve $\gamma(t)$ is autoparallel if the tangent vector at each point is equal to the parallel transport of the initial tangent vector $\frac{d\gamma}{dt}(t_0)$. Accordingly, every autoparallel curve turns out to be a straight line or its segment as widely known. Geodesics in $\mathbf{R}^3$ discussed here have the shortest-path property with respect to the Euclidean metric since the parallel transport (50) leaves the metric invariant. Note that parallel transports other than (50) can exist whose geodesics of course lose the shortest-path property with respect to the Euclidean metric.

3.3.2. *The m-parallel transport in the QIS*

We move on to the m-parallel transport in the QIS. Fortunately, the m-parallel transport can be described in a similar setting-up to that for the transport (50). Let us start with the space of $\ell \times \ell$ complex matrices, $\mathrm{M}(\ell, \ell)$, that includes the QIS, $\dot{\mathrm{P}}_\ell$, as a subset. The $\mathrm{M}(\ell, \ell)$ admits the matrix-entries as global (complex) coordinates like the Cartesian coordinates of $\mathbf{R}^3$. The tangent space $T_\rho\dot{\mathrm{P}}_\ell$ at $\rho \in \dot{\mathrm{P}}_\ell$ can be identified with the set of $\ell \times \ell$ traceless Hermitian matrices (see (23)), which can be dealt with in a smilar way to the Euclidean parallel transport setting-up. Indeed, in view of the definition (22), $\dot{\mathrm{P}}_\ell$ is understood to be a fragment of an affine subspace of $\mathrm{M}(\ell, \ell)$. Hence the tangent space $T_\Phi\dot{\mathrm{P}}_\ell$ at every $\dot{\mathrm{P}}_\ell$ admits the structure (23), which is looked upon as a linear subspace of $\mathrm{M}(\ell, \ell)$.

According to quantum information theory [25, 26], the m-parallel transport is written in a simple form

$$\Xi \in T_\rho\dot{\mathrm{P}}_\ell \mapsto \Xi \in T_{\rho'}\dot{\mathrm{P}}_\ell. \tag{51}$$

The geodesic from ρ_0 to ρ_1 with respect to the m-parallel transport is therefore characterized as an autoparallel curve,

$$\rho^{mg}(t) = (1-t)\rho_0 + t\rho_1 \quad (0 \le t \le 1), \tag{52}$$

which takes a very similar form to the Euclidean case. The parameter t in (52) can be chosen arbitrarily up to affine transformations; $t \to at + b$ $(a, b \in \mathbf{R})$.

A very important remark should be made here. From a very naive viewpoint, the geodesic $\rho^{mg}(t)$ in (52) looks like 'straight'. This is *not true*, however, since the QIS is not Euclidean due to the SLD-Fisher metric $\langle\langle\cdot,\cdot\rangle\rangle^{QF}$ endowed in the QIS. Precisely, $\rho^{mg}(t)$ has to be understood to be 'curved' in the QIS.

3.3.3. *The reduced search sequence is on a geodesic*

We are at the final stage to show that the search sequence $\{I_G^k(A)\}$ is reduced through $\pi^{(n,l)}$ on an m-geodesic, a geodesic with respect to the m-transport, of the QIS. We start with calculating the reduced sequence $\{\pi^{(n,l)}(I_G^k(A))\}$ explicitly. Though the initial states A for the search sequence $\{I_G^k(A)\}$ is out of the range $\dot{M}_1(2^n,\ell)$, we apply $\pi^{(n,l)}$ to $\{I_G^k(A)\}$ in the manner (38). Since we have

$$(W^\dagger W)_{jh} = \delta_{jh} \qquad (j,h = 1,2,\cdots,\ell), \tag{53}$$

$$(R^\dagger R)_{jh} = \frac{2^n - 2 + \delta_{jh}}{2^n - 1} \qquad (j,h = 1,2,\cdots,\ell), \tag{54}$$

$$(W^\dagger R)_{jh} = (R^\dagger W)_{jh} = \frac{1-\delta_{jh}}{\sqrt{2^n - 1}} \qquad (j,h = 1,2,\cdots,\ell), \tag{55}$$

$$(A^\dagger A)_{jh} = 1 \qquad (j,h = 1,2,\cdots,\ell), \tag{56}$$

the reduced search sequence $\{\pi^{(n,l)}(I_G^k(A))\}$ takes the form

$$\begin{aligned}
&\pi^{(n,l)}(I_G^k(A)) \\
&= \frac{1}{\ell}\left\{\left(\sin(k+\tfrac{1}{2})\theta\right)W + \left(\cos(k+\tfrac{1}{2})\theta\right)R\right\}^\dagger \left\{\left(\sin(k+\tfrac{1}{2})\theta\right)W + \left(\cos(k+\tfrac{1}{2})\theta\right)R\right\} \\
&= \frac{1}{\ell}\left(\sin^2(k+\tfrac{1}{2})\theta\right)W^\dagger W + \frac{1}{\ell}\left(\cos^2(k+\tfrac{1}{2})\theta\right)R^\dagger R \\
&\qquad + \frac{1}{\ell}\left(\cos(k+\tfrac{1}{2})\theta\right)\left(\sin(k+\tfrac{1}{2})\theta\right)(R^\dagger W + W^\dagger R) \\
&= (1-\tau_k)\left(\frac{1}{\ell}A^\dagger A\right) + \tau_k\left(\frac{1}{\ell}I_\ell\right) \qquad (k = 1,2,\cdots)
\end{aligned} \tag{57}$$

with

$$\begin{aligned}
\tau_k = 1 &- \frac{2^n-2}{2^n-1}\cos^2\left((k+\tfrac{1}{2})\theta\right) \\
&- \frac{2}{\sqrt{2^n-1}}\left(\cos(k+\tfrac{1}{2})\theta\right)\left(\sin(k+\tfrac{1}{2})\theta\right) \qquad (k = 1,2,\cdots),
\end{aligned} \tag{58}$$

where I_ℓ stands for the $\ell \times \ell$ identity matrix. The δ_{jh} in (53)-(55) indicates Kronecker's delta and θ is defined already to satisfy (15).

The expression (52) and (57) are put together to inspire us to consider the m-geodesic in the QIS of the form

$$\rho^G(t) = (1-t)\left(\frac{1}{\ell}A^\dagger A\right) + t\left(\frac{1}{\ell}I_\ell\right) \qquad (\varepsilon \le t \le 1) \tag{59}$$

where ε is a sufficiently small positive number subject to $0 < \varepsilon < \tau_1$ (see (58) with $k = 1$ for τ_1). Note here that the reduction $\frac{1}{\ell}A^\dagger A \in \mathrm{P}_\ell$ of the initial states $A \in \mathrm{M}_1(2^n, \ell)$ turns out to be placed as the limit point of the geodesic $\rho^G(t)$ in the sense that

$$\lim_{\varepsilon \to +0} \rho^G(\varepsilon) = \frac{1}{\ell}A^\dagger A. \tag{60}$$

Combining (59) with (57) and (58), we have

$$\pi^{(n,l)}(I_G^k(A)) = \rho^G(\tau_k) \qquad (k = 1, 2, \cdots, K_n) \tag{61}$$

where K_n is the integer nearest to $\frac{\pi}{4}\sqrt{2^n} - \frac{1}{2}$. The reduction of the initial state is placed at the limit point of the m-geodesic $\rho^G(t)$ in the sense (60). Thus we have the following outlined as Main Theorem in Sec. 1:

Theorem 3.3. *Through the reduction of the regular part, $\dot{\mathrm{M}}_1(2^n, \ell)$, of the extended space of ordered tuples of multi-qubits (ESOT) to the quantum information space (QIS), $\dot{\mathrm{P}}_\ell$, the reduced search sequence $\{\pi^{(n,l)}(I_G^k(A))\}_{k=1,2,\cdots,K_n}$ is on the m-geodesic $\rho^G(t)$ of the QIS given by (59).*

4. Concluding remarks

We have studied the Grover-type search sequence for an ordered tuple of multi-qubits. The search sequence itself is shown to be on a geodesic with respect to the Levi-Civita parallel transport in the ESOT. Further, the reduced search sequence in the QIS is shown to be on a geodesic with respect to the m-parallel transport in the QIS. The m-geodesics do not have the shortest-path property but they are very important geodesics in the QIS together with those with respect to e-parallel transport. The geometric reduction method applied this chapter is entirely different from the method in Miyake and Wadati [15].

A significance of this chapter is the discovery of a novel geometric pathway that connects directly the search sequence in the ESOT with an m-geodesic in the QIS. According to a crucial role of the m-geodesics and the e-geodesics together with their mutual duality, the pathway will be a key to further studies on the search in the ESOT from the quantum information geometry viewpoint. Further, since the QIS is well-known to be the stage for describing dynamics of quantum-state ensembles of quantum systems [2, 3], the pathway shown in this chapter will be of good use to connect the search in the ESOT with dynamics of a certain quantum system.

An direct application of the search in the ESOT is not yet found: However, if a problem with a strong relation to relative ordering of data (see around Eq. (39)) exists, our search will be worth applying to the problem.

On closing this section, three questions are posed below, which would be of interest from the viewpoint of the expected benefits listed in Sec. 1.

1 In view of the results in this chapter, we are able to clarify that the 'Grover search orbit' given by a continuous-time version of (16) is an m-geodesic. A question thereby arises as to 'Is it possible to characterize the m-geodesics by orbits of a certain dynamical system on $\mathrm{M}_1(2^n,\ell)$?'. To this direction, a variation of the free particle system on $\mathrm{M}_1(2^n,\ell)$ would be a candidate (Benefits 1 and 2).

2 Accordingly, another question would be worth posed: 'Is it possible to characterize the e-geodesics by orbits of a certain dynamical system on $\mathrm{M}_1(2^n,\ell)$?' (Benefits 1 and 2).

3 The celebrated fact on the duality between the m-transport and the e-transport (see [25] and [26]) may provide us with a further question: 'If there exists a pair of dynamical systems on $\mathrm{M}_1(2^n,\ell)$ whose reduced orbits characterize the m-geodesics and e-geodesics respectively, which kind of relation does it exist between those systems ?' (Benefit 3).

Acknowledgements

The author would like to thank the editor for his valuable comments to improve the earlier manuscript of this chapter. This work is partly supported by Special Fund for Strategic Research No. 4 (2011) in Future University Hakodate.

Appendices

Appendix 1. Glossary of symbols and notation

Acronyms

- ESOT: The abbreviation of the extended space of ordered tuples of multi-qubits, which is denoted by $\mathrm{M}_1(2^n,\ell)$ (see Eq. (5)).
- QIS: The abbreviation of the quantum information space, which is realized as $\dot{\mathrm{P}}_\ell$, the set of $\ell\times\ell$ positive definite Hermitian matrices with unit trace, endowed with the quantum SLD-Fisher metric $((\cdot,\cdot))^{QF}$ (see Eq. (22) for $\dot{\mathrm{P}}_\ell$, and (24)-(31) for $((\cdot,\cdot))^{QF}$).
- SLD: The abbreviation of the symmetric logarithmic derivative (see Eq. (24)).

Sets and spaces

- $\mathrm{Hor}(\Phi)$: The horizontal subspace of $T_\Phi\dot{\mathrm{M}}_1(2^n,\ell)$ (see Eq. (42) with (41)).
- $\mathrm{M}(\ell,\ell)$: The set of $\ell\times\ell$ complex matrices.
- $\mathrm{M}(2^n,\ell)$: The set of $2^n\times\ell$ complex matrices.
- $\mathrm{M}(2^n,2^n)$: The set of $2^n\times 2^n$ complex matrices.
- $\mathrm{M}(2^n-\ell,2^n-\ell)$: The set of $(2^n-\ell)\times(2^n-\ell)$ complex matrices.

- $\mathrm{M}_1(2^n,\ell)$: The subset of $\mathrm{M}(2^n,\ell)$ consisting of $2^n \times \ell$ complex matrices with unit norm referred to as the extended space of ordered tuples of multi-qubits (see Eq. (5)), which is abbreviated to the ESOT.
- $\dot{\mathrm{M}}_1(2^n,\ell)$: The subset of $\mathrm{M}_1(2^n,\ell)$ consisting of the elements of $\mathrm{M}_1(2^n,\ell)$ with the maximum rank equal to ℓ (see Eq. (32)).
- $\mathrm{M}_1^{OT}(2^n,\ell)$; The subset of $\mathrm{M}_1(2^n,\ell)$ consisting of $2^n \times \ell$ complex matrices whose columns are of unit length (see Eq. (7)).
- P_ℓ The set of $\ell \times \ell$ positive semidefinite Hermitian matrices with unit trace; the space of $\ell \times \ell$ density matrices (see Eq. (21)).
- $\dot{\mathrm{P}}_\ell$: The set of $\ell \times \ell$ positive definite Hermitian matrices with unit trace; the space of $\ell \times \ell$ regular density matrices (see Eq. (22)).
- $T_\Phi \mathrm{M}_1(2^n,\ell)$: The tangent space of $\mathrm{M}_1(2^n,\ell)$ at $\Phi \in \mathrm{M}_1(2^n,\ell)$ (see Eq. (17)).
- $T_\Phi \dot{\mathrm{M}}_1(2^n,\ell)$: The tangent space of $\dot{\mathrm{M}}_1(2^n,\ell)$ at $\Phi \in \dot{\mathrm{M}}_1(2^n,\ell)$ (see Eq. (40)), which is identical with $T_\Phi \mathrm{M}_1(2^n,\ell)$ if $\Phi \in \dot{\mathrm{M}}_1(2^n,\ell)$.
- $T_\rho \dot{\mathrm{P}}_\ell$: The tangent space of $\dot{\mathrm{P}}_\ell$ at a point $\rho \in \dot{\mathrm{P}}_\ell$ (see Eq. (23)).
- $\mathrm{U}(l)$: The group of $\ell \times \ell$ unitary matrices (see Eq. (27)).
- $\mathrm{U}(2^n)$: The group of $2^n \times 2^n$ unitary matrices (see Eq. (34)).
- $u(2^n)$: The Lie algebra of the group $\mathrm{U}(2^n)$ (see Eq. (43)).
- $\mathrm{U}(2^n-\ell)$: The group of $(2^n-\ell)\times(2^n-\ell)$ unitary matrices (see Eq. (37)).
- $\mathrm{Ver}(\Phi)$: The vertical subspace of $T_\Phi \dot{\mathrm{M}}_1(2^n,\ell)$ (see Eq. (41)).

Maps, operators and transformations

- α_g: The unitary transformation of $\mathrm{M}_1(2^n,\ell)$ associated with $g \in \mathrm{U}(2^n)$ (see Eq. (33)).
- $(\alpha_g)_{*\Phi}$: The differential map of α_g at $\Phi \in \dot{\mathrm{M}}_1(2^n,\ell)$. See also Appendix 2 for the definition.
- I_A: The unitary transformation of $\mathrm{M}_1(2^n,\ell)$ defined by (11).
- I_G: The unitary transformation composed of $-I_A$ and I_W (see Eq. (10)).
- I_W: The unitary transformation of $\mathrm{M}_1(2^n,\ell)$ defined by (12).
- $\mathcal{L}_\rho$: The symmetric logarithmic derivative (SLD) (see Eqs. (24) and (29)).
- $\pi^{(n,l)}$: The projection of $\dot{\mathrm{M}}_1(2^n,\ell)$ to $\dot{\mathrm{P}}_\ell$ (the QIS) (see Eq. (38)).
- $(\pi^{(n,l)})_{*\Phi}$: The differential map of $\pi^{(n,l)}$ at $\Phi \in \dot{\mathrm{M}}_1(2^n,\ell)$. See also Appendix 2 for the definition.
- ^{T}: The transpose operation to vectors and matrices.
- †: The Hermitian conjugate operation to vectors and matrices.

Metrics

- $((\cdot,\cdot))^{ESOT}$: The Riemannian metric of the ESOT and of its regular part (see Eq. (18)).
- $((\cdot,\cdot))^{QF}$: The quantum SLD-Fisher metric of the QIS (see Eqs. (25) and (30)).
- $((\cdot,\cdot))^{RS}$: The Riemannian metric of the QIS other than $((\cdot,\cdot))^{QF}$, that makes the projection $\pi^{(n,l)}$ a Riemannian submersion (see Eq. (47) with (46)).

http://dx.doi.org/10.5772/53187

Others

- A: The matrix expressing the initial state for the Grover-type search in the ESOT (see Eq. (9)).
- R: The matrix with which forms an orthonormal basis of the subspace consisting of all the superpositions of A and W (see Eq. (13)).
- W: The matrix expressing the target state (namely the marked state) for the Grover-type search in the ESOT (see Eq. (9)).
- $\sim$: The equivalence relation both on $\mathrm{M}_1(2^n, \ell)$ and on $\dot{\mathrm{M}}_1(2^n, \ell)$ (see Eq. (35)).

Appendix 2. Differential maps

We here give a detailed explanation of differential maps, $(\pi^{(n,l)})_{*\Phi}$ and $(\alpha_g)_{*\Phi}$. For any $X \in T_\Phi \mathrm{M}_1(2^n, \ell)$, we can always find a curve $\gamma(t)$ $(-\tau < t < \tau, \tau > 0)$ on $\mathrm{M}_1(2^n, \ell)$ subject to $\gamma(0) = \Phi$ and $\frac{d\gamma}{dt}(0) = X$. The differential map $(\pi^{(n,l)})_{*,\Phi}$ is defined to be

$$(\pi^{(n,l)})_{*\Phi}(X) = \left.\frac{d}{dt}\right|_{t=0} \pi^{(n,l)}(\gamma(t)), \tag{62}$$

which turns out to take the explicit form

$$(\pi^{(n,l)})_{*\Phi}(X) = \frac{1}{\ell}(X^\dagger \Phi + \Phi^\dagger X). \tag{63}$$

The differential map $(\alpha_g)_{*\Phi}$ of α_g at Φ is defined in the same way: On the same setting-up to the curve $\gamma(t)$ with $X \in T_\Phi \mathrm{M}_1(2^n, \ell)$, the $(\alpha_g)_{*\Phi}$ is defined by

$$(\alpha_g)_{*\Phi}(X) = \left.\frac{d}{dt}\right|_{t=0} \alpha_g(\gamma(t)), \tag{64}$$

which yields

$$(\alpha_g)_{*\Phi}(X) = gX. \tag{65}$$

Appendix 3. The standard parallel transport in S^2

In this appendix, the standard parallel transport is concisely reviewed. In particular, we present the fact that the transport depends on the choice of the paths connecting a pair of points in S^2. Below, S^2 is realized as the set,

$$S^2 = \{y \in \mathbf{R}^3 \,|\, y^T y = 1\}, \tag{66}$$

in the 3-dimensional Euclidean space $\mathbf{R}^3$.

Let us fix a pair of distinct points, y_0 and y_1, in S^2 arbitrarily, which connect by a smooth curve $\gamma(s)$, where s is the length parameter. Namely, the $\gamma(s)$ satisfies

$$\gamma(0) = y_0, \quad \gamma(L) = y_1 \quad (L : \text{the full curve length}). \tag{67}$$

Again, we remark that $\gamma(s)$ takes 3-dimensional vector form. To express tangent vectors of S^2 at $\gamma(s)$, we prepare the orthonormal basis $\{v_1(s), v_2(s)\}$ of $T_{\gamma(s)}S^2$ subject to

$$v_1(s) = \dot{\gamma}(s), \quad v_2(s) = \gamma(s) \times \dot{\gamma}(s) \tag{68}$$

where the overdot $\dot{}$ stands for the derivation by s and $\times$ the vector product operation. We note here that the vector, $\dot{\gamma}(s)$, tangent to $\gamma(s)$ is always of unit length since s is the length parameter. This ensures that the basis $\{v_1(s), v_2(s)\}$ is orthonormal. In terms of the basis $\{v_1(s), v_2(s)\}$, any tangent vector at $\gamma(s)$ can be expressed in the form of linear combination, $c_1 v_1(s) + c_2 v_2(s)$ $(c_1, c_2 \in \mathbf{R})$. Accordingly, the parallel transport along the curve $\gamma(s)$ is understood to be the way of connecting $\{v_1(L), v_2(L)\}$ at $y_1 = \gamma(L)$ to $\{v_1(0), v_2(0)\}$ at $y_0 = \gamma(0)$.

To express the parallel transport concisely, it is of good use to introduce the one-parameter rotation matrix,

$$\Gamma(s) = (v_0(s), v_1(s), v_2(s)), \quad v_0(s) = \gamma(s) \quad (0 \leq s \leq L). \tag{69}$$

On denoting by $P_\gamma(v_1(0), v_2(0))$ the parallel transport of $(v_1(0), v_2(0))$ along the curve $\gamma(s)$, we have

$$(v_1(L), v_2(L)) = P_\gamma(v_1(0), v_2(0)) \begin{pmatrix} \cos a & \sin a \\ -\sin a & \cos a \end{pmatrix} \tag{70}$$

with

$$a = -\int_0^L \frac{1}{2} \mathrm{trace}\,(N \dot{\Gamma}(s)\, \Gamma(s)^T)\, ds, \quad N = \begin{pmatrix} 0 & 0 & 0 \\ 0 & 0 & -1 \\ 0 & 1 & 0 \end{pmatrix}. \tag{71}$$

If we choose $\gamma(s)$ to be a big circle or its segment, the $\gamma(s)$ is autoparallel since a in (70) and (71) vanishes, so that big circles and their segments turn out to be geodesics.

Author details

Yoshio Uwano

Department of Complex and Intelligent Systems, Faculty of Systems Information Science, Future University Hakodate, Kameda Nakano-cho, Hakodate, Japan

References

[1] Nielsen MA., Chuang IL. Quantum Computation and Quantum Information. Cambridge: Cambridge U.P.; 2000.

[2] Benenti G., Casati G., Strini G. Principles of Quantum Computation and Information Volume I. Singapore: World Scientific; 2004.

[3] Benenti G., Casati G., Strini G. Principles of Quantum Computation and Information Volume II. Singapore: World Scientific; 2007.

[4] Mermin ND. Quantum Computer Science An Introduction. Cambridge: Cambridge U.P.; 2007.

[5] Shor PW. Algorithms for quantum computation: discrete logarithms and factoring. In: Proceedings of the 35th Annual Symposium on Foundations of Computer Science. Los Alamitos: IEEE Press; 1994.

[6] Grover L. A fast quantum mechanical algorithm for database search. In: Proceedings of the 28th Annual ACM Symposium on the Theory of Computing. New York: ACM Press; 1996.

[7] Grover L. Quantum mechanics helps in searching for a needle in a haystack. Phys. Rev. Lett. 1997; 79 p325-328.

[8] Gruska J. Quantum Computing. London: McGraw-Hill International; 1999.

[9] Ambainis A. Guest column: Quantum search algorithms. ACM SIGACT News 2004; 35 p22-33.

[10] Galindo A., Martin-Delgado MA. Family of Grover's search algorithms. Phys. Rev. A 2000; 62 062303 p1-6.

[11] Grover LK. A different kind of quantum search. arXiv:quant-ph/0503205v1 2005.

[12] Grover LK. Fixed-point quantum search. Phys. Rev. Lett. 2005; 95 150501 p1-4.

[13] Tulsi T., Grover LK., Patel A. A new algorithm for fixed point quantum search. Quantum Info. and Comput. 2006; 6 p483-494.

[14] arXiv: http://arXiv.org (accessed 5 September 2012).

[15] Miyake A., Wadati M. Geometric strategy for the optimal quantum search. Phys. Rev. A 2001; 64 042317 p1-9.

[16] Farhi E., Goldstone J., Gutmann S., Sipser M. Quantum computation by adiabatic evolution. arXiv:quant-ph/0001106v1 2000.

[17] Roland J., Cerf NJ. Quantum search by local adiabatic evolution. Phys. Rev. A 2002; 65 042308 p1-6.

[18] Rezakhani AT., Abasto DF., Lidar DA., Zanardi P. Intrinsic geometry of quantum adiabatic evolution and quantum phase transitions. Phys. Rev. A 2010; 82 012321 p1-16.

[19] Uwano Y., Hino H., Ishiwatari Y. Certain integrable system on a space associated with a quantum search algorithm. Physics of Atomic Nuclei 2007; 70 p784-791.

[20] Nakamura Y. Completely integrable dynamical systems on the manifolds of Gaussian and multinomial distributions. Japan J. Indust. Appl. Math. 1993; 10 p179-189.

[21] Nakamura Y. Lax pair and fixed point analysis of Karmarkar's projective scaling trajectory for linear programming. Japan J. Indust. Appl. Math. 1994; 11 p1-9.

[22] Nakamura Y. Neurodyanamics and nonlinear integrable systems of Lax type. Japan J. Indust. Appl. Math. 1994; 11 p11-20.

[23] Uwano Y., Yuya H. A gradient system on the quantum information space realizing the Karmarkar flow for linear programming – A clue to effective algorithms –. In: Sakaji A., Licata I., Singh J., Felloni S. (eds.) New Trends in Quantum Information (Special Issue of Electronic Journal of Theoretical Physics). Rome: Aracne Editorice; 2010 p257-276.

[24] Uwano Y., Yuya H. A Hebb-type learning equation on the quantum information space – A clue to a fast principal component analyzer. Far East Journal of Applied Mathematics 2010; 47 p149-167.

[25] Hayashi M. Quantum Information. Berlin: Springer-Verlag; 2006.

[26] Amari S., Nagaoka H. Methods of Information Geometry. Providence: American Mathematical Society; 2000.

[27] Bengtsson I., Życzkowski K. Geometry of Quantum States. Cambridge: Cambridge U.P.; 2006.

[28] Nakahara M. Geometry, Topology and Physics. Bristol: IOP Publishing; 1990.

[29] Montgomery R.A Tour of Subriemannian Geometries, Their Geodesics and Applications. Providence: American Mathematical Society; 2002.